Solar Concentrators

A Bibliography

Energy Series :
S B 01

S S Mathur
Head, Center of Energy Studies
Indian Institute of Technology
New Delhi, India

T C Kandpal
Center of Energy Studies
Indian Institute of Technology
New Delhi, India
&
R N Singh
Department of Physics
Indian Institute of Technology
New Delhi, India

Innovative
Informations
Incorporated

Library of Congress Catalogue Card Number : 82 - 82841

I S B N : 0 - 910661 - 01 - 4

Published 1982, First impression.

Innovative Informations Inc., P.O.Box 408, Greenbelt,
Maryland, 20770, U.S.A.

Printed in the United States of America

Preface

Solar energy has attained a very prominent position as a source
of energy for the future. Extensive efforts are being made the
world over to collect, store and to convert this energy to more
usable forms. Thermodynamically it is important that it be collected at higher temperatures so that its conversion to mechanical or electrical form can be possible at reasonable efficiencies. Development of solar concentrators, therefore, is a high
priority work. Large number of scientists are working in this
area and they have to scan the diffuse literature in this field
at a heavy cost of precious time. It is becoming more and more
difficult for a scientist or an engineer to keep track of the
relevant literature and to get adequate information about the
earlier work done on the subject.

At present there are a number of sources of literature and it is
not usually possible for any worker to have access to all those
sources. It will be a great help if a consolidated uptodate
bibliography along with the abstracts of the subject, the research papers and reports is made available. The present monograph has been prepared to fill this gap. It gives a bibliography on solar concentrators and contains over 1400 references with
abstracts, published upto mid 1982.

Solar concentrators can generally be classified as line focus or
point focus systems. Both the line focus and point focus systems
may be either (i) reflecting or refracting type, (ii) tracking or
non tracking type, (iii) imaging or non imaging type, (iii) single
stage or two stage, or (v) single piece or composite surface type

solar concentrators. The entries in the bibliography are
arranged in a classified manner representing different designs
and usage. Entries are arranged alphabetically according to
surnames of authors. Where there is more than one entry by a
single author, it is arranged chronologically. A single author
is placed earlier than the same author with coauthors.Where the
title of an entry is not in English, it has been translated
alongside into English. Abbreviations have been avoided to the
extent possible. An author index is also included.

We have made every effort to ensure that this bibliography is
as inclusive as possible. Although it is substantially complete, omissions are possible. We seek readers' indulgence.
We shall be grateful for any information reders may provide for
incorporation in the next edition.

The authors express deep gratitude to Professor M.S.Sodha, De-
puty Director, Indian Institute of Technology, New Delhi, for
his encouragement during the preparation of this work. Thanks
are also due to Mr. Ramesh Kumar Thareja for his efficient typ-
ing of the manuscript.

New Delhi   S.S.Mathur
                                               T.C.Kandpal
                                               R.N Singh

Publisher's Note

The grave consequences faced by a large number of countries as
a result of their dependence on a few countries for fuels to
meet domestic energy needs has speeded up research in solar
energy the world over. This research will continue and diver-
sify in various fields of specialization. Thus the need for
uptodate books of reference in each of these specialized fields
is growing. This bibliography is prepared by three scholars
to meet this need for a handy record of research done on solar
concentrators and some allied technologies.

The authors have culled from all available material to make this
bibliography as comprehensive as possible in the specialized
field. The book has been divided into chapters each covering
one aspect of the subject. The citations are sufficiently ex-
haustive to facilitate speedy location of relevant references.

An addendum, with citations numbered in continuation of the main
text, updates the bibliography with material published till mid
1982. The authors' index lists citations both in the text and
the addendum to facilitate search. A list of sources of material
is also given in two broad heads viz. publishers of journals and
periodicals and of proceedings of conferences, workshops, etc.

We have pleasure in placing at the service of scholars a comprehensive and compact book of reference in their specialized field.

Contents

Cylindrical Parabolic Trough Concentrator

001

1 KW Solar Pump System : Almanza R, Diaz A, Gonzalez A and Lopez S, " Proceedings ERDA Conference on Concentrating Solar Collectors ", Atlanta, Georgia (USA), p. 8 - 47 (1977).

The configuration concept of a solar power station for the power range of 1 KW using a Rankine cycle is presented. The main components of prototype plant with cylindrical parabolic collectors array are herein described.

002

A Stationary Solar Energy Concentrator : Archibald P B , " Solar Energy ", 3(4), 65 (1959).

A parabolic reflector having a short focal length will keep the sun's rays focussed within a small area for several hours without the employment of a heliostat mechanism. Its use as a steam generator and as a stove is described.

003

A Semifixed Solar Energy Concentrator: Archibald P B,"Lawrence Livermore Laboratory Report ",UCID-16877 (1975) .

Described herein is a solar concentrator, consisting of an array of short focal length parabolic reflectors, mounted in a panel. the focal line of the reflectors is oriented in an east-west direction. The concentration ratio is 8.8 : 1, and the absorber tube remains on focus for more than 6 hours a day. The concentrator follows the sun by tilting in a north-south direction, rather than by rotating about the focal line. This tilting motion, controlled by photocells is several degrees each day, over a total range of about 60° for the year. The concentrator would furnish heat at a temperature high enough to efficiently operate an absorption air conditioner, and on a unit area basis, it would be ten times as effective as a conventional flat plate collector. The construction of a concentrator, using presently available commercial materials is also described; the estimated labour and material cost, including the mechanism is less than $ 60.00/ sqm.

004

A Parabolocylindrical Concentrator for Photoimpulse Irradiation : Baibutaev K B & Kakharov M K, " Applied Solar Energy ", 5 (3), 57 (1969).

Concentrated photoimpulse irradiation on agricultural seeds is described. The design of a cylin - drical parabola concentrator is described. Calculated and experimental data are given.

005

Textile Drying Using Solar Process Steam: Bessing M E, "Mechanical Engineering",102(1), 40 (1980).

A solar energy collector system is providing process heat to a textile drying process, in a West Point Pepperell mill in Fairfax, Alabama. The solar collector system, uses 24 single axis tracking parabolic trough concentrating collectors, to heat water in a high temperature water loop.The high temperature water, fuels a steam generator to provide process steam. The process that was solarised, is a textile drying process using cylindrical can dryers. The dryers are ustilized in the slashing operation, a process in which the yarn is treated with sizing in preparation for weaving. The solar energy system supplies process steam at 70 psi (0.48 mPa) and 317 deg F (158 deg C)to the slasher manifold. Slasher operation at 5 to 60 psi (0.03-0.41 mPa) is insensitive to the steam source and allows displacement of fossil fuel when under solar operation. The system is expected to provide 1000 lbs/hr (450 Kg/h) of steam, under clear sky conditions. Based on 300 working days per year and representative weather tape data for Atlanta, Georgia, the system is expected to deliver 10^9 Btu/yr to the process.

006

Design Criteria and Thermal Performance of Cylindrical Parabolic Solar Collectors : Bhide V G , Vaishya J S, Tripathi T C and Sootha G D, " Proceedings Indian National Solar Energy Convention ", Bombay , p. 89 , (1979).

The performance of cylindrical parabolic solar collectors having metallic absorber tube enveloped by a glass tube, is analysed. The effects of concentration, collector physical parameters, such as the smoothness of the reflecting surface, span and focal length of the parabola etc. on the performance of the collectors, are discussed. The variation in collector efficiency with the absorber tube diameter, the collector aperture for different reflector contour errors, temperature of operation etc., are analysed and discussed. Effect of selective coating on the absorber tube and evacuation of the space in between the absorber tube and the outer envelope on the performance of the collector, are also discussed. It is found that selective coating, on the absorber tube, considerably improves the performance of the collector, whereas evacuation of the space between the outer glass envelope, and the metallic tube absorber, has relatively little effect on its performance. Parabolic reflectors of glass with copper absorber tube, have been designed and fabricated in the laboratory. The experimental results obtained with these collectors are discussed in the paper.

007

Method for Estimating Hourly Solar Radiation for Parabolic Trough Collectors : Bingham C H, Posner D M, and Bradley J O, " Proceedings Conference on Alternative Energy Sources ", Miami (USA), p. 329 (1978).

Performance simulation of focussing parabolic collectors, are hampered by a limited amount of useful insolation data. Hourly values of total radiation on a horizontal surface are the most common data available from the National Oceanographic and Atmospheric Administration.   This   paper develops a method to convert this total horizontal radiation to direct radiation on the surface of the tracking collector.

008

Utizzazione Dell' energie Solare; Analisi del Collectore di Raiazione / (Utilisation of Solar Energy : Analysis of a Radiation Collector): Boffa C & De Socio L," Termotechnica "(Milan) 29 (2), 671 (1975).

In this paper one of the most favourable solutions to the solar energy power plant design is considered. The system is constituted by a collector, an energy transfer loop and a storage system. The collector is formed by a movable specular parabolic cylinder that collects and focusses the radiation on a cylindrical absorber coaxial with the focal line of the reflector. Three coaxial cylinders constitute the absorber i.e. an inner heat pipe, a radiation shield and an outer transparent envelope. A slit is open along the shield, through which the radiation leaving the mirror can reach the heat pipe. The shield reduces radiative dissipation from the heat pipe.The present paper analyses the geometric characteristics of the collector from the point of view of the collector's efficiency. Also taken into account are other parameters related to the practical design of the system. Among them the reflector orientation, the angular standard deviation from the normal to the ideal parabolic surface, the energy losses in the absorber and the optical characteristics of the adopted material are considered.

009

Hexcel Parabolic Trough Concentrator : Branch G P, " Proceedings of the ERDA Conference on Concentrating Solar Collectors ", Atlanta, Georgia (USA) p. 2-11, (1977).

Hexcel has now been involved with the manufacture of concentrators for over two years. During that period of time, the Hexcel concentrator has gone through an optimization and design evaluation to improve the BTU/$ cost effectiveness and reduce the shipping and field installation costs.   Since the Hexcel concentrator module is rather large, a 165.5 sq. ft. aperture (8.5'x 19.5'), it is necessary to provide a few field assembled components as possible,and yet provide a compact high density stuffing configuration. This paper discusses some of these considerations and some of the solutions Hexcel has taken to solve these problems.

010

Suntec/Hexcel Parabolic Trough : Davison John H, "Proceedings of the ISES Silver Jubillee International Congress, Georgia, (USA), (1979).

In order to meet the industrial process heat need, Suntec Systems Inc., has begun marketing and further development of the Hexcel parabolic trough. This paper presents the technical data for the Suntec / Hexcel collector along with a review of current installations.

011

Textile Drying Using Solar Process System : Byfield D A & Mitchell P D. (Unknown).

This paper describes a solar energy collection system for providing process heat to a textile drying process. The solar collection sub-system uses 24 parabolic trough, single axis tracking concentrating collectors to heat water in a high temperature water (HTW) loop.   A system generator is fueled with the HTW to produce process steam. The process being solarized is textile drying using can dryers. The can dryers are part of a stirling operating in a West Point Pepperell mill in Fairfax, Alabama.

012

Design & Analysis of a Uniaxial Tracking Device with a Cylindrical Parabolic Solar Concentrator System : Carlton R J & Hewitt H C, " Proceedings of the ISES (American Section) Conference", Orlando (USA) (1977).

The results of thermal testing of four cylindrical parabolic solar concentrators with a uniaxial tracking device are presented. Exit temperatures of 200-260 deg.F were accomplished with 50 per cent efficiency. An inexpensive uniaxial tracking device with an adjustable accuracy as low as ± 1 along with programable lugstetesis is described. Construction of the concentrators with emphasis on durability, efficiency and cost is outlined.

013

Approaches to Fabrication of Parabolic Trough Reflectors :Champion R L, "National SAMPE Technical Conference ", 7 (Prt II), 35 (1975).

Design criteria and parameters of parabolic reflectors for focussing solar collectors are discused. The emphasis is on design concepts which offer potential for low cost mass production, include honey-comb/stressed skin structure, moulded plywood, stretch-formed metal skins & rigid foams.

014

Solar Production of Industrial Process Steam : Cherne J, Gelb G & Pinkeston D, "Proceedings of the 14th IECE Conference", Boston (USA) p. 124 (1979).

TRW is developing the application of solar technology to food processing at the Ore-Ida Foods . plant in Ontario, Oregon, under the auspices of the Department of the Environment. Steam generated at 417 deg.F, is used for frying, then cascaded to lower temperatures for other processes in the plant. This plant consumes all of the solar produced steam without the need for costly storage. The system selected consists of high efficiency, parabolic trough, concentrating collectors and a pressurized water heat transfer loop which generates steam in a flash tank. This system is applicable to many industries. The system interfaces with the industrial process by simple teeing into the existing process steam lines. This paper treats the trade-offs performed to select the system configuration.

015

Textile Drying Using Solarized Cylindrical Can Dryers : Curtner K L, Mitchell P D & Rausch R A ,
"Proceedings of the ISES (American Section) Conference", Orlando , (USA) (1977).

This paper describes a solar energy collection system for providing process heat to a textile drying process. The solar collection subsystem uses 773 sq.mtrs. of parabolic trough, single   axis
tracking, concentrating collectors to heat water in a high temperature water loop. A steam generator is fueled with the high temperature water to produce process steam. The process being   solarized is textile drying using cylindrical can dryers.

016

Performance Testing of the Hexcel Parabolic Trough Solar Collector : Dudley V E & Workhaven R M,
"Sandia Laboratories (USA) Report # 78 - 0381"(1978).

This report summarizes the testing which was performed on the Hexcel Parabolic Trough solar   collector at the Solar Total Energy Facility. Test objectives are defined, test procedures are described and results and conclusions are given.

017

Performance Analysis of a Cylindrical Parabolic Focussing Collector & Comparison with Experimental
Results : Edenburn M W, " Solar Energy " 18, 437 (1976).

The efficiencies for focussing collectors which consist of a cylindrical parabolic reflector and
a collector tube surrounded by a transparent envelope and which heats a fluid flowing through the
collector tube have been predicted using heat transfer analytical methods. The analysis considers
visible radiation transfer, IR radiation exchange, conductive and convective losses and   energy
transferred to a fluid flowing through a collector tube. The collector may have a tilted north to
south axis, an east to west axis or it may fully track the sun and geometric parameters with tracking the sun. Predicted results are in excellent agreement with recent collector performances measured using the Sandia Laboratories collector test facility.

018

On the Performance of Cylindrical Parabolic Solar Concentrators with Flat Absorbers : Evans D L ,
"Solar Energy", 19, 379 (1977).

An integral relationship is developed for evaluating the intensity distribution on flat absorbers
used with cylindrical parabolic solar concentrators. Calculations are presented for perfect cross
section concentrator using various modes, rim angles, off axis angles and defocussing   amounts.
Peak concentration ratios are shown to vary as the sine of the rim angle. Off axis and defocussed
operations are shown to result in considerably reduced intensities. The effect of surface   slope
errors is also investigated. Normally distributed surface slope errors with a standard deviation
of 0.25 degree are shown to reduce peak intensities by more than a factor of 3.

019

Concentrating Systems for Small Solar Power Stations : Feustel J E & Maryhofer O, "Proceedings of
the Commission of the European Community Workshop on Concentrators for Solar Energy Applications",
Heverlee, (Belgium), (1978).

This paper describes after a survey on concentration systems, collector influence factors   and
some practical application examples for concentrators of the parabolic trough and the paraboloid
type. These concentrators seem to be most promising for the use in solar thermal systems of the
solar farm type and for the use in photovoltaic systems.

020

Incidence Angle Modifier and Average Optical Efficiency of Parabolic Trough Collectors : Gaul H &
Rabl A ,"Proceedings of the ISES Silver Jubilee International Congress", Georgia, (USA) , (1979).

The incidence angle modifier for parabolic troughs is investigated in order to clarify the connection between collector tests & prediction of long term energy delivery by collector arrays .
The optical efficiency of a parabolic trough collector decreases with incidences   for several
reasons (i) decreased transmission and absorption; (ii) increased width of solar image on receiver and (iii) spillover of radiation in troughs of finite length without end reflectors. In order
to be able to apply test results from a (usually short) collector module to collector arrays of
arbitrary length, it is necessary to separate analytically the end loss from the first two effects.
This analysis is applied to several collectors which have been tested at Sandia Laboratories and
at the Solar Energy Research Institute (SERI). For improved accuracy the measurements of the incidence angle modifier at SERI were carried out at low temperatures with an open water test loop.
The results are presented in two forms, first as a polynomial fit to data, and second as a single
number, the all day average optical efficiency for typical operating conditions.

021

Small Solar Power Plant with a Freon Turbine : Gehrke H ,"Proceedings of the ISES Conference" ,
New Delhi (India), p.1719 (1978).

Small solar power plants might be used mainly to provide electricity to small villages in regions
which are poorly inhabited but rich in solar insolation. A prototype of a small solar power plant
with an electrical output of 10 KWe is presented. Thermal energy collected by flat plate collectors and parabolic troughs is converted into mechanical and subsequently electrical energy by means
of the freon - Rankine process.

022

Parabolic Trough Collectors with Thermal or Photovoltaic Receiver : Giuseppe B, Ramano F & Gugli-
elms T, " Proceedings of the ISES Silver Jubilee International Congress", Georgia (USA) , (1979).

The present paper deals with the development of the ANSALDO parabolic trough collector   which is
suitable to concentrate the incident solar radiation onto both a strip of solar cells or on a lin-

ear cavity receiver. The design of the reflector, reflecting material, mounting structure, tracking, receiver design etc. have been discussed. The necessary changes in the shape of parabola or the positioning of the receiver so as to produce a uniform flux onto the strip of solar cells are brought out.

023

Development and Evaluation of A Mid Temperature Concentration Collector : Gupta B P, "Proceedings of the ERDA Conference on Concentrating Solar Collectors", Atlanta, Georgia (USA), p.2-41, (1977).

A cylindrical parabolic solar collector was designed and fabricated and is undergoing tests to obtain performance data at operating temperatures upto 400 deg.F. The concentrating collector is designed to track the apparent motion of the sun in one axis. The tracking axis can be aligned in any orientation as the design allows for 270° of rotation during the day. The concentrator consists of one half of a parabola reflecting the sunlight onto a cylindrical absorber tube.The parabolic structure is of a honeycomb sandwich construction to combine high strength and stiffness with light weight. The selectively coated absorber tube is insulated within a metal housing with a high transmittance glass window forming the receiver aperture. The receiver and the concentrator are arranged such that the receiver does not block any part of the concentrator aperture.

The modular construction in 20 feet lengths allows convenient collector row lengths upto 120 feet with a compatible motor/gear bone drive mounted in the middle of the row. The design allows storage of the collector in a low profile and low drag position with the mirror facing down to enhance the life of the second surface aluminized acrylic reflective material.

Simulation analysis result of the thermal loss from the receiver and the overall collector efficiencies are presented. The results of the experimental evaluation with pressurized water as the heat transfer fluid is also presented here.

024

Thermal Performance of Cylindrical Parabolic Concentrators : Gupta R K, Patel R C, Barve K M and Jain B C, "Proceedings of the Indian National Solar Energy Convention", Bombay, p.83 , (1979).

The thermal performance of cylindrical parabolic concentrators has been studied, both theoretically as well as experimentally for 200-300 deg.C temperature applications. The effects of variations in direct solar insolation, selective and non - selective coatings, absorber tube material & configuration, sectorial insulation, and orientation of concentrators etc. on collector perform - ance have been analysed. A number of cylindrical parabolic concentrators with different features have been fabricated and tested. In all cases, a fixed absorber has been used with a tracking reflector. Thermal performance of a concentrator array having 80 m^2 area has been studied with both east - west and north - south orientations. Different tracking arrangements to follow the sun movement have been identified and a tracking system has been developed which uses an electric motor, for continuous tracking of the array.

025

Mid-Temperature Solar Thermal Systems Design Handbook Parabolic Troughs : Harrigan R W, "Proceedings of the ISES Silver Jubilee International Congress", Georgia, (USA), (1979).

This paper outlines ongoing efforts at Sandia Laboratories (SLA) to develop a solar thermal systems Design Handbook. The handbook deals at this time only with parabolic trough collector systems operating upto 315 deg.C (600 deg.F). While useful to anyone doing solar thermal design involving parabolic troughs the handbook application is primarily aimed at commercial architectural and engineering firms. Updating of the handbook is expected to occur continuously as the rapidly developing solar technology advances.

026

Theoretical Performance of Cylindrical Parabolic Solar Concentrators : Hassan K E & El-Refaie M F, "Solar Energy", 15, 219 (1973).

This is a theoretical study of the performance of cylindrical parabolic solar concentrator under ideal conditions. Particular attention is given to the application of this type of concentrator in the intermediate temperature range, where wide targets are used. Suitable indices that describe the performance are defined and calculated and the effects of the different parameters are studied. In particular the energy available at the focal plane and the distribution of its intensity are determined. All ranges of rim angles are studied: less than 90°, more than 90° and the whole range of 0 - 180°. The performances obtained are the best possible and can, therefore, be used as a standard of reference for the appraisal of actual concentrators of the type studied.

027

Performance of a Solar Thermal Collector : Higa W H, "NASA (USA) Technical Memorandum 33-748"(1975).

Many evolutionary steps are required before a solar thermal power system can be moved from the research laboratory phase to the field application area. It is the objective of this paper to discuss some possible means for achieving the required technology. Simplifications in construction techniques as well as in measurement techniques for parabolic trough collectors are described . Actual measurement data is also given.

028

High Temperature Solar Review : Hildebrandt A F, "Chemical Engineering Progress",75 (11),20 (1979).

This paper describes parabolic trough collectors or linear concentrators which are utilized for intermediate temperature process heat production. By focussing the sun onto a long tube and using selective surfaces of high absorptivity and low emissivity, 60 percent of the solar beam energy can be captured in a fluid. Single axis tracking results in efficient concentration ratios of 50 and temperatures of upto 200 deg.C. The linear concentrator can be used to supply process heat as well as some electricity, its low physical profile makes it suitable for installation on building roofs. For higher temperatures - 300°and above a point focus reflector is required. Chemical engineering solutions are needed in the development of suitable receivers with heat transfer fluid and chemical reactions in the 300-1000°C and higher range in resolving problems with a variety of materials, heat transfer and chemical processes. A pilot plant is described which has been constructed in California to define the operating problems of the solar system, called solar tower. In this system, the energy from one square kilometer can be continuously reflected on the elevated receiver through computer control of the heliostat. The energy can be used by pumping water to a central receiver, converting this to steam, and piping the steam back down the tower into a conventional steam electric turbine.

029

Thermal Analysis of a Black Liquid Cylindrical Parabolic Collector : Huang B J, Wung T Y & Nieh S, " Solar Energy " 22, 221 (1979).

In the present paper a simple theoretical analysis and an experiment are carried out for a modified concentrating collector which consists of a cylindrical parabolic reflector, a transparent glass tube centered along the focal line, and a highly absorbent black liquid which flows in the glass tube to directly absorb the concentrated solar beam radiation. The analytical results are presented in normalized form and proved to be in very good agreement with the experimental results.

030

Results of Development of Linear Parabolic Trough : Hutchison G, " Proceedings of the ERDA Conference on Concentrating Solar Collectors",Georgia, (USA), p. 2-93, (1977).

After more than two years of development and testing, Solar Kinetics Inc. has completed the T-500 solar collector. Initial design considerations included short focal length for reasonable mirror accuracy requirements, monocoque construction for maximum strength per weight ratio and angular misalignment compensation for minimum maintenance and installation expense. Black chrome was chosen for the selective surface and the receiver insulated with pyrex glass, suspended in Silicone'0' rings. Hydraulics are used for tracking due to the many benefits which include superior tracking accuracy. Finally, testing was reduced to various performance curves.

031

Geometrical Aspects of a Cylindrical Parabola Collector : Kandlikar S G & Vij S K, " Proceedings of the ISES Conference ", New Delhi (India) p.1254 (1978).

This paper deals with a practical configuration of a parabolic reflector (axis north-south) with a circular absorber located at the focus. For this case, the geometric relationships among various parameters are analyzed. The results should be useful for selecting different parameters of the parabolic collector system for a given application.

032

An Experimental Study on Horizontal Parabolic Cylinder Type of Collector for Solar Cooling: Kimura K, Udagawa M, Miyazaki T & Yagino T, " Proceedings of the ISES Conference,Los Angeles(USA), (1975).

The performance of horizontal parabolic cylinder type of collector with a focal line along the east-west axis is examined by experiments combined with 1 RT of water/lithium bromide adsorption refrigeration machine.

033

Der Parabolische Zylinderspiegel als Reflector für Solarkollektoren Wirkungsgrade und Optimierungsbetrachtungen / (The Cylindrical Parabolic Mirror as Reflector for Solar Collectors, Efficiencies & Optimisation : Kohne R, " Dentsche Luft-und Raumfahrt, Forschnungsbericht ", 76 55, (1976).

After introducing the concentration ratio and intercept factor of focussing collectors with parabolic cylinder mirrors the energy balance equations were derived to determine the efficiencies under steady state conditions. The components of the collector were varied and optimized with respect to maximum efficiency. Furthermore the dynamic behaviour of the collector was calculated and the average efficiency compared with the efficiencies in the steady state conditions.

034

30/50 KW Thermal Power Plant of the Distributed Type : Kohne R & Kraft M, "Proceedings of the ISES Silver Jubilee International Congress", Atlanta, Georgia (USA), (1979).

A joint project between M.A.N. (Germany) and AUXINI (Spain), is presented which describes a small solar thermal power plant of the distributed type delivering a maximum power output of 30 KW and 50 KW respectively, if a storage is used additionally to the collector field. The plant built up at Getafe near Madrid (Spain) consists presently of a first collector field with parabolic trough collectors mounted rigidly on a platform which is tracked azimuthally. Collector system efficiencies have been measured and are compared with theoretically predicted values from a transient simulation model. Different temperature spreads and inlet temperatures (between 60° to 240° C) are investigated and some results of the measured collector system efficiencies are also given.

035

Theoretical and Experimental Results of a Small Solar Power Plant with Concentrating Collectors : Kohne R & Kraft M, " Proceedings of the International Symposium on Solar Thermal Power and Energy Systems", (France), (1980).

The solar thermal power plant at Getafe (Spain) with an electric peak output of 50 KW is described which consists of three different collector fields CTF 1....3 with parabolic trough collector which have a total reflective area of 572 m². Results of the thermal efficiencies of CTF 1 and one collector module of CTF 3 are given. The values are 60 and 50 percent respectively for temperatures between 180 and 280 deg. C for the collector system CTF 1 & 2 and 56 percent respectively for the non collector module for CTF 3. Effect of dust covers as well as the transient behaviour of CTF 1 are treated. Extended Computer simulations are presented which deviate not more than 6 per cent from the experimental values.

036

Study of Thermal Characteristics of Parabolic Cylindrical Solar Installations : Kolos Ya G, "Teploenergetika ", 5(3), 73 (1958) .

Formulas and graphic relations for determination of optimal operating conductions,for solar installations, for various temperatures and pressures in boilers are presented.

037

Optimum Design Parameters of Horizontal Coaxial Cylinders for a Solar Energy Collector: Kunitomo T,
and Aizawa K, " Proceedings of the ISES Conference",New Delhi (India), p. 1259 (1978).

In this paper, the optimum combinations of design parameters of a solar collector system,for thermal use, of horizontal coaxial cylinders, with a cylindrical parabolic mirror are discussed, from
the standpoint of energy, collector efficiency and outlet temperature of a fluid. Heat   balance
calculation is carried out, using the exact relations of simultaneous radiative, convective and
conductive heat transfer in the system.

038

Linear Concentrating Solar Collector in an Air-Supported Enclosure : Preliminary Design S t u d y,
Final Report : Laakso J H & Zimmerman D K,"Sandia Laboratories Report SAND 78-7022" (1979).

A preliminary design, for a low cost linear parabolic concentrating solar collector, in a pneumatically stabilized cylindrical plastic film enclosure, analysis of its theoretical performance
and projected cost are described. Potential applications for the concentrator, are in heating
fluids to the mid temperature range for use in industrial process heating. The study objective
was to develop an innovative design concept, having pneumatically stabilized plastic film enclosure as an approach to achieving low collector cost. Both, circular film and rigid parabolic reflector concepts, were investigated; the concept fulfilling the study objective, has a light weight
aluminium honeycomb sandwich, parabolic trough reflector with an aluminized polyester film reflective surface. Key design features, for a collector with a weathered polyester enclosure 2.8m
diameter by 30.5m long (9.33 x 100 ft.),and an aperture area of 69.3 sqm. (745.9 sft.) are given.
For this collector configuration, structural analyses, thermal performance modeling, mass produced component costs, field assembly methods and maintenance requirements are discussed. Daily efficiencies in excess of 45% are predicted, with estimated installed field collector costs of $90
per sqm. ($8.34/sft.). Results of cost and performance studies indicate that the collector has
potential for low cost, and offers attractive cost/performance figures - of merit with further
development.

039

An Axial Temperature Differential Analysis of Linear Focussed Collectors for Solar Power: Lee D O
& Schimmel Jr. W P, "Proceedings of the Ninth IECE Conference", San Francisco, p.347,   (1974).

Collection of solar energy over a moderate temperature range has been analytically studied. Specifically, Therminol 66 was used as a collector fluid, in this study. The effects upon performance
of receiver absorptance and emittance, silvering of a collector envelope, and diameter of the envelope are considered. Convective and radiative transport, between the receiver and envelope, and
from the envelope to the environment are modeled. However, wind losses are not considered. Radiative exchange processes are modeled, with an approximate two wavelength band model.   Calculated
results are presented for (i) Varying values of specular solar input for Albuquerque, New Mexico
on the statistical June 21, (ii)Fluid flow rate , (iii) Initial fluid temperature for both,silver-
ed and unsilvered envelopes, and (iv) both evacuated and unevacuated envelopes. The effect of varying envelope diameter is also considered. This study indicates, that the best extraction of energy
and the best temperature rise in the fluid, occurs in the turbulent flow regime of the collector
fluid. The maximum, however, for any given flow rate with a given inlet temperature, occurs just
after transition from laminar to turbulent flow.

040

Energy Balances on a Parabolic Cylinder Solar Collector : Lof G O G, Fester D A & Duffie J A ,
"Journal of Engineering for Power (Trans. ASME)", 84, 24 (1962).

A technique for optimizing the design of focussing solar collectors, was developed through a detailed study of the energy balances, for a parabolic cylindrical reflector with tubular receivers
of three diameters. Experimental data for concentration ratios of 10 to 22.5 and surface temperatures of 88 to 353 deg.F are presented. Receiver temperature, meteorological variables (including
solar radiation), and distribution of reflected radiation in the focal zone of the reflector, were
measured and corelated, so as to permit optimization of concentration ratio.

041

Optimization of Focusssing Solar Collector Design : Lof G O G & Duffie J A, "Journal of Engineering for Power (Trans. ASME)", 85, 221 (1963).

A set of general graphical relationships, to be used for establishing the receiver-reflector area
ratio, which will provide the maximum useful heat delivery from a focussing collector is presented.
The method, applied to paraboloids and parabolic cylinders, gives the optimum area ratio, in terms
of incident radiation intensity, optical properties (including precision of reflector), & thermal
loss rate. Use of the method for design, and for evaluation of existing reflectors, is illustrated; its use in economic optimization will be dependent on availability of adequate cost data .

042

Theoretical and Experimental Study of Parabolic Troughs & Paraboloidical Dish Concentrators w i t h
Plane Absorbers : Lucifredi A, Bisagni M, Ravina E & Castellazzi P, " Proceedings of ISES Silver
Jubilee International Congress", Georgia, (USA), (1979).

Chosen for a distributed components power plant, the parabolic troughs or the paraboloidal d i s h
collectors with plane absorber, many problems remain to be solved, e.g. the mounting selection,the

drive and tracking system, the supporting structure, the absorber width, the reflecting materials etc. Many factors have been taken into account, in order to evaluate the efficiency of various kinds of collectors, in particular the enlargement of the focal effective strip, the thermal flux or focal mean concentration change, the non uniform distribution of the flux intensity on the absorber due to the inclination of the sun's rays, the manufacture and tracking errors, the shadows caused by the absorber or by the adjacent collectors. The influence of each factor is evaluated through mathematical models, whose composition gives a value of efficiency, which is variable depending on the design choice, and on the latitude of the site of the installation of the power plant.A methodology for the choice of the type, and for the design of a distributed elements power plant is reported. The methodologies which are used, have been developed only theoretically, but have also been checked from the point of view of experimental results deduced from a series of tests, performed on orientable modules (fully orientable or uniaxial) directly to the sunlight or from laboratory measurements, on various types of reflecting surfaces. In addition, strength and alternating fatigue tests were performed on sandwich structures with foamed or honeycomb material. From the results which are presented, it becomes possible to form an optimal choice of the reflecting material and of its supporting structure.

043

Performance and Cost of Solar Industrial Process Heat Systems Utilizing the Parabolic Cylinder Collector : Ludwig D, "Proceedings of ERDA Conference on Concentrating Solar Collectors",Georgia, (USA) , p. 9 - 31 (1977).

Intertechnology Solar Corporation, has evaluated the technical performance and life cycle cost, of a solar thermal system to produce industrial process heat, utilizing the parabolic cylinder. A comparison of the performance and cost factors, for flat plate, compound parabolic, low concentration Fresnel lens, parabolic cylinder,and solar pond collectors, showed the parabolic cylinder collector to be the most cost effective, for solar supplied industrial process heat applications above 175 deg.F. For applications below 175 deg.F, a well designed single glazed, black chrome coated flat plate collector, is the most economical system, even though the usable energy is less than that produced by the parabolic cylinder. The system studies were performed at six locations, representative of six constant performance solar regions, covering the USA. Performance and cost factors were computed, for a wide range of solar fractions, of load and process temperature requirements, ranging from 125 deg.F through 500 deg.F.

044

On Heat Exchsngers Used with Solar Concentrators : Lumsdaine E & Chering J C,   " Solar Energy", 18, 157 (1976) .

The possibility of using an elliptical cylinder, with a parabolic cylinder mirror, is explored. It is shown that for an aperture ratio of 4, an elliptic cylinder will attain a higher concentration, than parabolic or circular pipe targets.

045

Design Analysis of an Asymmetric Solar Receiver : McCulloch W H & Treadwell G W, " Sandia Laboratories. (USA) Report SAND 74-0124 (1974).

This study considers in detail, the effects of varying the receiver design, with focal length. As the rim angle is decreased, an increasing portion of the receiver and its envelope, receive little or no reflected solar flux, and this portion may be designed, to minimise thermal losses. The result is a design which is not symmetric about the receiver axis. The purpose of this analysis, is to examine the performance of such an asymmetric receiver design, and to determine the optimum rim angle. The results of the study show, that for the receiver configuration and operational conditions described, the optimum has a 90 rim angle. The impact of varying a number of design parameters is also evaluated.

046

Concentrating Collectors Applied to the Industrial Process of Textile Drying : Mitchell P D and Gupta B P, "Proceedings ERDA Conference on Concentrating Solar Collectors",Georgia(USA),p.8-29,('77)

The first phase of this ERDA Solar Industrial Process Heat Program, has resulted in the detailed design of a solar energy collection system, for providing process heat to a textile drying process. The solar collection subsystem, uses 700 sqm. of parabolic trough, single axis tracking, concentrating collectors, to heat water in a high temperature water (HTW) loop. At the system design point (clear day), the solar collectors nominally generate 198 deg.C water with the HTW loop at 2×10^6 Pa. A steam generator is fueled with the HTW, and produces 490 Kg/hr. of process steam at the nominal design point conditions. The generated process steam is at 0.5×10^6 Pa and 160 deg C. The solar system will provide 1.2×10^6 MJ/yr. to the process. This is 41% of the direct insolation available to the collector field, during the operational hours (300 days/yr) of the Fairfax mill. The process being solarized is textile drying using cylindrical can dryers. The can dryers are a part of a slashing operation in a West Point Pepperell mill in Fairfax, Alabama. Over 50 % of all woven goods are processed through slashers and dried on cylindrical can dryers.

047

Theoretical Behaviour During the Day & the Year of a Solar System Using Cylindrical Parabolic Collectors : Mora J L & Almanza R, "Proceedings International Symposium on Solar Thermal Power and Energy Systems", Marsulls (France) (1980).

In this paper a theoretical model is presented, which was developed for prediction purposes, and allows the study of the behaviour of a proposed experimental 1000 sqm solar concentrator system, as a function of time, during any day of the year. The solar system considered, consists of a set of cylindrical parabolic collectors and a storage system. The results are represented graphically and also discussed.

048

The .Acurex Cylindrical Parabolic Solar Collector : Muller T K, " Proceedings ERDA Conference on Concentrating Solar Collectors ", Georgia, (USA) p.2 - 11 (1977).

Acurex Corporation has designed and currently manufactures, a parabolic trough , concentrating solar collector of the single axis tracking type. This system has eight modules each with a 6'x 10' aperture coupled torsionally to a common rotational axis. Each row of eight modules, is positioned with the rotational axis oriented either east-west or north-south, and tracking horizon to horizon. The collectors are designed to heat water, transfer oils, and other fluids to temperatures in the range of 140 deg F to 600 deg F. Operation at these temperatures is required for the following applications : process hot water and steam, space heating, single and two stage absorption cooling and refrigeration, drying operations and solar thermal and photovoltaic power generation. Design trade studies were conducted, to determine individual module size, means of sun tracking ,   and specific details of the fluid passage-way design. The trough concept provides sufficient design flexibility to integrate this collector, into any given solar energy system.

049

Experimental Evaluation of a Cylindrical Parabolic Solar Collector : Ramsey J W, Gupta B P   and Knowles G R, "Journal of Heat Transfer (Trans. ASME)", 99, 163 (1977).

Results are presented for a series of solar collector experiments, in which the incident   solar flux was concentrated by a single axis tracking, parabolic trough mirror. The concentrated solar flux, was directed onto an absorber tube whose axis coincided with the focal axis of the concentrator. The performance of the collector was evaluated, using three different absorbers, a black painted tube designed to operate near ambient temperature, a heat pipe which had a selective solar absorber coating applied to its surface, and a heat pipe, which had its surface coated with a non-selective black paint. The peak efficiency for the collector in the absence of heat losses, is approximately 62% when the incoming solar energy is normal to the collector aperture. The h e a t losses which occurred at elevated temperatures (300 deg C) decreased the peak efficiencies to 50 and 30% respectively, for the selectively coated and black painted tubes. The experimental result established the technical feasibility, of using parabolic trough collectors for application   requiring thermal energy, at temperatures upto 300 deg C.

050

Solar Thermal Electric Power Generation Using a System of Distributed Parabolic Trough Collectors: Ramsey J W, Sparrow E M & Eckert E R G, "Proceedings AICHE Symposium", San Francisco, 74 (174), 271 (1978).

The system employs water as the working fluid, and steam at 60 bars pressure and 276 deg C, is generated locally within a field of distributed parabolic trough solar collectors. A transfer loop conveys the steam to a central site, where the power plant is situated. The design of the collector and the operating characteristics, of three transfer loop configurations are described. Results of experiments performed at a desert test site, using a scale model of a solar collector module   are presented. The data establishes the efficiency of the collector, both in the absence of heat losses and under normal operating conditions. The findings of life test being performed, on samples o f candidate solar reflector surfaces are reported. A parametric study of the performance of a parabolic trough collector, using a Monte Carlo digital simulation of the collector is described. Examples of the effects of collector orientation on performance are given. Predictions o f power absorbed, versus time of day for a typical trough collector are also presented.

051

Steady-State Wind Loading on Parabolic Trough Solar Collectors : Randall D E, McBride D D and Tate R E, "ASME paper No. 80-C2 /Sol - 20 (1980).

Two wind tunnel force and moment tests have been conducted on parabolic trough solar collector configurations. The two tests were conducted in different flow field environments, first a uniform flow infinite air-stream, the second a simulated atmospheric boundary layer flow, with the models simulating a ground mounted installation. The influence of various geometric design parameters for collector modules, and arrays has been established. Data indicate.that f o r c e s   and moments increase with mounting height, and with trough aspect ratio.

052

Receiver Assembly Design Studies for 2-m 90° Parabolic Cylindrical Solar Collectors : Ratzel A C, "Sandia Laboratories (USA) Report", SAND 79 - 1026 (1979).

Results are presented from a parametric study of the Sandia Laboratories second generation, 2 - m 90° parabolic-cylindrical solar collector design. A computer simulation was developed, to provide cumulative all day performance results or instantaneous solar noon results, for three a n n u l a r solar receiver assemblies; 2.223, 2.54 and 3.125 cm O.d.tubes with concentric glass jackets. Representative clear spring, summer and winter conditions for Albuquerque, New Mexico, were modeled. Design problems considered in the analysis included misalignment of the receiver assembly from the focal line, reflector trough tracking bias, variation in receiver tube operating temperature, and variation in the reflector trough one-dimensional slope errors, and two-dimensional m i r r o r errors. Changes in collector material radiative properties and wind effects, are also summarized, and comparative performance results for evacuated, versus non-evacuated annular receivers are given. Summarized performance results, for all studies are provided graphically. For operating receiver tube temperatures $<$475°K the 3.175 cm receiver tube provides the best overall collector performance results. For higher operating temperatures, where detrimental receiver heat losses become more significant, the smaller 2.54 cm tube is more effective for solar energy collection.

053

Structural Optimization of SMC Line Focussing Collectors : Reuter Jr., R C and Allred R E, "Proceedings National SAMPE Symposium ", San Francisco, (USA), p. 866 (1979).

One of the primary structural components, of a line focussing solar collector, is the reflector support. In this design, it must be a stiff, light-weight, parabolic trough, which provides and maintains, correct optical shape for the reflective surface, and protects it from d e t r i m e n t a l environmental effects. A viable candidate for the reflector support, is a moldable fiber reinforced rib stiffened structure, which has a low cost, high production rate potential. This p a p e r covers material and analytical considerations, made to ensure that the sheet molding compound panel design meets specified structural requirements. A detailed analysis is provided and optimization is achieved through steps taken to minimize panel weight.

054

Concentration Characteristics of Parabolic Trough Mirrors : Sakuta K, Tani T, Sawata S & Tanaka T, "Bull Electrotech. Laboratory ", Tokyo, (Japan), 43 (1 - 2) 13 (1979).

An investigation was conducted of the relationship between the concentration characteristics and curvature accuracy, of parabolic trough mirrors used as collectors and concentrators of energy, in distributed type, thermal electrical power systems. Errors of curvature of several parabolic trough mirrors were measured, with a new equipment, developed for the purpose. Measurement results make clear, the relationships between the concentration ratio and the collection efficiency.

055

Experimental and Theoretical Investigations of an East - West Oriented Parabolic Concentrator : Samuel A A, Bhaskaran K A, Sriramulu V and Gupta M C,"Proceedings Indian National Solar Energy Convention ", Bhavnagar (India), p. 131, (1978).

An east - west oriented parabolic concentrator has the advantage, that it does not require diurnal tracking. In view of the fact that the reflector is stationary, focussing is not so efficient as in a tracking device, but its efficiency is greater to a flat plate collector. The paper discusses the theoretical and experimental results obtained during the course of investigation of such a concentrator.

056

Conceptual Design and Analysis of a 100 MWe Distributed Line Focus Solar Central Power Plant Final Report, Volume 1 - Executive Summary : Semmens M M, Fong A, Collaros G J, Dascher R E, Grarsberger R E and Griego D B, " The BDM Corporation (USA) Report ", (BDM / TAC - 79 - 375 - TR), (1979).

The conceptual design and analysis of a nominal 100 MW solar central power plant is performed.The applicability of line focus technology for central power plants, when operating at temperatures in excess of 700 deg F (370 deg C) is assessed. It is concluded that the line focus technology can be fabricated, implemented and maintained in the field, at a far less capital cost than the more complicated point focus central receiver technology.

057

Performance and Optimization of a Cylindrical Parabola Collector : Singh P & Cheema L S, " Solar Energy ", 18, 135 (1976).

This paper includes, an analysis of the performance, of a cylindrical-parabola collector regarding the amount of energy collected, optimization of its various parameters, and experimental verification of the theoretical prediction. Theoretical concentration calculated by M H Cobble for various absorber shapes have been recalculated, with relatively more logical assumptions. The various parameters have been theoretically optimized for the maximum energy collection. To corroborate the theoretical predictions, a cylindrical parabola collector, capable of having two axes steering was constructed. Provision was made for changing the relative focal length, theoretical concentration and flow rate of heat transfer fluid.

058

Two Dimensional Composite Solar Concentrator : Singh R N, Mathur S S & Kandpal T C, " Proceedings Indian National Solar Energy Convention", Bhavnagar, p.123, (1978).

Geometrical concentration characteristics, of a composite parabolic trough with a flat plate and circular cylinder receiver are presented in this paper.

059

Using a Fin with a Parabolic Concentrator : Singh R N, Mathur S S and Kandpal T C,"International Journal of Energy Research, 3, 393, (1979).

The consequence of using a fin receiver, in focussing solar collectors, is examined in this paper and is found to have merit. An axially situated flat plate (fin) receiver will get concentrated solar radiation on both its sides. Therefore in this case, it will not be necessary to insulate either side of the receiver. Further the use of fin receiver, obtaining nearly equal temperatures on its two surfaces, may provide additional benefits from the point of view of thermal design.

060

Assembly and Alignment Errors of a CPT : Singh R N, Mathur S S & Kandpal T C, " Proceedings Indian National Solar Energy Convention", Bombay, p.99, (1979).

The effects of assembly and alignment errors, of a typical composite parabolic trough (CPT) on its geometrical concentration characteristics have been studied. The results are plotted graphically and interpretted to be helpful in the assessment of the tolerances of the concentrator.

061

Concentration Characteristics of Composite Parabolic Concentrators : Singh R N, Mathur S S and Kandpal T C , " Applied Energy " , 6,   315, (1980).

The variation of the geometrical concentration ratio, for cylindrical and flat plate receivers, with the strip width of a composite cylindrical parabola concentrator, has been examined numerically, in cases where the edges of the constituent strips lie on the parent parabola.

062

Design Stage Ray Optical Assessments of a Composite Parabolic Trough : Singh R N, Mathur S S and Kandpal T C, "Optica Applicata", 10 (3), 189 (1980).

A ray - optical evaluation, of the constructional and concentration characteristics, of a composite parabolic trough (CPT) with flat receiver is presented. The region of non-zero intensity in a focal plane, the intensity in the central solar image and the receiver intercept fraction for a typical CPT are studied, and the significance of the results at the design stage is pointed out.

063

Foam - Film Parabolic Cylindrical Solar Energy Concentrator : Sobirov O Yu & Gafurov A M, "Applied Solar Energy ", 15 (1), 25 (1979).

A method for fabricating a parabolic cylindrical concentrator surface, from high molecular and foam-polymer materials is considered. Experimental results indicate, that a foam-film parabolic - cylindrical concentrator, is equal in optical energy characteristics to glass and metal structures.

064

Research Applied to Solar Thermal Power Systems : Sparrow E M, Ramsey J W and Wehner G K, " Minnesota University, Minneapolis (USA) Report , TID - 26961",(1975).

The research and development efforts of the Minnesota/Honeywell team during the period July 1,1974 to December 31, 1974 are documented in this report. Experiments were conducted on the scale model parabolic trough collector module at a desert test site. The low temperature absorber tube used in the prior experiments, was replaced by a selectively coated heat pipe absorber tube. Collector performance was measured for absorber tube operation over a temperature range of 210 - 300 deg C. Auger electron spectroscopy studies, of the diffusion phenomenon in various solar absorber coatings, were continued. An experimental model of a solar boiler / heat exchanger has been designed and is being fabricated. The apparatus will serve to simulate a boiler / heat exchanger that would be a part of a solar thermal collector module. Various types of power and flow transients are to be investigated. Measurements were made of the thermal conductivity of a candidate pipe - line insulation. A computer program was written, and applied for determining the heat transfer characteristics of phase change heat storage media. Preliminary data runs were made for a single phase heat storage system.

065

A Terrestrial Solar Thermal Electric Power System - Development of Basic Model System : Tanaka T, Sawata S, Tani T, Sakuta K & Horrigome T, "Proceedings Conference Physics of Solar Energy", Tripoli (Libya), 115, (1977).

A model system of a distributed collector power was constructed at the Electrotechnical Laboratory, Tanashi, Tokyo, (Japan), in the summer of 1975. The model system consisted of a cylindrical parabolic concentrator, heat transfer loop and storage - type heat exchanger, and has been used for evaluating the system performance, the dynamic characteristics of each subsystem and a whole system. This paper presents the performance of each subsystem used in the model system and summarizes results obtained from tests made with the whole system.

066

Analysis of a Cylindrical Parabolic Focussing Collector for Distributed Collector Power System : Tanaka T, Tani T, Sawata S, Sakuta K and Horrigome T, "Policy Anal. Inf. Syst. 2(1), 249 (1978).

A solar collector used in a solar thermal electric power system, is one of the most important components because it has a significant influence upon the efficiency and construction cost of the plant. The analysis of a cylindrical parabolic focussing collector, to collect solar energy is studied in this paper. The effects of various factors such as reflectance of reflector, mass flow rate, spectral characteristics of selective surface etc. on the collector performance, are presented.

067

Performance Analyses of a Cylindrical Parabolic Focussing Collector for Distributed Collector Power System : Tanaka T, Tani T, Sawata S, Sakuta K & Horrigome T, " Proceedings ISES Conference", New Delhi (India), (1978).

A solar energy collector, used in a solar thermal electric power system is one of most important components because it has a significant influence upon the efficiency and construction cost of the plant. Therefore, the analysis of cylindrical parabolic focussing collector to collect solar energy in a high quality state is numerically studied in this paper to provide the detailed design of a 1000 KWe pilot plant promoted in Japan, which would be operational by the late 1980. The results , the effects of various factors, i.e. spectral characteristics of a sensitive surface, etc. on the collector performance were explained.

068

Design Considerations for Parabolic - Cylindrical Solar Collectors : Treadwell G W, " Sandia Laboratories (USA) Report , SAND 76 - 0082 ",(1976).

The significant factors which influence the design of parabolic - cylindrical solar collectors are presented in some detail.

069

Test Results From a Parabolic Cylindrical Solar Collector : Treadwell G W, McCulloch W H & Rusk R S, "Sandia Laboratories (USA) Report ", SAND 75 - 5333, (1975).

This report represents and analyzes results from the testing of receiver tubes using a focussing parabolic cylindrical reflector, north - south axis, which was fully tracking. Various receiver tube design differences have been compared experimentally and analytically at temperatures upto 600 deg F (315 deg C). Included are results for an asymmetric receiver tube design, and data showing the influence of an evacuated glass vacuum jacket. The asymmetric design consists of an insulating strip on the surface of the glass jacket closest to the sun. Also tested,have been receiver tubes coated with Pyromark, black chrome on a dull nickel - plated tube, black chrome on a bright nickel - plated tube. Pyromark is a non - selective coating. The test results show significant improvement through the use of a low-order vacuum, the use of selective coatings, and the use of an as-symmetric design for high emmitance coating. The analytical results are within the range of test results and the model will be a valuable tool for further studies.

070

Selection of Parabolic Solar Collector Field Arrays : Treadwell G W,"Sandia Laboratories (U S A) Report", SAND 74 - 0375, (1975).

A technique has been developed whereby the number of columns and rows of parabolic solar collectors required to achieve a given task can be determined for steady state operational conditions.

071

Theoretical Study on the Performance of Horizontal Parabolic Cylinder Type Collector f o r Solar Cooling : Udagawa M & Kimura K, "Heat Transfer Jap. Research ", 4 (4), 1 (1975).

Studies are made on the performance of a solar heat collector for a house cooling system with an absorption refrigeration machine. The horizontal parabolic cylinder type collector is considered suitable to achieve the high temperature level(about 100 deg C) of heating media for operation of the absorption refrigeration machine using LiBr-water. Theoretical studies on the setting, the shape of the parabola, and the required area of the collector are described.

072

Design of a Tubular Heat Collector for a Solar Powered Installation with a Parabolo - cylindric Concentrator : Umarov G Ya, Abidov T & Trukhov V S, "Applied Solar Energy",10(1), 33 (1974).

A method for the determination of the basic thermal parameter of a smooth-wall tubular heat ex-changer for use as the heat collector in a solar power installation with a parabolocylindric con-centrator is developed. Results of some numerical calculations are plotted g r a p h i c a l l y and discussed.

073

Honeywell General Offices Solar System : Waters Donald E & Block Rojer F, "Proceedings ERDA Conf-erence on Concentrating Solar Collectors",Georgia (USA), p. 8 - 21 (1977).

This paper presents a description of an advanced solar energy HVAC system presently under constr-uction by Honeywell Inc. The solar system will provide 82% of the cooling energy, 53% of the heat-ing energy and 100% of the domestic hot water energy for a 100,000 sft office building in Minnea-polis, Minnesota. The solar system components are described including the concentrating collectors, the heating and cooling subsystems and the solar system controls. The technical rationale for the selection of the collector, the Rankine cycle turbine, the heat transfer fluids and the thermal storage system is discussed. Annual and peak load system performance predictions are also presented.

074

Transient Energy Removal in Cylindrical Parabolic Collector Systems : Wijeysundera N E, "Proceed -ings 13th IECE Conference ", San Diego, (USA), p. 1570 (1978).

This paper describes two heat transfer models for the transient energy removal in cylindrical para-bolic collectors. The models are used to study the transient behaviour of cylindrical p a r abolic collectors and their respective accuracies of prediction are compared. A single design parameter, namely, the collector time constant on other collector design variables is studied using the heat transfer models refered to above.

075

Solar Parabolic Trough Forming Process : Final Technical Report : Williams O G and Skaggs R L , " Report ALO - 6158 - 1, Contract EG - 77 - G - 01 - 41587, May 3,(1978).

Eight feet and six feet long solar parabolic shells were formed from thin aluminium sheet by using this entirely new forming process, exceeding 81, concentration. Expanding on work previously done many variables were investigated. Using one foot long trough samples, a high degree of slope accu-racy was obtained. One laser beam scan sample showed a 75/1 concentration ratio based on receiver diameter and a 24/1 concentration ratio based on receiver circumference. The forming process eli-minated the need for ribs, bones or other devices now used to maintain parabolic shape of a para-bolic shell. The process in brief involves (1) Alloy aluminium (2) Moments (3) Temperature (4)Time (5) Thickness (6) Bearings (7) Flange angle bend (8) Hardness (9) Lengthwise sag (10)Material flat-ness and (11) Aperture control. A large, forming fixture was built. Although it was sufficient to prop the shells needed, it should be modified and improved. The process allows for large manufac-turing tolerances which were helpful with this fixture. The small, sample testing fixtures were satisfactory. Flanges add strength and provide a flat surface across which the end and receiver support is attached. This simplifies and reduces the cost of the end supports.

076

Thermal Conditions in the Cavity Radiation Receivers of Mirror Type Solar Energy Equipment, As De-rived from the Structure of the Radiation Field : Zakhidov R A & Teplyakov D I, " High Temperature' (USA), 4 (2), 249, (1966).

A method is given for calculating and measuring the distribution of the energy incident on the wall of a three dimensional receiver located near the focus of a parabolic concentrator. Experimental results and the optimal thermal conditions in cylindrical cavity receivers are given.

Fresnel Lens and Reflector Systems

077

<u>Distortion Properties of Fresnel and Conventional Opthalmic Prisms</u> : Barkan E and Kapash R J ,
" Journal of Optical Society of America " , 61, 686A, (1971).

The relative distortion characteristics, of Fresnel and conventional opthalmic prisms were studied by means of a computer ray trace. The optical configuration, was chosen to simulate as nearly as possible, the actual condition of a prism used by a human patient. By use of the set of five co-effients defined by Ogle [Kennetu N. Ogle, JOSA 41, 1023 (1951)] the distortion of both p r i s m types were computed, as a function of both, prism power and front surface (base) curvature. The two types of prisms were similar with respect to the three un-symmetric distortions (e.g. curvature of lines normal to the base - open meridian). The conventional prism, was found to be slightly better in two cases, and slightly worse in the third, the differences, however, should prove to be clinically insignificant. With respect to the two remaining distortions, horizontal and vertical symmetrical magnification, the Fresnel prism proved far superior. The improvement provided b y Fresnel prisms is shown to increase rapidly with front surface curvature. Prescription o f such prisms for monocular use without fear of incurring large differential magnification errors , is possible.

078

<u>Design & Construction of Fresnel Optics for Photoelectric Receivers</u> : Boettner E A & Barnett N E ,
" Journal of Optical Society of America ", 41, 849, (1951).

The design and construction of Fresnel type optics, particularly suitable, for use with area type photoelectric receivers are described. Such optics have several advantages, o v e r conventional lens and mirror systems ; large aperture systems may be constructed cheaply and are light in weight the radiant energy can be distributed uniformly over the sensitive surface to avoid local satura-tion, and the optics can be economically custom - made for each application. The design procedure is explained, and a method given for the construction of these Fresnel type lenses from plastic,in a modestly equipped machine shop. The design prediction and experimental performance d a t a a r e compared for lenses designed for three different applications.

079

<u>A Novel Square Fresnel Lens Design for Illuminating Circular Solar Cell</u> : Burgess E L, Marchi D L,
& Walter H, "Optical Engineering", 17, 299 (1978)

This new square Fresnel lens can focus radiation at high densities, on many photovoltaic arrange-ments, and can decrease system costs to a minimum. This new design uses a focal point, related to the corner of the Fresnel lens, which is different from that originating from the central part. An acrylic lens has been produced, by compressing a suitable material in a metallic mould. This lens has a transmission efficiency of 79.4 % within the spectral range utilized by solar cells.

080

<u>Design Construction and Test of a Collector System Using a Linear Asymmetric Fresnel Reflector</u> :
Butz J R, Fukuta N & Armstrong J A, "Proceedings of ISES (American Section),Conference" Orlando, (USA), (1977).

The concept of a linear asymmetric Fresnel reflector, to be used in a concentrating collector was explored in a research and development program. A design handbook which presents detailed   geo-metric relationship, in a parameterized form was produced. A prototype reflector a n d receiver were fabricated and tested.

081

<u>Theoretical Aspects of Solar Energy Collection with a Cylindrical Step Lens</u> : Cheema L S,
" Proceedings of ISES Conference", Los Angeles, (USA), (1975).

The present study involves the theoretical investigation, of various aspects of solar energy con-centration, by refraction through a cylindrical step lens. It has been found that for the concen-tration of solar energy by a lens, having a non-zero negative coma,combines the advantage of apla-natic lens with simplicity of design. A procedure for finding the optimum size of t h e receiver has also been discussed.

082

<u>Fresnel Lens Concentrating Collector</u> : Claxton R J, "Proceedings of ERDA Conference on Concentrat-ing Solar Collectors, Georgia, (USA), p. 2 - 57, (1977).

The Northrup Concentrating collector is a polar axis collector with equitorial tracking capabili-ty. The collector consists of a steel trough housing, with a black chrome plated, copper absorber tube and a curved Fresnel lens. Applications include heating, cooling and domestic hot water,some of which are among the largest solar installations in the world. Recent engineering advancements have overcome some of the initial difficulties with the suntracking system, and all installations are being upgraded with this new hardware. Meanwhile, research and development continues o n t h e next generation collector. Major advances to be expected include a more efficient Fresnel lens and an advanced receiver assembly.

083

<u>Efficiency of Fresnel Lens Solar Collector</u> : Cobble M H, Smith P R and Boyes J D,"Proceedings 23rd Institute of Environment Science Annual Technical Meeting",Los Angeles,California(USA),p.147,(1977)

A theoretical analysis is made, of a series of Fresnel lens collectors, in parallel c o n n e c t i o n that track the sun to produce wet steam. Equations are developed for the boiler efficiency, the steam quality and the fluid temperature distribution to be expected in the fluid. A series of experiments done at three mass flow rates and three pressures, verify the theoretical predictions.

084

The Linear Fresnel Lens Solar Concentrator Transverse Tracking Error Effects : Cosby R M , " NASA Contractor Report", No. NASA CR - 2889, (1977).

This report details an analysis of the solar concentration performance of a line focussing flat base Fresnel lens in the presence of small transverse tracking errors. Simple optics and ray tracing techniques, are used to evaluate the lens' solar transmittance and focal plane imaging characteristics. Solar transmission losses by Fresnel reflection and material absorption are included & an analysis of groove edge losses is presented. Computer - generated example data are presented for lenses with parameters corresponding to two NASA test articles : a 0.56 m wide, f/1.0 lens and a 1.83 m wide, f/0.9 lens.

085

Concentration Characteristics of a Fresnel Solar Strip Reflection Concentrator : Cosby R M, " NASA Report", NASA - CR - 120336, (1974).

An analysis of the concentration characteristics of an ideal Fresnel solar strip reflection concentrator is presented. Required reflection surface angles are calcualted as a function of position and concentrator focal length. The intensity distribution, of the concentrated solar radiation in the focal plane, is determined and characterized by a maximum intensity, a central solar image width, a full width at half maximum and a maximum beam spread. The fraction of the total concentrated solar flux intercepted by a target, placed in the concentrator focal plane is determined as a function of target width.

086

Solar Concentration by Curved Base Fresnel Lenses : Cosby R M," Proceedings of ERDA Conference on Concentrating Solar Collector, Georgia, (USA), p. 2-61, (1977).

The solar concentration performance idealized curved base line focussing Fresnel lenses is analyzed in this paper. A simple optical model, for studying the solar performance effects of base curvature, f - number of the lens, small transverse tracking errors and slight defocussing, is des-cribed. Computer generated example data, for a lens with a 36 in (91.4 cm) wide aperture, indicates that substantial improvements occur in the solar concentration characteristics over the flat lens case, when base curvature is introduced. Required target receiver widths are significantly reduced. Performance sensitivities to small transverse tracking deviations ($0°-2°$) improve when the lens base is curved, in contrast to the increased degradation of the image profile characteristics, with slight defocussing - (± 2% of the focal length).The computed lens solar transmission is generally in the 85 - 88% range. In comparison to an f/1.0 flat lens, selection of a curved base Fresnel lens with $0.8 < f$ number < 1.0 and curvature radius $\leq$ focal length, is predicted to improve the overall optical performance, while decreasing support structure and tracking system requirements.

087

Primary Aberrations of Fresnel Lenses : Delano E, "Journal of Optical Society of America ", 64(4), 459 , (1974).

Mathematical expressions derived for the primary aberrations, of a thin flat Fresnel lens (grooved on both surfaces, in general) as functions of the constructional parameters, are convenient for the analytic designs of optical systems, containing one or more Fresnel lenses. In general, such systems have five primary monochromatic aberrations of the Seidel type, plus a sixth that is identically zero for ordinary lenses. The new aberration (called line coma, to distinguish it from ordinary circular coma), bears the same relation to sagittal and tangential coma, that Petzval curvature bears to sagittal and tangential astigmatism. Moreover, line coma is independent of stop position, whereas Petzval curvature varies with the stop position, unless the line coma is corrected. The primary chromatic aberrations and the aberration contribution, of Fresnel aspherics, are the same as for ordinary lenses. Specific applications of the theory are discussed.

088

Primary Aberrations of Meniscus Fresnel Lens : Delano E, "Journal of Optical Society of America ", 68 (10) , 1306 , (1978).

Mathematical expressions, for the primary aberrations of a thin meniscus Fresnel lens (grooved on both surfaces), are derived and compared to the corresponding for the case of a thin flat Fresnel lens. It is shown that the only aberration that differ, are the spherical aberration and the line coma, both of which are independent of stop position. Moreover, a meniscus Fresnel lens has an additional degree of freedom available, for aberration correction. Some applications of the theory are discussed.

089

Primary Aberration Contributions for Curved Fresnel Surfaces : Delano E, " Journal of Optical Society of America ", 68 (10), 1306 , (1978).

Mathematical expressions are derived, for the six monochromatic primary aberration contributions , of a curved Fresnel surface. Some special cases are noted and a numerical example given .

090

Performance Testing of the McDonnell Douglas Fresnel Lens Solar Collector : Dudley V E & Workhoven R H, " Sandia Laboratories (USA) Report ", SAND 78 - 0625, (1979).

This report summarizes the results of test, performed on the McDonnell Douglas Fresnel Lens rotating array solar collector, at the Mid. Temperature Solar Systems Test Facility. Test objectives are defined, test procedures are described and test results and conclusions are given .

091

Design Construction and Performance of a Fresnel Lens for Solar Energy Collection : Garg H P, Mathur M L and Verma V K,"Proceedings of ISES Conference ", New Delhi (India), p. 1328, (1978).

This paper describes a study on circular type of Fresnel lens with steps on front for collection of solar energy. The effects of step width and plate thickness on design parameters, have been analysed and based on this, a design curve relating the radius to step angle has been drawn. The relationship between focal length and maximum attainable radius has been investigated. A method of manufacturing the Fresnel lens on a lathe is described. The experimental results with the lens manufactured are given.

092

Superheated Steam Generation in a Fresnel Lens Concentrating Collector : Glass R E & Smith P R,"ASME Paper No. 79 - WA/Sol. - 21",(1979).

Concurrent analytical & experimental studies were carried out on linear Fresnel lens and solar collectors to determine their potential for producing superheated steam. Nine tracking Northern collectors were connected in parallel as a single boiler. The results of the experimental study were used to verify the computer model developed to study Fresnel lens collectors. The computer model closely predicted the temperature profiles through the collector in all cases including the occurance of the critical heat flux point.

093

Solar Optical Analysis of a Mass Produced Fresnel Lens : Harmon S, "Solar Energy, 19, 105, (1977).

In this study experimental work was done to characterize the intensity distribution of the 3M Co lens. The data from this work was used with computer method, to produce a simulation of the intensity field anywhere within the theoretical focal cone. The experimental work consisted of two parts. In the first part, the intensity as a function of radial distance was plotted for several radii varying the parameters of polar angle and focal distance. These plots were constructed using a silicon photovoltaic cell and solar irradiation. The second part of the experimental work involved using a laser device to determine the deviation in the convergence of normal incident radiation to a single focal point.

094

Analytical & Experimental Evaluation of the Plano-Cylindrical Fresnel Lens Solar Concentrator : Hastings L J, Allums S L & Cosby R, " Proceedings of ISES Conference, Winnipeg (Canada) - S h a r i n g the Sun", 2, 275, (1976).

An analytical and experimental evaluation of line focussing Fresnel lenses with application p o t e n tial in the 200 - 300 deg C range is in progress. Analytical techniques are formulated to a s s e s s the solar transmission and imaging properties of a grooves down lens. Experimentation is based primarily on a 56 cm wide F/1.0 lens. A sun tracking heliostat provides a non moving solar s o u r c e. Measured data indicate more spreading at the profile base than analytically predicted resulting in a peak concentration 18% lower than computed peak of 57. The measured and computed transmittances are 85 and 87% respectively. Preliminary testing with a subsequent lens (1.85) indicates that modified manufacturing techniques correct the profile spreading problem and enable improved analytical - experimental correction.

095

Analytical and Experimental Investigation of a 1.8 by 3.7 meter Fresnel Lens Solar Concentrator : Hastings L J, Allums S L & Jensen W S, "Proceedings of ISES (American Section) Conference, Orlando, (USA), (1977).

Line focussing acrylic Fresnel lenses with application potential in the 200 - 300 deg C range are being analytically and experimentally investigated. The measured solar characteristics of a 1.8 by 3.7m lens and its utilization in a solar collection mode are summarised in this paper. A peak concentration ratio of 64, with 90% of the transmitted energy focussed into a 5 cm width, was achieved and demons - trated the feasibility of the Fresnel lens solar concentrator concept.

096

An Analytical and Experimental Evaluation of a Fresnel Lens Solar Concentrator : Hastings L J, Allums S L and Cosby R M, "NASA Technical Memorandum No NASA TMX - 73333", (1976).

An analytical and experimental evaluation, of line focussing Fresnel lenses with application p o t e n tial in the 200 - 370 deg C range is in progress. Analytical techniques have been formulated, to assess the solar transmission and imaging properties of a grooves-down lens. Experimentation was based on a 56 cm wide, f/1.0 lens. A sun tracking heliostat provided a non moving solar source. Measured data indicated more spreading at the profile base than analytically predicted, resulting in a peak concentration 18 percent lower than the computed peak of 57. The measured and computed t r a n s m i t tances were 85 and 87 per cent respectively. Preliminary testing with a subsequent lens indicated that modified manufacturing techniques corrected the profile spreading problem and should enable improved analytical experimental correlation.

097

Performance Characteristics of a 1.9 x 3.7 m Fresnel Lens Solar Concentrator : Hastings L J & Allums S L,"Journal of Energy", 3(2), 65 (1979).

Line-focussing acrylic Fresnel lenses, with application potential in the 200 - 370 deg C range h a v e been investigated analytically and experimentally, compared to other concentration concepts,the technique is relatively unexplored. However, the acrylic lens is durable and adaptable to mass production techniques. Initially an analytical model was developed and used in conjunction with subscale examination to define optical performance potential of the Fresnel lens concept. The present investigation was conducted in three phases. First, the lens transmission and focussing characteristics were analytically modeled and bench tested at the component level. Second, the lens was assembled in a full scale configuration and its optical performance checked to assess the impact of the assembly process. Third, the lens was interfaced with the receiver tube assembly and operated in a solar collection mode.

098

High Efficiency Fresnel Reflectors for Solar Energy Concentration : Heath A R,"NASA Contractor R e p-
ort NASA TND - 1358," (1962).

A study has been made of the Fresnel reflector, and two variations of this reflector, for use as solar
energy collectors. One variation is the conical Fresnel reflector, in which the serrations are located on the inner surface of the cone. The second variation consists of a reflector plane that is not
perpendicular to the incoming light rays. Segments of this reflector can be used to form a pyramidal
collector, which combines the desired flatness of the basic Fresnel reflector, with the high effici -
ency of the conical Fresnel reflector. Each of the variations can have an efficiency greater than the
basic Fresnel reflector, when the proper apex angle is chosen.

099

Zur Definition und Berechnung des Astigmatismus von Fresnel - Flachën/(Defining & Tracing the Astigmatism of Fresnel Surfaces): Hoffman C & Klebe J, "Optik", 25, 389, (1967).

The Astigmatic ray tracing formulae, for any desired Fresnel surface are derived, with the aid of the
definition, of effective principal curvatures. It is found, that the meridional curvature of field
is a direct function of the obliquity σB of the principal ray. This formulæ thus determined   are
used to the astigmatism of a Fresnel surface with a plane background.

0100

Efficient Fresnel Lens for Solar Concentration : Kritchman E M, Friesem A A and Yekutieli G, " Solar
Energy ", 22, 119, (1979).

This paper presents a new design, for an efficient linear Fresnel lens, capable of providing h i g h
concentration for a given acceptance angle. The predicted performance of the lens is comparable to
that of the "ideal" reflector, while providing greater reliability at a lower cost.

0101

Highly Concentrating Fresnel Lenses : Kritchman E M, Freisem A A and Yekutieli G, " Applied Optics",
18, 2688, (1979).

A new type of Fresnel lens is described, that is capable of concentrating solar radiation very near
the ultimate concentration limit. The differential equations that describe the lens, are solved to
provide computed solutions, which are then checked by ray tracing techniques. The performance (efficiency and concentration) of the lens is studied and compared, with that of the flat Fresnel l e n s
showing that the new lens is preferable for concentrating solar radiation.

0102

Cast Acrylic Fresnel Lenses for Solar Energy Concentration : Lingle C P, " Proceedings Seminar on
Testing Solar Energy Materials and Systems", Gaithersburg, (USA), p. 229, (1978).

The physical and mechanical characteristics of cast acrylic, with emphasis on its use in Fresnel lens
concentrators are addressed. Experimental evidence is cited which confirms that cast acrylic is resistant to long term degradation (optical and physical) when subjected to field weathering condi -
tions. Various configurations of acrylic Fresnel lenses that have been cast, are discussed i n t h e
context of their most common system applications. Other applications of cast acrylic are cited which
also entail prolonged outdoor exposure and in which, degradation would have impaired the material ac-
ceptability. Mounting configurations of acrylic lenses are discussed. The economic importance of
weatherability is analyzed on a system life cycle basis, to demonstrate the significant advantages of
cast acrylic over alternative materials.

0103

Prospects for Using Fresnel Lenses for the Concentration Systems : Lidorenko N S, Zhukov K V; Nabiul-
lin F Kh & Tver' yanovich E V, " Applied Solar Energy ", 13(4), 17 (1977).

The state of development work on Fresnel lenses is reported and the possibility of using t h e m i n
solar installations is analyzed.

0104

Practical Solar Concentrator : Lowery G W, Northrup L L & Hancock O, " ASHRAE Transcripts ", 82 ,(2),
416 , (1976).

A new solar energy collector has been developed that utilizes a specially designed, linear Fresnel
lens for concentrating sunlight on to an absorber heat exchanger. A key to both the performance and
the manufacture of this efficient collector, is the technique for extruding the ultra violet-resist-
ant acrylic lens. As the solar radiation passes through the lens it is refracted by a computer designed array of prisms that concentrate the rays on to a copper absorber tube. The present concen -
tration ratio is approximately 6 : 1.

0105

Thin Sheet Plastic Fresnel Lens of High Aperture : Miller O E, McLeod J H and Sherwood W T,"Journal
of Optical Society of America", 41, 807, (1951).

Fresnel type lenses of high precision and excellent surface quality, have been made in thin plastic
materials. The prismatic elements are very fine - about 0.003 to 0.005 in and are not visible to the
average unaided eye. A high degree of correction for spherical aberration has been achieved. Moulded
from high precision molds, lenses have been made in diameters of 2 to 15 in and focal lengths of $2\frac{1}{2}$
to $22\frac{1}{2}$ in. Relative apertures in excess of f/1.0 have been made. These lenses have found many applications as light collecting elements where weight and space are limited. Such applications include uses for large condensers, large field lenses in finders, camera viewing screens and translucent
screens for projection.

0106

Linear Fresnel Lens Concentrators : Nelson D T, Evans D L & Bansal R K, " Proccedings ISES Confer - ence ", Los Angeles, (USA), (1975).

The use of the linear Fresnel lens, as a seasonally adjusted or one axis tracked solar concentrator is investigated. The lens is considered to be a series of parallel strip prisms, each of triangular cross section, molded into a single element. For seasonally adjusted operation, the axis of the concentrator system is oriented in an east - west direction, with the lens normal considered to be aimed towards the sun at noon. For one-axis tracking, rotation about a longitudinal east - west axis is permitted. Ray tracing techniques applied to each prism are used to show that this type of lens provides very good collection efficiencies, within one hour of solar noon. Detailed calculations were carried out every half hour from noon to 6 p.m. on the 21st day of each month for a lens having a relative aperture of 1.0. Individual prism widths were calculated so that the bundle of light through each prism at noon would, after refraction, just fill a width of 1.0 cm at the absorber.

0107

Cast Acrylic Fresnel Lens Solar Collector: Nixon G, " Proceedings ERDA Conference on Concentrating Solar Collectors ", Georgia, (USA), p. 5 - 33, (1977).

A description of Fresnel lens solar collectors, prepared for solar thermal and photovoltaic applica - tions is presented with arguments for this refracting method of concentration, compared with that obtained by reflection. Acrylic material used in a casting process has been utilised successfully by SWEDLOW INC. in producing such Fresnel lenses and arguments for both, this selected material and the noted process are delineated. A status report of lens production and field experience is made along- with lens price projections. Design techniques, performance characteristics and the next generation of Fresnel lens development are described.

0108

A Practical Concentrating Solar Collector : Northrup Jr. L L and O'Neill M J, " Proceedings ISES Conference Los Angeles (USA), (1975).

A new concentrating solar energy collector has been developed to provide excellent collection efficiency in the temperature range of 100 - 800 deg F. The efficiency of the new concentrator, and the daily heat collected, have been measured. An approximate cost analysis has also been made to assess the economic viability of the system. The collector uses a special computer designed Fresnel lens, to concentrate incoming solar radiation upon the small copper absorber tube, which is coated with a selective optical coating. Standard collector modules utilized a one foot wide frontal aperture and are ten feet long. The lenses are laser - checked for optical performance, the reflectors aid in collecting diffused solar radiation, fibreglass insulation is enclosed within a painted steel housing . The overall collector unit is simply efficient and attractive.

0109

Compound Lens Solar Energy System : Northrup Jr. L L, " US Patent ", 4, 022, 186/May 10 (1977).

A   solar collector includes an elongated heat exchange element onto which the sun's rays are to be directed. A compound concentration lens system is mounted to direct the sun's rays o n t o the heat exchange element and includes a center refractor, extending the length of the element for refracting the sun's rays onto the portion of the heat exchange element, facing the sun. A pair of reflectors extends on each side of the refractor, said reflectors, each having a plurality of parallel   prisms arranged one above the other for reflection of light onto the sides of the heat exchange element.

0110

A Unique Fresnel Lens Solar Concentrator : O'Neill M J, "Proceedings of ISES Silver Jubilee International Congress", Georgia (USA), (1979).

A new Fresnel lens solar concentrator has recently been developed, which offers higher transmittance, higher concentration and shorter focal length than prior lenses. This superior performance is the result of optimizing the design of each prism within the lens. Each prism achieves maximal transmittance by utilizing equal angles of incidence, with the solar rays at both the front and the rear prism faces. The theory, configuration and performance of the new concentrator is described. Direct comparisons of optical performance with prior lenses are presented. Prototype collector test results , which have confirmed theoretical performance of the new concentrator, are also presented.   Finally, the economics of a thermal energy collection system (scheduled for production in 1980) based upon the new concentrator is presented, which indicates competitiveness with fossil fuel energy costs.

0111

Step Lenses and Step Prisms for Utilization of Solar Energy : Oshida I, " Proceedings U N Conference on New Source of Energy ", Rome (Italy), 4 , 598 , (1964).

In this paper, the possible applications of Fresnel lenses, as well as cylindrical step lenses a n d step prisms, are investigated.

0112

The Effects of Defocus on a Cylindrical Fresnel Lens : Robertson C E & Banker J H, "Sandia Laboratory (USA) Report ", SAND 75 - 0358, (1975).

This report presents a graphic demonstration of image defocus and movement in case of a fixed cylindrical Fresnel lens with the diurnal apparent motion of the sun. The effects of defocussing by changing the lens image plane distance are also illustrated. Sensitometric techniques are employed to show relative intensity changes in the image due to focus changes.

0113

A New Hybrid Fresnel Lens for Solar Concentration : Sala G & Lorenzo E, " Proceedings I S E S Silver Jubilee International Congress", Georgia, (USA), (1977).

A new type of hybrid Fresnel lens is proposed which consists of a sheet glass acting as a support for

a thin film of transparent silicone, which acts as the refractive layer. This new lens has the fol -
lowing advantages : high rigidity, and high resistance to abrasion and to radiation. Another signi -
ficant advantage is its low cost which is about half of the cost of acrylic lenses of the same size .
This is so although transparent silicones are not mass produced in the same quantities as a r e t h e
methacrylates. Finally, the equipment necessary to fabricate the hybrid lenses is quite simple and
cheap, because the curing of silicones at atmospheric pressure gives an excellent reproduction of the
mold. Fresnel lenses using the silicone resin - Sylgard 182, of Dow Corning have been fabricated and
tested. A square hybrid Fresnel lens of 0.40 x 0.40 m² intended for photovoltaic conversion has also
been designed. The estimated efficiency of the lens without considering the dispersion loss is 86%.

0114

Thermostatically Controlled Non-tracking Type Solar Energy Concentrator : Shimada K, " U S Patent",3,
915, 148/Oct. 28, (1975).

A solar energy concentrator composed of an array of cylindrical Fresnel lenses, all of which are fix-
edly aligned, in the east - west direction was developed. Each lens concentrates the sun rays a n d
forms a line image, which extends in the east - west direction and are spaced apart in the n o r t h-
south direction. Each line image, focusses on to not more than two of the channels, w h i c h absorb
heat of the concentrated sun rays. Each channel has a thermostatically controlled valve, which controls fluid flow through the channel to take place,only when the channel temperature or t h e fluid
temperature exceeds a threshold level.

0115

Some Geometrical Design Aspects of a Linear Fresnel Reflector Concentrator : Singh R N, Mathur S S &
Kandpal T C, " International Journal of Energy Research ", 4, 59, (1980).

A somewhat new approach to the design of solar concentrators, of Fresnel reflector geometry is   outlined. The constituent mirror elements of the concentrator surface are characterized by three para -
meters shift,tilt and width. The evaluation of these parameters and the concentration characteristics
are investigated on the basis of a simple ray optical model.

0116

Defocussing and Transverse Tracking Error Effects in the Case of a Vee Type Linear F r e s n e l Lens :
Singh R N, Singhal A K, Mathur S S and Kandpal T C, "Proceedings of National Solar Energy Convention"
Annamalai (India), p. 237, (1980).

It has been found advantageous to approximate a linear curved base Fresnel lens, by a double convex
inverted Vee type Fresnel lens. We have earlier studied the concentration characteristics of a perfectly tracking inverted Vee type Fresnel lens, but errors of various types were not taken into ac -
count. However, in actual practice, the concentration characteristics are affected by the presence
of transverse tracking errors and also due to receiver defocussing errors. In this paper the effects
of both these errors have been presented for a typical concentrator receiver system.

0117

A Linear Solar Concentrator System   ; Spitzberg L A and Williams J K, " Proceedings of ISES Conference ", Los Angeles (USA), (1975).

A water heating and storage system with a linear solar concentrator has been designed and constructed.
This system uses a large Fresnel lens array with one dimensional tracking to focus the direct insolation on a linear absorber. The heat transfer equations have been analysed and solved to predict the
system efficiency and are compared with field results.

0118

A Solar Strip Concentrator : Szulmayer W, " Solar Energy ", 14, 327, (1973).

A solar concentrator, capable of focussing solar heat on a linear target viz. a rod or a water pipe,
consists of a transparent extruded plastic strip 6 in wide and 0.04 in thick. A pattern of prismatic
parallel grooves and ridges produces the required beam refraction. Temperatures between 140-290 deg F
were produced in exploratory heating experiments, which indicate the possibility of using the strips
for water heating, steam production, regeneration dessicants (silica gel) as well as thermo-electric
power generation. Some aspects of design, production and aperture of the strips are discussed.

0119

Fresnel Condensers for Cylindrical Solar Concentrators :Tyagi R C, "Proceedings of National S o l a r
Energy Convention ", Bhavnagar (India), p. 148, (1978)

Based on the principle of Fresnel condenser, a one dimensional concentrating system having a width of
one meter has been designed for a geometrical concentration ratio of one hundred. One prototype concentrator has been fabricated in 'Perspex" plastic and in glass by pressing a master, cut with optical accuracy on a mild steel roller, at an elevated temperature. The design of the master and t h e
technique of transferring the pattern even on to glass, makes this process extremely economical, as
it can be directly adapted for rolling out any length of the condenser lens, like the figured window
glass. Apart from its low cost, the glass condenser will be radiation and abrasion resistant and ideal for out door applications.

0120

Linear Fresnel Lenses : Umemoto Y, "Sun World", No. 6, p. 10, (1977).

This article describes a new extra-large, low cost Fresnel lens which can be used in a variety of thermal solar energy applications. The lens is made of methacrylic resin, has 80% transmission a n d is
only 7-10 mm thick. It is expected that the lens will sell for about $ 20 to $ 40 per square meter.

0121

Linear Echelon Refractor/Reflector Solar Concentrators : Vanderwerf D F, Anderson R H & Appledorn R H,
"Proceedings SPIE Conference", 161, 23 (1978)

This paper describes some new configurations for linear incremental or Fresnel refractors and reflectors for solar energy concentration. The first system, a reflex type lens, uses crossed linear echelon elements and has convergence power in two dimensions. The refractor/reflector concentrates solar radiation to a spot focus. Large area solar concentrators can be sectionally constructed to provide high power solar flux concentration. The second type of echelon reflector forms a linear focus of incident solar radiation. The reflector is selectively tilted with respect to incoming solar radiation such that the design eliminate all riser step blockage of radiation at the reflecting echelons. This results in a higher efficiency concentrator than is achievable with a normally oriented linear focus concentrator. Design parameters and a ray-trace analysis are presented for both concentrator systems.

0122

Fresnel Lenses for Collecting Solar Energy : Weinz E A,"Industrial Diamond Review ",p.119,Apr.(1979).

Some suggestions as to how gratings and Fresnel lenses and mirrors might be used in soalr energy technology are presented in the first part of this article. The advantages of such components derive from their slender configuration, the use of cheap, optically transparent plastics, and the simplicity of automatic reproduction. The concluding part of the article presents practical guidelines for the diamond machining of brass Fresnel lens moulds and for the direct machining of perspex and o t h e r lens materials.

0123

Collection and Concentration of Solar Energy Using Fresnel Type Lenses : Wilson R F," N A S A (USA) Report No. NASA - CR - 14294", (USA),(1975).

The efficiency of collecting solar energy using a Fresnel type lens, was measured for two different collectors. A flow collector utilizes the temperature difference and heat capacity in water measurements, to determine the amount of absorbed energy retained from sun rays passing through the Fresnel lens. A static collector is a hollow copper box filled with vegetable heating oil, for absorption of focussed solar radiation.

Russell's Fixed Mirror Solar Concentrator

0124

Fixed Mirror Solar Concentration for Electrical Generation : Backus C E , " Proceedings of 10th IECE Conference ", (1975).

The fixed mirror solar concentrator (FMSC) is a recent discovery that has a unique geometry that is attractive for solar electric generation systems needing concentration. Its geometry is such that the mirrors can be built in a fixed position and supported directly on the ground. The absorber tube is moved to stay in the focal line, which is a sharp focus, regardless of the sun's direction . The reflector is a series of long narrow strips of flat mirrors along a cylindrical trough. T h e edge losses due to its faceted nature are found to be less than 10% of the total energy. The concentration ratio expected in a practical FMSC is about 30 - 50 .

0125

Fixed Mirror Solar Concentrators for Central Power Generation : Backus C E, Bansal R K & Russell Jr. J L , " Proceedings of ISES Conference ", Los Angeles (USA), (1975).

A theoretical analysis has been made of the FMSC to determine its characteristics. A major objective was to minimize the losses inherent in this geometry due to the light impinging directly on to t h e "edges" between the slats and the light reflected onto these edges. Parametric analyses have. been performed to determine edge losses, adjacent mirror shadowing and hourly, daily, monthly and annual variation in collected energy. The considerations of edge losses indicated that the window for efficient light collection with FMSC's is only 60° to 90° wide depending upon the FMSC parameters and orientation.

0126

Solar Collector Field Subsystem Program on the Fixed Mirror Solar Concentrator ; Final Report March 28, 1976 to Sept. 30, 1976 : Eggers G H, Housman J J, Openshaw F L and Russell Jr. J L, " G e n e r a l Atomic Company (USA) Report GA - A 14209 (Rev.) UC - 62", (1976).

This report describes the work done by   General Atomic Company (GA) to complete the preliminary design of a fixed mirror solar concentrator (FMSC). The objective of this work was to establish a n FMSC preliminary design, based on a General Atomic patent, that would deliver approximately 2×10^6 Btu/day to the Sandia Solar Total Energy System. The complete design requirements and the preliminary design is given and the transient performance of the system is discussed. Specifications of the sensors, instrumentation requirements and data process interfaces were prepared to meet the requirements established for control, performance measurements and system/component diagnosis. Some results of an economic analysis of the FMSC projected into a commercial sized facility of 10 MW are also presented.

0127

Solar Collector Subsystem Program on the Fixed Mirror Solar Concentrator Design & Tests : Eggers G H, Housman J J, Openshaw F L & Russell Jr. J L, "General Atomic Co.(USA) Report GA-A14595(UC-62)",(1977).

The first part of this report describes the additional testing and design work performed in Phase - II of the Solar Collector Field Subsystem Program on the Fixed Mirror Solar Concentrator (FMSC). The second part summarizes the test and test results achieved during the test period Sept.30,1976 to Jan.31,1977.

0128

Characteristics of the Concentrated Solar Flux Produced by the FMSC Prototype : Harmon S Y, Backus C E and Pinon R, " Proceedings of ISES Conference ", Winnipeg (Canada), ' Sharing the Sun', 2, 291, (1976).

The present paper deals with the characterisation of the concentrated flux field produced by a prototype model of a Fixed Mirror Solar Concentrator (FMSC) constructed by General Atomic Co. The intensity profiling showed the focal zone of the FMSC to have a fairly well defined symmetrical focus. The intensity in the focal zone varied a maximum of 30% along the z-axis of the concentrator. The radiometric analysis indicated a large part of the flux in the focal zone to be within a width of 2 cm.   Work with solar cell arrays indicates the FMSC to be capable of producing concentration ratios of around 20.

0129

Fluid Selection for a 100 MW(e) Line Focus Solar Central Power Station : Neill J M & Schuster J R, " Proceedings AIChE Symposium ", 75 (189), (1979).

A system evaluation study was conducted to enable selection of a heat transport and storage fluid for 100 MW(e) line focus solar central power station. The study based on the General Atomic Co. Fixed Mirror Solar Concentrators, addressed and resulted in the selection of draw salt over T h e r m i n o l 88 and sodium.

0130

Evaluation of a Fixed Mirror Solar Concentrator : Pinon R and Wood B D, " Proceedings of ISES (American Section) Conference", Denver, Colorado (USA), 2.1, 398, (1978).

A simplified analysis is presented for calculating the reflected energy distribution across the focal line for a given linear fixed mirror solar concentrator (FMSC). An expression is given for evaluating the effect of angular variations in the mirror slat alignment. A technique for m e a s u r i n g the energy distribution across the focal line is presented along with the results for a particular F M SC. The effects of dust on mirrors and mirror slat angular alignment are shown to be critical in the design of a FMSC.

0131

Central Station Solar Power Making a Viable System Economics : Russell Jr. J L, " Power Engineering", (November 1974).

Model of solar system with fixed mirror concentrator and movable heat collector pipe for e c o n o m i c solar power generation is presented.

0132

Principles of the Fixed Mirror Solar Concentrator : Russell Jr. J L, DePlomb Eugene P & Bansal R K, " General Atomic Co. (USA) Report GA-A 12902 (Rev.) ", (1977).

A new concept for a fixed mirror solar concentrator is described. The fixed, stepped surface of the proposed cylindrical mirror is designed to produce a sharply focussed line image regardless o f t h e incident sun direction. This is in contrast with the severe, off axis aberration (hence poor concentration factor) of a parabolic mirror for all sun directions other than on axis. It is shown t h a t (i) the heat absorber follows a circular path to remain in focus as the sun moves, (ii) the annual light losses from self shadowing of the "stepped" structure of the mirror surface can amount to less than $\sim 10\%$ for appropriate mirror orientation, (iii) by facing the mirror in a southerly direction (in the northern hemisphere), the winter deficit in sun input can be significantly compensated a n d (iv) practical concentration factors of around 50 can be achieved with a single stage of concentration.

0133

Preliminary System Analysis of a Fixed Mirror Solar Power Central Station (Final Report) : RussellJr. J L, Eggers G, Emrich W, Openshaw F & Walker W, "General Atomic Co(USA) Report GA-A13974(Rev.),"(1977).

A preliminary systems design and evaluation of the Fixed Mirror Solar Concentrator (FMSC) is presented. A computer program was written to model a preliminary system definition of a power   plant design for an FMSC. Analytical models were developed to simulate the performance and to model the cost of subsystems using both heat transfer salt (marketed as Hitec) and Carbon dioxide (CO_2) as the primary heat transfer fluids. It was found that the cost of the Hitec system was about one-half that of the Carbon dioxide system. The Computer code was exercised to obtain preliminary plant cost estimates & data on the sensitivity of the plant cost to changes in design parameters.

0134

Demonstration Model of Solar Power Concentrator : Russell J and Potthoff R, " General Atomic Company (USA) Report GA-A 1352", (1975).

This report presents the work performed by General Atomic Company and Arizona State University during calender year 1974 on the design and construction of a small scale (3.61 ft by 10 ft aperture) Fixed Mirror Solar Concentrator. A summary of the work done, a discussion of the design and construction of the model, idealized performance calculations and preliminary operating experience is described .

0135

Development Status of the Fixed Mirror Solar Concentrator : Russell Jr. J L, Schuster J R and Eggers G H, " General Atomic Co (USA) Report GA-A 14375 ", (1977).

This paper presents a brief technical background of the Fixed Mirror Solar Concentrator(FMSC) project and summarizes the technical status. The design and construction techniques of FMSC concrete modules is presented and the results of some preliminary testing result are reported.

0136

Development Status of the Fixed Mirror Solar Concentrator : Russell Jr. J L, Schuster J R & Eggers G H, " Proceedings of 12th IECE Conference ", 2, 1141, (1977).

The General Atomic Company fixed mirror solar concentrator employs a fixed mirror trough that produced a sharp line focus regardless of the position of the sun. The receiver is moved in a circular arc of track and focal line. General Atomic is fabricating FMSC modules of precast concrete and glass mirror to supply a 260 m^2 (2800 ft^2) mirror field of Sandia Laboratories Water Contract 05 - 4569 for their solar energy demonstration facility.

0137

The Fixed Mirror Solar Concentrator : Russell Jr. J L and Schuster J R , " Proceedings E R D A Conference on Concentrating Solar Collectors ", Georgia (USA), p. 3 - 1, (1977).

The General Atomic Co.(GA) Fixed Mirror Solar Concentrator (FMSC) employs a fixed mirror trough that produces a sharp line focus regardless of the sun position. The heat receiver, which employs a compound parabolic (Winston) secondary concentrator, is moved in a circular arc to track the focal line. General Atomic is fabricating FMSC modules of precast concrete and glass mirrors to supply a 260 m^2 (2800 ft^2) mirror field to Sandia Laboratories for their Solar Total Energy Demonstration Facility . These modules will be constructed on site at Sandia using local labour. The combination of concrete substrate and glass mirrors yields a rugged, low-maintenance concentrator that has good potential for low cost, high volume, on-site construction. Use of a secondary concentrator makes possible the design of a practical FMSC to efficiently generate steam at modern steam plant operating c o n d i t i o n s. Plans are under way at GA to measure heat losses from various high temperature heat receivers and to operate a small solar hardware demonstrative unit using the heat transfer salt Hitec.

0138

Fixed Mirror Solar Concentrator for Application to a 100 MW(e) Electric Generation Plant : Schuster J R, Neill J M and Bars J, " Proceedings IECE Conference", Boston(USA), p. 15, (1979).

A design study was performed of a 100 MW(e) power plant that uses a Fixed Mirror Solar Concentrator (FMSC) to supply energy for steam generation. Various heat transport fluids were considered and draw salt was selected over Therminol 88 and sodium. The complete plant was modeled in a cost performance optimization code which automatically performed trade-off such as reduced blocking versus increased piping costs, field piping heat loss versus overall system efficiency. Various collectors and receiver designs were studied and evaluated for their cost effectiveness. Operational requirements impacting the component designs were identified.

0139

Operating Experience with the General Atomic Fixed Mirror Solar Concentrator : Schuster J R, Eggers G H and Russell Jr. J L, " Proceedings ISES (American Section), Conference",Denver, Colorado (USA), 2.1, 863 (1978).

The General Atomic Co. (GA) Fixed Mirror Solar Concentrator (FMSC) employs a fixed mirror circular trough that produces a sharp line focus regardless of sun position. The heat receiver, which employs a compound parabolic (Winston) secondary concentrator, is moved in a circular arc to track the focal line. With secondary concentration the FMSC has a theoretical upper concentration limit of 206 suns, and the secondary concentrator makes possible the design of a practical FMSC to efficiently generate steam at modern steam plant operating conditions. Because the mirror trough of the FMSC is stationary it can be made using low cost type construction approaches.

0140

Design Construction and Testing of a Fixed Mirror Solar Concentrator Field : Schuster J R, Russell Jr. J L and Eggers G H, " General Atomic Co. (USA) Report GA-A 15003 ", (1978).

General Atomic has fabricated Fixed Mirror Solar Concentrator (FMSC) modules of precast concrete and glass mirrors to supply a 260 m^2 collector field to Sandia Laboratories for their Solar Total Energy Demonstration Facility. These modules were constructed on site at Sandia using local labour. The combination of concrete substrate and glass mirrors yields a rugged, low maintenance concentrator that has good potential for low cost, high volume on site construction. Scientific Atlanta Inc., has developed another construction approach based on supporting the glass mirrors with inexpensive m e t a l stampings. These structures can be quickly assembled in the field.

0141

Design & Predicted Performance of Scientific Atlanta's Fixed Faceted Mirror Concentrator : Shelton S V, Andrews Blackshaw and Hutchins Samuel F, " Proceedings ERDA Conference on Concentrating Solar Collectors ", Georgia (USA), p. 3 - 9, (1977).

A high volume manufacturing design has been developed by Scientific Atlanta Inc., for a linear east-west fixed faceted mirror solar concentrator. This concept utilises fixed east - west flat mirror slats orients around a circular arc such that the resulting focal line is an east - west linear bond in front of the mirror slats which moves north - south on another circular arc. Both the arc on which the mirrors are located, and the arc on which the focal band moves, are on a common circle lying in a vertical plane. The maximum energy is generally collected annually if the normal to the aperture is tilted from the vertical towards the equator, an amount equal to the latitude of its location.

The developed design incorporates 28 rear silvered mirror slats each of which are 7.3 cm w i d e and spaced over 103° of a 127 cm radius arc. The total collector aperture width is 2 m. The east-west aperture length can be any integral numbers of 3 m. The moving receiver at the focal band incorpor - ates a secondary trapezoidal concentrator (groove type) with a 2 to 1 concentration ratio. Its aperture is 12.7 cm wide concentrating onto a 6.4 cm wide flat absorber surface. The mechanical d e s i g n utilizes die stamped and roll formed galvanized sheet metal parts for the mirror slat supporting structure. This ensures low manufacturing costs with the necessary accuracy of mirror location. The mirror slats are held onto the sheet metal brackets by spring clips. The receiver is made up of a flat single glazed absorber consisting of multiple side by side tubes and coated with black chrome. The sides and back of the absorber are insulated.

0142

Measurement of Heat Loss from a Heat Receiver Assembly of a Fixed Mirror Solar Concentrator : Walker W E, Housman J J and Russell J L, " Proceedings 13th IECE Conference", San Diego (USA),p.1554,(1978).

This paper reports on an experimental measurements program on a heat receiver of the type currently used by the Fixed Mirror Solar Concentrator (FMSC) at the Sandia Solar Energy Test Facility. The objective of the measurement was to obtain a reliable data base suitable for comparison with, and evaluation of, engineering design methods of heat loss from a class of linear solar heat receivers.

0143

Conceptual Design of a Demonstration Fixed Mirror Solar Concentrator : Walker W E, " General Atomic Company (USA) Report GA-A-13926 ", (1976).

This report presents a conceptual design and discusses the features of a Fixed Mirror Solar Concentrator Demonstration Prototype Unit (FMSC Demo.). The FMSC Demo. is a length of FMSC typical of what could be used for a solar power plant. The FMSC Demo. features a mirror unit, 200 ft long with a 21 ft wide aperture, and a traversing Heat Receiver Assembly (HRA), suspended from a beam mounted on five support arms.

0144

Development of a 540 ft^2 Prototype Fixed Mirror Solar Concentrator : Williams J R & Hutchins Samuel F, " Proceedings IECE Conference ", San Diego (USA), p.195, (1975).

A new type of solar collector has been assembled and tested at Georgia Tech. which promises to provide heat at about the same cost as commercially available flat plate collectors, but at a much higher temperature. The higher temperature permits higher COP air conditioning and heat pump operation, with a resulting decrease in the collector area required and lower air conditioning equipment costs as compared with systems using flat plate collectors. In addition, this focussing collector can supply low grade steam and toher industrial process heat at temperatures to several hundred degrees Centigrade. Using air as the heat transfer medium, collection temperatures in excess of 400 deg C have already been achieved using the 540 ft^2 FMSC at Georgia Tech. These data indicate that an average collection efficiency of at least 50% at the design temperature of 260 deg C should be achieved using an improved heat exchanger currently under construction.

0145

Performance Testing of the General Atomic Fixed Mirror Solar Concentrator : Workhoven R M & Dudley V E " Sandia Laboratories (USA) Report SAND 780624 ", (1978).

This report summarizes the testing which was performed on the General Atomic Fixed Mirror Solar Concentrator at the Mid Temperature Solar Systems Test Facility. Test objectives are defined, test procedures are described, and results and conclusions are given.

Hemispherical   Bowl / SRTA   System

0146

Optical Study of Fixed Spherical Solar Collectors : Authier B,"Journal of Optics",(France),8,331(1977).

Study of geometric optics of a spherical solar collector permits a graphical determination of the radiation incident along the sagittal line. In the tropical regions, the addition of a mobile mirror element (visor) considerably increases the intercepted radiation.

0147

Optical Simulation for a Fixed Spherical Solar Collector : Authier B, Hill L, Duban M, Trarieux P, Sarazin M and Nadeau P "Applied Optics", 18, 3081 (1979).

To calculate the absorber dimensions for a fixed spherical solar collector, an optical simulation of the ray tracing type is proposed. The physical quantities, which have an effect upon these dimensions, are described as well as the measurement methods. Once the dimensions are determined, the incident flux on the absorber surface can be calculated by the same program in terms of different zenith distances. These calculations can be checked by comparing the calculations flux on the surface of the absorber with the measured flux at different points along with the absorber aimed at the full moon instead of at the sun.Through the data obtained from the measurements, fluctuating points of high flux and permanent zones which receive double and triple reflection rays have been studied.

0148

High Concentration Solar Collector of the Stepped Spherical Type ; Optical Design Characteristics : Authier B and Hill L, " Applied Optics ", 19, 3554, (1980).

An analysis of the optical design characteristics of a new high concentration solar collector is presented in this article. This type of collector consists of spherical segments that are sections of a spherical cap by planes perpendicular to its axis. These ring shaped spherical segments are so arranged along their common axis that the planes of their circles of least confusion are superimposed. The optical characteristics and simulation of this system are developed to provide information for the engineering design of this type of solar energy collector system. The calculations are checked by a laser scanning onto a breadboard mock up.

0149

Performance of Evacuated Concentrating Spherical Collector : Bayazitoglu Y and Pearson S, " Proceedings of ISES (American Section), Conference ", Denver, Colorado (USA), 2.1, 640, (1978).

This paper describes the mathematical model to simulate the quasi steady state performances of the evacuated-concentrating spherical collector. The evacuated-concentrating spherical collector consists of a nontracking evacuated glass sphere internally reflecting at the lower hemisphere and a fixed receiver absorber of an annular sector contained inside it. The entire sphere is evacuated, thus eliminating the convection loss. A heat transfer fluid is circulated through the absorber to collect the concentrated solar energy.

0150

Pressure Stabilised Solar Collector (PSSC): Brantley Jr. L W," Proceedings ERDA Conference on Concentrating Solar Collectors ", Georgia (USA), p. 4-21, (1977).

Presented here is a concentrating, two axis tracking, spheroidal collector designed to minimize the cost of the tracking system and the reflector. The cavity absorber has been designed to minimize the tracking and reflector surface accuracy requirements while maintaining high performance efficiency. Working fluid temperatures of 200-1000 deg F are achievable at efficiencies of 50-70% of the direct solar radiation at a cost projected in mass production to be as low as $ 5/ft^2 for units 25-50 ft in diameter.

0151

Potential of a Solar Collector with a Stationary Spherical Reflector and a Tracking Absorber for Electrical Power Production : Clausing A M, "Sandia Laboratories (USA) Report SAND 76-8039", (1976).

A fixed segment of a concave spherical mirror can be used to concentrate beam radiation onto a tracking absorber which pivots about the center of curvature of the mirror. A possible economic advantage of this system over concentrating collectors with tracking mirrors, is reduced mirror cost. The characteristics of this system are analytically studied. The strong influences of the geometrical parameters of the stationary reflector/tracking absorber are established. The penalty caused by the fixed reflector and the influence of the absorber temperature is also determined. The daily and yearly variations in the power produced by the system are presented.

0152

Optical and Thermal Characteristics of a Solar Collector with a Stationary Spherical Reflector and a Tracking Absorber : Clausing A M, " Sandia Laboratories (USA) Report SAND 76 - 8663 ", (1976).

A fixed segment of a concave spherical mirror can be used to concentrate beam radiation into a tracking absorber which pivots about the centre of curvature of the mirror. A possible economic advantage of this solar collector over concentrating collectors with tracking mirrors is reduced mirror cost. The objective of the investigation is to determine the potential of the system for electrical power production. Special emphasis is given to identifying the penalty associated with the fixed reflector. The results showed appreciable cosine losses even at the best time of the year. The over all system efficiency was found to be strongly dependent on the rim angle, the optical efficiency, the absorber temperature, and the degree of selectivity of the absorber surface.

0153

Design Considerations for the Energy Receiver in a Fixed Mirror Distributed Focus (FMDF) Solar Energy System : Clements L D, " Proceedings International Conference on Alternative Energy Source, Miami, (USA), p. 295, (1977).

This paper describes the factors which go into the design of a receiver for a particularly attractive solar power concept the fixed mirror distributed focus system. In this system the solar collector is a fixed spherical segment mirror giving a line focus which moves within the mirror region during the day. The receiver is a heat absorber specially matched to the geometry of the focal region, which track the line focus. The problem of maximizing heat absorption efficiency while minimizing heat losses, choice of heat transfer medium, choice of flow geometry, and materials and manner of construction are detailed.

0154

Optical-Thermal Performance Analysis for a Fixed Mirror Distributed Focus Solar Thermal Electric Power System : Clements L D, " Proceedings IECE Conference ", Boston (USA), p 11, (1979).

This paper is concerned with both optical and thermal analysis for a Fixed Mirror-Distributed Focus (FMDF) system. This work has arisen as part of the Crosbyton Solar Power Project at Texas Tech. University sponsored by the U.S. Department of Energy. The paper addresses both detailed and approximation modeling of receiver performance for a once-through steam boiler receiver for the FMDF concept. The final performance results are given in terms of enthalpic power absorbed as a function of solar inclination angle, fluid inlet conditions and wind speed.

0155

Development of a Spherical Reflector/Tracking Absorber Solar Energy Collector : Grossman G & Fruchter E, " Israel Journal of Technology ", 17 (1), 5 (1979).

A concentrating solar collector based on the Stationary Reflector/Tracking Absorber (SRTA)concept was built and tested. The collector consists of a 2.5 m diameter fixed spherical mirror, which concentrates solar radiation on a cylinder shaped tubular absorber tracking the sun. The system was exposed to weather conditions for more than a year and tested at different operating temperatures. The design of the mirror, the absorber, their method of fabrication, the automatic tracking mechanism and the heat removal system are described. Experimental results are given and compared with a theoretical performance prediction. Total efficiency of 42% has been obtained for a wide range of outlet temperatures upto 150 deg C. Higher temperatures, upto 300 deg C, have been reached but could not be maintained at no-boiling condition. The results of this study indicate the possibility of obtaining higher efficiencies with a better mirror surface, and the practical independence of the efficiency with regard to the operating temperature for a wide range of the latter. The system is hence capable of providing heat at high temperatures sufficient for space heating and air conditioning, and for compact heat storage.

0156

Determination De La Concentration Geometrique D'un Capteur Solaire A Mirror Spherique / (Determination of the Geometrical Concentration of a Solar Collector Using a Spherical Mirror) : Imbert B and Pasquetti R,"Journal of Optics", (France), 9, 25, (1978).

This paper is a presentation of a study of a focussing collector of solar radiation. The focussing reflector is a mobile spherical dish, which follows the path of the sun and t h e energy convertor opening is a disc, the radius and exact position of which is calculated. A relationship which permits the optimization of the collector has been established between the rate of geometrical concentration and the angle of the reflector opening, the influence of the imperfections of this reflector, or of the system for tracking the sun is then brought out. According to the degree of these imperfections, a comparison is finally made with the solar collector having a parabolic reflector. The solar collector of "spherical type" seems well adapted to the various uses which require geometrical concentration in the hundreds. One of these applications could be for example, the thermodynamic production of electricity.

0157

Thermal Performance Analysis of the Stationary Reflector/Tracking Absorber (SRTA) Solar Concentrator: Kreider J F, "Journal of Heat Transfer (Transcripts ASME) ", 97, 451, (1975).

The performance of a solar energy concentrating system consisting of a fixed concave spherical mirror and a suntracking cylindrical absorber is analysed in detail. The effects of mirror reflectance,concentration ratio, heat transfer fluid flow rate, radiative surface properties, incidence angle , an evacuated absorber envelope, and insolation level upon thermal performance of the concentrator are studied by means of a mathematical model. The results of the study show that the high temperature heat energy can be collected efficiently over a wide range of useful operating conditions. The analysis indicates that mirror surface reflectance is the single most important of the principles governing system performance.

0158

A Modular Fixed Mirror Brayton-Cycle Solar Power System : Meinel A B, McKenny D B and Meinel M P, " Proceedings International Conference on Solar Electricity ", Toulouse,(France), p. 867, (1967).

A quasi-hemispherical fixed mirror provides a distributed focus of the sun within the bowl. A cylindrical absorber tracks the diurnal motion of the sun, heating compressed air which then expands in the turbine stage of a Brayton cycle engine which is attached to the moving absorber structure, The fixed mirror and moving structure are below the land surface providing excellent protection o f t h e system from storm damage. Elimination of water for cooling makes this modular system well adapted for desert areas.

0159

Optical Analysis of the Fixed Spherical Mirror/Tracking Linear Receiver Solar Concentrator Including Statistically Distributed Reflector Slope Errors : O'Neill M J, " Proceedings ISES Silver Jubilee International Congress",Georgia, (USA), (1979).

This paper presents an optical analysis which treats randomly varying slope errors over the reflector of the concentrator so as to calculate the receiver flux distribution for imperfect reflection. An actual reflective concentration will have surface normal errors due to manufacturing, alignment, deflection etc. These errors are well described by a two-dimensional circular normal (Gaussian) probability distribution. Such an error distribution can be combined with a cone optics solution to the flux profile problem to define the probable flux profile for an imperfect concentrator and a receiver of any desired size and shape. The paper will present the mathematical formulation and solution of the combined statistical error cone optics problem as as parametric results for various error levels, receiver sizes and receiver shapes. Since the same techniques can be applied to other linear concentrators, the paper should be of interest to investigators of other collector types as well.

0160

Optical Analysis of the Fixed Mirror/Distributed Focus (FMDF) Solar Energy Collector :   O'Neill M J, " Proceedings ISES (American Section) Conference ", Orlando, (USA), (1977).

A collector utilises a large stationary mirror concentrator of spherical geometry to focus incident sun light upon a tracking linear receiver. This paper describes the optical characteriatics o f t h e FMDF collector. A closed form analytical solution based upon cone optics is presented for the f l u x concentration distribution over the linear receiver. Results of this cone optics solution are compared with the numerical solutions of other investigators.

0161

The Crosbyton Solar Power Project : Fixed Spherical Mirror/Tracking Receiver : Reicher J D, " Proceedings ERDA Conference on Concentrating Solar Collectors ", Georgia (USA), p. 3-61, (1977).

The Crosbyton Solar Power Project at Texas Tech. University deals with the solar Gridiron Concept , often called the FMDF (fixed mirror distributed focus) concept. Basically the concept involves t h e use of large spherical segment mirrors fixed in position with the symmetry axis declined about 15 deg south of vertical. The support for the mirror involves excavation near the south lip and space frame near the north rim. A conical receiver, typically operating with a water steam cycle, is the tracking element. The tracking requirement is for the receiver cone symmetry axis to lie, at all times, on the line determined by the center of the sun and the center of curvature of the spherical segment. A conceptual design for a nominal 5 MW(e) solar thermal electric system has been developed matched to t h e load and climatic conditions of Crosbyton, Texas. The solar Gridiron concept offers reasonable high temperature performance at relatively low cost and there is some indication that such systems m a y be cost effective in the near term market.

0162

Stationary Concentrating Reflector cum Tracking Absorber Solar Energy Collector - Optical Design Characteristics : Steward W G and Kreith F, " Applied Optics ", 14, 1509, (1975).

This article presents an analysis of the optical design characteristics of a stationary reflector / tracking absorber solar energy collector. This type of collector consists of a segment of a spherical mirror, placed in a stationary position facing the sun, but having a linear absorber that can track the image of the sun by a simple pivoting motion about the center of curvature of the reflector. The optical characteristics of this system and the axial variation of its concentration ratio are developed to provide information for the engineering design and sizing of this type of solar energy collector system.

0163

Fixed Mirror/Distributed Focus Solar Thermal Electric Power System Development : Walters R R, O'Neill M J & Gupta Y P, " Proceedings ISES (American Section) Conference ", Orlando (USA), (1977).

This paper presents the current development in the overall design, performance and cost analyses of a solar thermal power system concept which uses large fixed spherical mirror segments for the concentrator. This concept is referred to here as the Fixed Mirror/Distributed Focus Solar Thermal Electric Power System (FMDF - STEPS) where a large multimegawatt power facility would use an array of FMDF collector modules. Results of optical, thermal, structural and economic studies utilizing detailed computer modeling of the FMDF-STEPS are presented along with cost performance data in terms of dollars per kilowatt of installed costs. These data are compared with that for other solar system concepts currently under consideration by the Energy Research and Development Administration.

0164

Distribution of Radiant Vector in Radiation Field of Spherical Concentrators : Zakhidov R A, Vainer A A and Gafurov A M , " Applied Solar Energy ", 12 (4), 7 , (1976).

The concentration of solar energy by a spherical mirror has been studied. It was found that the maximum irradiance of an ideal concentrator is as much as 0.277 of the focal irradiance of an ideal paraboloidal concentrator with 60° flare angle and identical coefficient of specular reflection. It is shown that it is desirable to replace a long focus paraboloidal mirror [$U_m \approx 30°$] by a spherical mirror.

Tabor — Zeimer Circular Cylinder

0165

An Inflated Cylindrical Concentrator for Producing Industrial Process Heat : Gerich J W, "Proceedings ERDA Conference on Concentrating Solar Collectors", Georgia (USA), p. 2-103, (1977).

The use of industrial process heat below 170 deg C accounts for five percent of this country's total energy consumption. The authors have developed a concentrating solar collector to produce hot water and steam in this temperature range. The collector structure consists mainly of an inflated thin film plastic cylinder which is clear on the upper portion and is an aluminized reflector on the lower portion. The reflector concentrates on sun light on a receiver tube which is jacketed with a heat transfer suppressing thin film plastic cylinder. Because of its simplicity this collector will be cost effective relative to fossil fuels such as oil at $ 115/per barrel. We have written computer codes to analyze the optical and thermal properties of this collector. Results indicate that weekly tilting of the collector provides over 90% of the energy available from continuous tracking. A selective surface on the receiver tube (α=0.84, ϵ=0.20) increases the useful energy gain by over a factor of five at 170 deg C. The stagnation temperature at the outer receiver tube surface is calculated to be below 350 deg C. Combining calculated optical and thermal efficiencies gives an overall collector efficiency of 20% for 170 deg operation. This is based on an aperture of the full diameter times the length and or beam radiation. The first experimental collectors are being constructed to verify our computer code studies. These units have an outer diameter of one meter and are four meters long.

0166

Inflated Cylindrical Solar Concentrator : Gerich W, " Proceedings ISES (American Section) Conference ", Denver, Colorado (USA) 2.1, 889, (1978).

A concentrating collector is developed to produce pressurized hot water upto a temperature of 170 deg C. The collector structure consists mainly of an inflated thin film plastic cylinder that is clear on the upper portion and is an aluminized reflector on the lower portion. The reflector concentrates sun light on a receiver tube which is jacketed with a heat transfer suppressing thin film plastic cylinder. Computer codes are presented to analyze the optical and thermal properties of this collector. Results indicate that weekly tilting of the collector provides over 90% of the energy available from continuous tracking.

0167

A Non-Tracking Inflated Cylindrical Solar Concentrator : Gerich J W, "Proceedings ISES Silver Jubilee International Congress", Georgia (USA), (1979).

A concentrating solar collector to produce pressurized hot water upto a temperature of 170°C is being developed. The collector structure is an inflated thin film plastic cylinder that is clear in the upper portion and is an aluminium reflector on the lower portion. The reflector concentrates sunlight on a receiver tube, which can be either circular or triangular in cross-section, and has a diameter or side dimension of seven to ten per cent of the outer cylinder diameter. Surrounding this receiver tube is a thin film plastic jacket which significantly reduces the convection and radiation heat transfer losses. The axis of the cylinder lies horizontal in an east-west direction, and the collector is designed so that it can be manually tilted in the north-south plane. Computer codes have been written to analyze the optical and thermal properties of this collector. Results indicate that the collector has an acceptance angle of five to ten degrees,depending on the geometric concentration ratio, and therefore will need manual tilt adjustments 10 to 16 times a year. A selective surface on the receiver tube increases the useful energy gain by more than a factor of five at 170 deg C. The stagnation temperature at the outer receiver tube surface is calculated to be 220 to 320 deg C depending again, on the geometric concentration ratio. Collector efficiency is influenced by many geometric and material properties. However, at 170 deg C operating temperature, we expect typical daily efficiencies to be 20 to 25 per cent. Several experimental collectors have been constructed. These units are one meter in diameter and upto ten meters in length. Efficiency tests have verified the calculated optical and thermal efficiencies.

0168

The Circular Cylinder Reflector : Application to a Shallow Solar Pond Electricity Generating System : Kooi C F, " Solar Energy ", 20, 69, (1978).

The circular cylindrical reflector is shown to be a means of effectively tilting a solar energy collecting plane. This is particularly useful for swimming pools and solar ponds with free water surfaces that cannot be tilted. Such reflectors are applied to an electricity generating system which is driven by shallow solar pond at a 40° latitude. Thereby the annual electrical energy production can be increased by 40%. To a large extent power production can be levelled on an annual basis, and the turbine inlet temperature can be maintained at a constant level through the year.

0169

Low Cost Focussing Collector for Solar Power Units : Tabor H & Zeimer H,"Solar Enery",6(2),55, (1962).

A focussing collector for producing heat from solar radiation for power use consists of an inflated cylinder of plastic film 12 m long and 1.5 m in diameter. The cylinder is made in two segmencs, the portion exposed to the sun being clear plastic to permit radiation to enter and fall upon the rear segment, which is an aluminized plastic that acts as a mirror. The circular section is shown to be superior to a parabola for overall focussing, in addition is both easier and less expensive to form. A collector, including end rings, weighs only 1.2 Kg/sqm of solar aperture. Being deflatable, it is portable. Cost is estimated as $ 20/sqm of solar reception. Collection efficiency is about 40%.

0170

Theoretical Analysis and Performance Evaluation of a Semi-Cylindrical Solar Concentrator : Wen-Hsiung Huang, "Proceedings ISES Silver Jubilee International Congress ", Georgia (USA), (1979).

This paper sheds light on certain basic optical and thermal characteristics of a semi-cylindrical solar concentrator and, presents mathematical derivatives which are to function under ideal conditions for the solar radiation incidence at any arbitrary angle on the separate absorbers of the concentrator.

Mirror Boosters / V—Trough / Polygonal Troughs

0171

Moderately Concentrating (Not Focussing) Solar Energy Collectors : Bannerot R B,"Proceedings Workshop on Solar Collectors for Heating and Cooling of Buildings ", New York (USA), p.206, (1974).

It is illustrated how the new concept differs from the conventional flat plate collector. Moderate concentration (possibly as high as 10 to 1 but more probably 3 to 1) is achieved by installing mirror like surfaces above the actual absorber plate to redirect incident radiative energy to the plate. The feasibility of stationary, low cost, low maintenance collectors of this type was investigated.

0172

The Radiative Characteristics of Non-Tracking Moderately Concentrating Solar Energy Collectors : Bannerot R B, " Proceedings ERDA Conference on Concentrating Solar Collectors ", Georgia (USA), p. 3-71 , (1977).

The ideal (geometric) and effective (actual) concentration ratios are used to compare the radiative performances of various booster mirror configurations and nontracking, east - west aligned grooved collector designs. The groove designs are represented by various side wall shapes, including a single straight facet (trapezoidal), a double facet and a range of curved wall designs including the compound parabolic concentrator or Winston collector.

0173

Performance Comparison Between Flat-Plate and Moderately Concentrating Solar Energy Collectors : Bannerot R B, Elliott D G and Reber J, " Journal of Energy", 1, 329, (1977).

The object of this work is to evaluate the potential of trapezoidal grooves as thermal energy collectors. The work discussed represents the first step in analyzing the grooves success in reducing the thermal energy loss from the receiver significantly.

0174

The Effect of Non-Direct Insolation on the Radiative Performance of Trapezoidal Grooves Used as Solar Energy Collectors : Bannerot R B and Howell J R, "Proceedings ISES Conference", Los Angeles(USA),(1975).

Moderate concentration of solar energy can be achieved by redirecting (but not focussing) incident beams over a given area on to a smaller one. Since focussing is not required, both direct and non-direct insolation can be utilized. A simple,easily fabricated collector of this type consists of a series of long, trapezoidal, parallel stationary grooves. The side walls are highly reflective while the base is highly absorptive of solar radiation. The radiative performance of such a collector exposed to direct insolation has previously been reported. In this paper the radiative performance of trapezoidal grooves used as moderately concentrating solar energy collectors and exposed to non-direct insolation is examined.

0175

Predicted Daily and Yearly Average Radiative Performance of Optimal Trapezoidal Grooves Solar Energy Collectors: Bannerot R B & Howell J R, "Proceedings ISES Conference", Winnipeg (Canada), "Sharing the Sun", 2, (1976).

In this paper the daily and annual radiative performance of moderately concentrating east-west aligned, parallel trapezoidal grooved collectors with depth to base ratios ranging from 1.1 to 5 are presented. The results are based on assumed tilting routines chosen to match the collectors capabilities.

0176

The Effect of Nondirect Insolation on the Radiative Performance of Trapezoidal Grooves Used as Solar Energy Collectors : Bannerot R B & Howell J R, "Solar Energy", 19, 549, (1977).

In this paper the radiative performance (the effective solar absorptivity and concentration ratio) of moderately concentrating collectors, composed of east - west aligned parallel, trapezoidal grooves is examined. Both direct and nondirect insolation performance is discussed. Results are presented indicating performance degradation due to nondirect insolation. Finally, nomographs are presented which allow one to determine the radiative performance of the grooved collectors when exposed to various combinations of direct and nondirect insolation.

0177

Analysis, Design, Fabrication and Testing of Moderately Concentrating Solar Energy Collectors : Bannerot R B & Howell J R, "3rd Annual Solar Heating & Cooling R&D Contractor's Meeting",(USA), (1978).

This paper presents the development of low cost,moderately high temperature, efficient solar energy collectors utilizing non-tracking concentration in east - west aligned groove like geometries. The concentration is achieved with reflective side walls which redirect the insolation onto a receiver located at or near the bottom of the groove. The design concept, a brief summary of the work performed and technical accomplishments are presented in this paper.

0178

Natural Convection in Groove-Like Geometries : The Overall Heat Transfer Coefficient : Bannerot R B & Nabavi H, "Proceedings 2nd AIAA/ASME Thermophysics & Heat Transfer Conference", Palo Alto,California (USA), (1978).

Groove-like geometries are being used to form non-tracking concentrating solar energy collectors. Experimental determinations of the overall heat loss coefficients from a trapezoidal groove, heated from the short side and insulated on the sloping side are reported. The effects of tilt and internal baffling are investigated. The relative importance of the various modes of heat transfer is analysed.

0179

Flat-Sided Rectilinear Trough as a Solar Concentrator ; An Analytical Study : Burkhard D L, Strobel G L & Burkhard D G , "Applied Optics", 17, 1870, (1978).

Formulas are derived for the concentration factor and irradiance distribution at the base of the flat - sided linear trough. Performance is affected by the number of reflections the solar rays undergo before reaching the base, the cone apex angle, and the coefficient of reflection. Results are presented in such a way that one can choose the optimum configuration, which is the minimum material required, to achieve a given concentration factor. Practical concentration factors range from 1 to 4 depending on the geometry and coefficient of reflection.

0180

Solar Concentrating Properties of Truncated Hexagonal, Pyramidal and Circular Cones : Burkhard D G, Strobel G L and Shealy D L, " Applied Optics ", 17, 2431, (1978).

The concentrating properties of specularly reflecting pyramids, hexagons and circular cones are examined. The concentration factor is determined as a function of the coefficient of reflection and the shape and orientation of the incident sun light. Reflector designs allowing multiple reflections for both normal and oblique incidence are considered.

0181

Solar Collector Cost Reduction with Reflector Enhancement : Clausing A M & Edgecombe A L," Proceedings ISES (American Section), Conference", Orlando (USA), (1977).

This study is concerned with the use of fixed specular reflectors in combination with conventional flat plate solar collectors. The objective is to reduce the high capital cost of solar collector systems through the use of inexpensive specular reflectors. A mathematical formulation of a general reflector-collector system is given which encompasses a wide range of geometries and is applicable to any sun position. Two new key parameters are introduced, the reflector effectiveness and the reflector

efficiency which provide a meaningful and simple means of evaluating the performance and cost effectiveness of reflector-collector systems. Instantaneous and time averaged performance data are   given
for reflector enhanced collector systems and comparisons are provided with conventional collector systems which clearly show the benefit of reflector enhancement.

0182

Analysis of a Flat Mirror Solar Concentrator : Collier R K & Matthew G K,"ASME Paper No.76-WA/HT-11(1976)

The theoretical performance of a flat mirror solar concentrator was analysed where the mirrors formed
flat arrays in two planes. Mirror spacings were minimised by a generally coded computer program for
a number of operating conditions. Design variables included the height of the collector tube and the
elevation angle associated with V trough style configurations. The results show that concentration
ratios of the order of 100 to 1 are attainable for   reasonable sizes and a wide range of sun angles.

0183

A General Model for Predicting the Performance Characteristics of Planar Concentrating Systems :
Edgecombe A L & Clausing A M, " Proceedings ERDA Conference on Concentrating Solar Collectors", Georgia
(USA), p. 3 - 109, (1977).

The specular reflector enhancement of flat plate solar collectors is analysed in this study. A mathematical model is developed and two key geometrical parameters are introduced in the analysis of the
spatially averaged energy flux over the collector surface. The model allows these parameters to be
determined as a function of the collector size and tilt, and the position of the sun. The key parameters, when time averaged over a period of interest, yield a relative measure of the direct beam component and the specularly reflected component of the total energy flux during the given period. The
relative values of these two flux components are used in optimizing the reflector collector system.
Data generated by computer simulation is presented as an example of determining an optimum system with
respect to one of the system variables.

0184

New Design for an Efficient Low Concentrating Solar Collector : Eldifrawi A A, "Proceedings ERDA Conference on Concentrating Solar Collectors", Georgia (USA), p.3 - 115, (1977).

A design of a low concentrating collector has been developed. The design incorporates two reflectors,
one mounted above and the other mounted in front of a flat plate collector. This assembly is enclosed inside an air-supported transparent membrane structure. Besides the increase in solar insolation
received by the collector absorber due to the reflectors, the temperature around the flat plate collectors increases either due to the green house effect or by utilizing the building exhaust air in
supporting the structure. This results in a significant increase in system efficiency. The membrane
inflated structure provides protection against weather conditions and the structural integrity of the
system. It is possible, therefore, to use an inexpensive and thinner glazing material for the enclosed flat plate collectors. The preliminary analysis indicated that the flat plate collector's performance , when incorporated inside the structure, would increase over 100% under extreme cold weather .
An output of 200 deg F, required for solar absorption air conditioning, is possible at a reasonable
efficiency. The economic analysis of the utilisation of this concentrator indicates that it would be
cost effective.

0185

Fixed Concentrating Flat-Plate Collectors for Heating and Hot Water Applications : Espy P N,"Proceedings ERDA Conference on Concentrating Solar Collectors", Georgia (USA), p.3 - 55, (1977).

An alternative to simple flat plate collector used for solar heating and hot water systems is a fixed
concentrating collector utilizing low cost flat plate collectors. In this paper it is shown that
the least expensive collectors augmented by single or double faceted reflectors perform at high efficiency and at a low total collector array cost. Additionally the system operating requirements for
summer for domestic hot water at about 60 deg C are adequately met. The design proposed has a reflector oriented at a right angle to a flat plate collector. The results of this design approach is
to provide a concentration ratio of about 1.5 during the winter season, and a concentration ratio of
1.0 during the mid summer period. Winter performance of an inexpensive and relatively inefficient
collector may thus be elevated to more than twice the collected energy of conventional configurations,
the summer efficiency need only be that which is required for meeting hot water heat loads. The overall system may thus be seen to better match the load requirement of heating and hot water applications.

0186

Use of Planar Reflectors for Increasing the Energy Yield of Flat Plate Collectors : Grassie S L and
Sheridan N R, " Solar Energy ", 19, 663, (1977).

A mathematical model to simulate the performance of flat plate collector reflector systems is presented. First the collector energy balance is modified to account for the reflected energy. Then the
exchange area for a diffused reflector is obtained by integrating over both reflector and collector
surfaces. For the specular reflector, the collector area exposed to reflected radiation is calculated
from geometrical relations. Shading effects are also found from the system geometry. The model is
used to predict the annual performance of a water heating system with several values of the reflector
angle.

0187

Augmented Solar Energy Collection Using Various Planar Reflective Surfaces ; Theoretical Calculations
and Experimental Results : Grimmer D P, Zinn K G, Herr K C & Wood B E, " Los Alamos Scientific Labora -
tory (USA) Report LA- 7041", (1978).

This report presents the results of theoretical calculations and experimental tests on the use of
different types of flat reflective surfaces to increase the collection of solar energy by flat collectors. Specular,diffuse and combination specular/diffuse reflective surfaces are discussed. An attempt has been made to describe the reflective properties of surfaces in more generalised terms than
simple direct or simple diffuse. Most real surfaces possess a combination of specular like & diffuse

like reflective properties. A computer model has been generated to describe surfaces as a combination of specular and diffuse like reflectivities. The reflective properties of a given surface can be measured in the laboratory as a function of incident and reflected angles, and these measured reflective properties can be used in the computer model to predict the increase in collector performance with such a reflector. Predictions of system performance were made for various collector/reflector configurations and compared with the performance of an optimally oriented collector without a reflector.

0188

Enhancement of Flat Plate Solar Collector Performance Through the Use of Planar Reflectors: Hill John M & Perry E H, " Proceedings ISES (American Section) Conference", Orlando (USA), (1977).

A study was undertaken to determine insolation and collector performance enhancement factors from the presence of specular planar reflectors in front of flat plate solar energy collectors. The study involved both, a mathematical analysis of the problem and experimental measurements of the enhancement factors. Among the variables included in the analysis are the collector and reflector tilt angles, the reflector to collector weight ratio, and the collector aspect ratio. The study shows that such reflectors can significantly improve the performance of flat plate solar collector arrays.

0189

A Concentrator for Thin Film Solar Cells : Hollands K G T, " Solar Energy ", 13, 149, (1971).

This paper presents a theoretical study of a novel concentrator reflector specially suited to thin film solar cells. The concentrator consists simply of a V trough with side walls which are specularly reflecting and a base to which the solar cells are attached. Mounted with its axis in the east - west direction, the concentrator is designed to track the seasonal but not diurnal motion of the sun, by adjustments to its tilt made several times per year.

0190

Trapezoidal Grooves as Moderately Concentrating Solar Energy Collectors : Howell J R & Bannerot R B , " Proceedings 10th AIAA Thermophysics Conference", Denver, Colorado (USA), (1975).

The radiative behaviour of trapezoidal grooves used to provide moderate concentration of direct beam incident energy is examined experimentally. With highly reflective side walls and a highly absorbant base, such a configuration is a candidate for a solar energy collector to provide temperatures above those practically attainable with conventional flat-plate designs. The base in this case would include a means of removing the collected energy, e.g. a finned tube. The results obtained are presented as concentration ratio versus incident zenith angle for various geometric configurations. These results are compared with those obtained from a Monte Carlo ray tracing computer simulation. Good agreement is shown. Side wall material tested include, back surfaced glass, plexiglass mirrors and aluminium foil. From results shown here and from previous analytical results, trapezoidal grooves are v i a b l e candidates for achieving moderate concentration of solar energy and for enhancing the performance of solar collectors at elevated temperatures (upto 150 deg C).

0191

Evaluation of a Surface Geometry Modification to Improve the Directional Selectivity of Solar Energy Collectors ; Progress Report,1 Jan. 1975 - 31 March 1975 : Howell J R & Bannerot R B, "University of Houston Report TID - 26838",(1975).

The performance of a flat plate collector can be greatly enhanced with the addition of a spectrally (wavelength) and/or directionally selective surfaces. This document reports on the progress in examining, proposed moderately concentrating collectors composed of parallel, east-west oriented stationary, trapezoidal grooves. Computer simulations of radiative behaviour with direct beam solar incidence have been verified by model testing. Based on an optimization study, two prototype collectors have been designed and are currently under construction. Work is continuing on a more realistic solar input in the computer simulation as well as on other candidate geometries.

0192

Experimental Evaluation of the Reflector Collector System : Kaehn H D, Geyer M, Fong D, Vignola F and McDaniels D K, " Proceedings ISES (American Section) Conference", Denver, Colorado(USA),21, 654, (1978).

The experimental arrangement utilized to measure the useful heat collected with a reflector-collector combination,is described and the results obtained from the first few months of operation are given. The theoretical analysis of reflector-collector systems are reviewed. The key conclusion from these calculations is that the useful heat output of the reflector system averaged over an entire heating season is typically about 40-50% greater than that obtained with a standard flat plate collector combination.

0193

Non Tracking Solar Energy Collector System : Lovelace A M & Selcuk M K, US Patent 4,122,833/Oct.31, 1978.

A solar energy collector system is disclosed characterized by an improved concentrator directing incident rays of solar energy on parallel strip-like segments of a flat plate receiver and a plurality of individually mounted reflector modules of a common asymmetrical triangular crossectional configurations,supported for independent reorientation,and defining a series of asymmetric V trough concentrators for deflecting incident solar energy toward the receiver.

0194

Mirror Enclosures for Double Exposure Solar Collectors : Larson D C, " Solar Energy ", 23, 517,(1979).

Several flat mirror collector panel configurations are evaluated and optimum configurations are determined for different solar energy applications at latitudes 35, 40 and 45° .The various mirror configurations are evaluated theoretically by calculating direct beam and diffuse solar radiation enhancement factors. The enhancement factors are defined as the ratio of the solar flux absorbed by both sides of a double exposure panel to that absorbed by an identical single exposure panel tilted at the latitude angle from the horizontal. The enhancement factors are calculated using the method of images and take account of the variation of glazing transmittance with incidence angle. Optimal mirror configurations were determined for direct beam solar radiation for both fixed mirror configurations and adjustable mirror configurations with semi-annual mirror rotations. Optimal fixed mirror configurations were obtained for both winter space heating and year round applications.

0195

Optimal Geometries for One and Two-Faceted Symmetric Side-Wall Booster Mirrors : Mannan K D,"Proceedings Flat Plate Solar Collector Conference ", Orlando, Florida (USA), p. 207, (1977).

Moderate concentration for east - west aligned non tracking, trough like solar energy collectors is examined. Two designs are evaluated in detail. They are the one and two-facet plane side wall configurations. The maximum performance designs are shown not to be the most practical designs since they tend to require disproportionately large reflectors. A significant reduction in reflector area can be made with only a small degradation of performance. At solistice a 9° collector acceptance angle is necessary for 8 hours of collection. Under this restriction the practical concentration ratios are limited to about 2 and 2.6 for the one and two-facet designs respectively.

0196

Performance of Optical Geometry Three Step Compound Wedge Stationary Concentrator ; Mannan K D, "Proceedings ISES Conference ", New Delhi (India) p.1218, (1978).

The concentration ratios including maximum ones obtainable with east - west aligned non - tracking three step compound wedge solar energy collector has been examined. It has been shown that maximum concentration design is not the most practical design because of its requirement of disproportionately large reflecting surfaces. It has been shown further that significant saving of reflecting surface can be effected with only small reduction in concentration ratios available. The experimental performance of a practical design has also been investigated.

0197

Compound Wedge Cylindrical Stationary Concentrator : Mannan K D and Cheema L S, " Solar Energy ", 19, 751, (1977).

This paper describes a non-tracking cylindrical concentrator of a normal design using flat reflecting panels. Series of flat reflecting panels are assembled in such a way that each ray of solar energy striking the reflecting panels is beamed down to the absorber largly after a reflection, w h e n averaged over the energy collection hours. Each successive panel is positioned at a different angle, thus forming a stepped or compound wedge. The concentrator is oriented with its length along east-west direction and requires seasonal arrangements only. Concentration ratios of about 6 to collect solar energy for 8 hrs. a day are possible with such a design.

0198

Enhanced Solar Energy Collection Using Reflector Solar Thermal Collector Combinations : McDaniels D K, Lowndes D H, Matthew H, Reynolds J and Gray R, "Solar Energy", 17, 277, (1975).

The amount of direct light gathered by a combination of reflector plus flat plate collector is analysed. The calculations are made allowing variable reflector and collector orientation angles, variable latitudes and arbitrary sun hour angle away from the solar noon. The effects of reflection and transmission losses and of polarization of the incident light and included correction is also made for the finite size of the reflector. It is found that the optimum orientation has the collector plane almost perpendicular to the plane of the reflector. The optimum reflector angle is found to be between 0° and 10° above the horizon for winter solar conditions and the enhancement is by a factor of 1.4 - 1.7 . The overall enhancement in the useful heat output of diffused solar radiation is found to be by a factor of 1.5 . Comparison with the preliminary analysis of the performance of the Coos Bay - Oregon solar house shows substantial agreement with the predictions of the present analysis.

0199

Design Evaluation and Testing of a Moderately Concentrating Non-Tracking Solar Energy Collector : Olvera A and Bannerot R B, " ASME Paper No. 79 - WA/Sol - 3, (1979).

The thermal performance of a moderately concentrating, non-tracking solar energy collector is predicted, based on a series of experimental evaluations of its components. Four reflector designs were constructed and tested. Six simple tubular receiver designs were tested. A collector utilising one of the reflector designs and one of the receiver designs was constructed and tested. The predicted performance closely approximated the actual thermal performance of the collector. The component evaluations are discussed in such detail that the analysis could easily be extended to other designs by the reader.

0200

Effects of Boosters on the Performance of Flat Solar Collectors : Pande P C, Garg H P & Thanvi K P , " Proceedings Indian National Solar Energy Convention ", Bhavnagar (India), p.61, (1978).

The effect of boosters (plane reflectors) on the performance of flat plate collectors inclined at an optimum tilt is worked out for different seasons. The integrated enhancement in the energy gain b y the collector due to the boosters is calculated considering variation of transmittance of glass cover with angle of incidence, absorptance by the collector plate and the total area of the collector ex - posed by the reflected radiations. Different collector booster systems viz booster at top edge, booster at lower edge, booster on east - west edges of a south facing flat plate collector are considered and a comparative study between these systems is carried out. Results of an experimental study p e r-formed at Jodhpur (India), are also described in detail.

0201

Measurements on the Effect of Planer Reflectors on the Flux Received by Flat Plate Collectors : Reid R L, Chilcoat M and Yako M J, " Proceedings ISES Conference (American Section)", Orlando(USA),(1977).

The effect of planer reflectors on the flux received by flat plate collector was determined b o t h through experiments and analytical model. The particular application studied was an inverted V-roof solar house configuration. Experimental data were taken for mirror, plexiglass and aluminium. These results were compared with the analytical model written for specular reflection. Results from the model were generated for flux received both with and without the reflector, for an yearly period.

0202

Computer Simulation Results for Planer Reflectors and Flat Plate Solar Collectors : Rudloff F A, Swanson S R and Boehui R F, "ASME Paper No. 79 – WA / Sol – 37, (1979).

The benefits attributed to reflectors for increasing performance of flat plate collectors vary quite widely in the literature. In the present study, a detailed computer simulation is carried out to investigate the increase in performance. The results show that collectors should be placed at a higher tilt angle than usual; reflectors placed in front of the collectors can be tilted up 5 or 10 degs ; and performance has a broad peak with respect to tilt angle. The system performance enhancement in Salt Lake City, Utah, ranges from 1.16 for domestic hot water to 1.28 for space heating for equal reflector collector areas, and the break even ratio of per area reflector collector cost is from 16 to 28 % of these systems.

0203

Double Exposure Collectors with Mirrors for Solar Heating System : Savery C W, Anderson P M & Larson D C, " ASME Paper No. 76 – WA/HT – 16", (1976).

A retrofit solar water heating system has been designed for a three storey building at Drexel University. The system will use two conventional collector banks mounted at 40 degs from the horizontal & one bank of double exposure collectors mounted vertically ; the latter bank is mounted in a fixed mirror enclosure which is designed to provide additional collector insolation, and to reduce convective heat losses from the collectors. The performance of the collector mirror system has been evaluated in comparison with an identical reference collector mounted at the latitude angle from the horizontal. The solar energy collector enhancement was determined for both direct beam and diffuse radiation. The heat loss coefficient of the collector mirror system was also determined relative to a reference collector when consideration was given to the effect of the enclosure upon the convective film coefficient. The useful heat output of the collector-mirror system relative to the reference collector was then estimated.

0204

Fixed Flat-Plate Collector with a Reversible V-Trough Concentrator : Selcuk M K, "ASME Paper #76-WA/HT-12 (1976).

An asymmetrical reversible Vee trough concentrator with a modest optical concentration factor of about three is suggested in order to improve the performance of a fixed flat plate collector. The reversible Vee trough can maintain a year round average concentration factor of two or better for the most significant collection period by simple reversal of the Vee-trough. The performance of a Vee-trough reflector is formulated and year round variation of the concentration factor is analytically determined. By using an enhanced solar flux condition, the performance of collectors with evacuated and non-evacuated receivers is predicted. Thermal energy costs using non evacuated receivers are predicted.

0205

Solar Energy Collection System : Selcuk M K," US Patent ", 4, 044, 753/August 30, 1977.

An improved solar energy collection system, having enhanced energy collection and conversion capabilities is delineated. The system is characterised by a plurality of receivers suspended above a heliostat field comprising a multiplicity of reflector surfaces , each being adapted to direct a concentrated beeam of solar energy to illuminate a target surface for a given receiver. A magnitude of efficiency suitable for effectively competing with systems employed in collecting and converting energy extracted from fossil fuels is indicated.

0206

Experimental Evaluation of a Fixed Collector Employing V-Trough Concentrator and Vacuum Tube Receivers: Selcuk M K, "Proceedings Winter Annual Meeting of ASME", Atlanta (USA), p. 33–37, (1977).

A test bed for experimental evaluation of a fixed solar collector which combines an evacuated glass tube solar receiver with a flat plate, black chrome-plated copper absorber and an asymmetric V trough concentrator was designed and constructed. Earlier predictions of thermal performance were compared with test data acquired for a bare vacuum tube receiver, and receiver tubes with Alzak aluminium, aluminised FEP Teflon film laminated sheet-metal and second surface ordinary mirror reflectors. Owing to its high temperature capabilities (150 to 250 deg C), the proposed scheme could also be used for power generation purposes in combination with an organic Rankine conversion system.

0207

A Fixed Moderately Concentrating Collector with Reversible Asymmetric V-Trough & Vacuum Tube Receiver: Selcuk M K, " Proceedings ERDA Conference on Concentrating Solar Collectors",Georgia(USA),p.3-93,(1977).

A vacuum tube receiver combined with an asymmetrical Vee-trough is being developed as an efficient solar heat collector for a range of about 100 – 200 deg C. An improvement in the efficiency of the vacuum tube receiver and a reduction in collector cost result with the use of the Vee-trough reflector. A test bed for experimental evaluation of a fixed solar collector which combines an evacuated g l a s s tube solar receiver with a flat plate/black chrome-plated copper absorber and an asymmetric V - trough concentrator was designed and constructed. Analytical predictions of thermal performance were compared with test data acquired for a bare vacuum tube receiver and receiver tubes with Alzak aluminium,aluminized FEP Teflon film laminated sheet metal and second surface ordinary mirror reflectors. Test results and system economics are discussed.

0208

Non-tracking Solar Energy Collector System : Selcuk M K, " US Patent " 4, 122, 833/Oct. 31, (1978).

A solar energy collector system is described. It is characterized by an improved concentration for directly incident rays of solar energy on parallel strip like segments of a flat plate receiver. Individually mounted reflector modules of a common asymmetrical triangular cross sectional configuration supported for independent orientations are asymmetrically included with Vee trough concentrators for deflecting incident solar energy towards the receiver.

0209

Low Cost Vee Trough/Evacuated Tube Collector Module : Selcuk M K, " Proceedings ISES Silver Jubilee International Congress ", Georgia (USA), (1979).

A low cost solar collector capable of operating at 150-200 deg C is described. An evacuated tube receiver is combined with asymmetric Vee trough concentrators. Peak efficiencies of about 40% at 120 deg C and 30% at 180 deg C are expected. Predicted future collector cost is $ 70/sqm which yields an energy cost of $ 4.20/GJ at 120 deg C. During the development of the Vee trough/evacuated tube collector (DVTETC) both mathematical models to predict thermal and optical performance were developed & test run to verify the theory.

0210

Analysis and Two Years of Testing of the Vee Trough Concentrator/Evacuated Tube Solar Collector : Selcuk M K & Aghan A, " Proceedings ISES Silver Jubilee International Congress", Georgia(USA), (1979).

The Vee Trough/Evacuated Tube Collector (VTETC) was analysed and tested. Black silvered mirror, Alzak, Aluminized Teflon and Kinglux reflectors were used during the tests run in spring/summer of 1977 - 78 at temperatures ranging from 80 to 190 deg C. Efficiencies were about 40% at 120 deg C and 30 %   at 175 deg C respectively.

0211

Analysis Development and Testing of a Fixed Tilt Solar Collector Employing Reversible Vee Trough Reflectors and Vacuum Tube Receivers : Selcuk M K, " Solar Energy ", 22, 413, (1979).

The Vee Trough/Vacuum Tube Collector (VTVTC) aimed to improve the efficiency and reduce the cost of collectors assembled from evacuated tube receivers. The VTVTC was analysed rigorously and a mathematical model was developed to calculate the optical performance of the Vee trough concentrator and the thermal performance of the evacuated tube receiver. A test bed was constructed to verify the mathematical anaylsis and compare reflectors made out of glass, Alzak and aluminized FEP Teflon. Tests were run at temperatures ranging from 95 to 180 deg C during the months of April, May, June, July and August 1977. Vee trough collector efficiencies of 35 - 40 % were observed at an operating temperature of about 175 deg C. Test results compared well with the calculated values. Test data covering a complete day are presented for selected data throughout the test season. Predicted daily useful heat collection and efficiency values are presented for a year's duration at operation temperatures ranging from 65 to 230 deg C. Estimated collector costs and resulting thermal energy costs are presented. Analytical and experimental results are discussed along with an economic evaluation.

0212

Performance Studies on a Flat Plate Collector Array with Booster Mirrors and Associated Hot Water Thermal Storage Tank : Sukhatme S P, Kandlikar S G, Sharma G K and Arun Kumar S, " Proceedings Indian National Solar Energy Convention ", Bombay (India), p.36 (1979).

Initial results of an experimental study on the performance of a flat plate collector array w i t h booster mirrors and an associated hot water thermal storage tank are described. The energy collected by the collector, cum storage system of a 0.5 ton cold storage unit. Based on the experiments , the maximum temperature attained by the water in the storage tank, the heat loss characteristics of the storage tank and the efficiency curve for the collector array are obtained.

0213

Mirror Boosters for Solar Collectors : Tabor H, " Solar Energy ", 10, 111, (1966).

This paper deals with the use of flat mirrors as boosters to flat-plate collectors and shows the considerable improvement in output particularly for collectors used at elevated temperatures. The Shuman case of side mirrors on the north and south edges of a collector is discussed in detail; the yield is increased but is very peaky. An alternative system uses an east facing mirror placed on the w e s t wedge of the collector in the forenoon, which is transferred (manually) at noon to be west facing on the east edge. This system produces about the same amount of boost as the Shuman case but the output is approximately rectangular. Several configurations and transposition systems are given: one h a s been used in a turbine power installation.

0214

Effect of Off-South Orientation of Optimum Conditions for Maximum Solar Energy Absorbed by Flat Plate Collector Augmented by Plane Reflector : Taha I S & Eldighidy S M, " Proceedings ISES Silver Jubilee International Congress ", Georgia (USA), p. 332, (1979).

Complete analysis of the off-south orientation, simple flat plate collector, augmented by flat sheet specular reflector, is developed. The enhancement of heat flux absorbed by solar collector due   to the use of reflector is calculated as a function of solar altitude and azimuth angles, off-south orientation angle of collector, and relative sizes and tilt angles of both collector and reflector. The shading effect due to the presence of the reflector is also considered in the analysis. The collector and reflector variables are optimized for maximum solar energy flux absorbed by the collector during a prespecified period of time. The Hooke and Jeeves optimization technique has been used in the analysis.

0215

Optical Performance of the Compound Trapezoidal Collector (An Optimized Non-Focussing Concentrating Collector) : Truong H V & Villanueva J, "Proceedings ERDA Conference on Concentrating Solar Collectors ", Georgia (USA), p.3 - 103, (1977).

This paper is concerned with the theoretical evaluation of the operating characteristics of an optimized non-focussing Compound Trapezoidal Collector (CTC). A CTC consists of two successive trapezoidal grooves, the dimensions of which are optimized to accept all the solar insolation impinging upon it when sun rays are directed along the optical axis of the collector. Theoretical concentration ratio as high as 5 can be achieved, however a daily average concentration of 3.5 can be obtained with this optical geometry. The computer simulations presented in this paper include the optical performance of the CTC for various reflective wall reflectivities and receiver surface absorptivities. Only direct solar radiation has been considered in the theoretical estimation so far.

0216

Compound Trapezoidal Collector (An Optimized Stationary Concentrator) : Villanueva J and Truong H V, " Proceedings ISES Conference (American Section)," Orlando (USA), (1977).

This paper discusses the design evaluation and optimization of a nontracking compound trapezoidal groove collector (CTC). The proposed collector is compared to the R. Winston's compound parabolic collector studied by J R Howell and R B Bannerot. The proposed collector geometry consists of two successive trapezoidal grooves, the relative dimensions of which are optimized to accept (with no more than one reflection) all the solar energy impinging upon it when the sun's rays are directed along the optical axis of the collector. Computer simulation of the proposed collector shows that instantaneous concentration ratios as high as 5(five) can be achieved with the geometry, while time averaged concentrations of about 3.5 (during a complete collecting period of 8 hours) are possible.

0217

Design and Experimental Evaluation of a V-Groove Solar Collector : Wiebelt J A and Thatree N, "ASME Paper No. 76 - HT - 53 ", (1976).

A V-groove (trough type) collector was considered for a solar absorption system. Solar insolation is concentrated on a base which acts as the energy receiver thereby increasing the absorptance with no increase in emittance. Design of the collector, insolation measurments, and power collection technique are described. Comparison of the experimentally measured optical concentration ratio with analytically determined concentration ratios indicated very good agreement. Experimental efficiency values are compared to analytically determined efficiencies for flat plate collectors operating under similar conditions. This comparison indicated eight to ten per cent efficiency for the V-groove collector.

Compound Parabolic Concentrator / Non – imaging Concentrator

0218

Natural Convection in Inclined Two-Dimensional Compound Parabolic Concentrators : Abdel-Khalik S I & Li H W, [Preprint-Department of Nuclear Engineering, Univ. of Wisconsin, Madison(USA),p.23-28]

Finite element techniques are used to determine natural convection heat transfer coefficients between absorber surfaces and cover plates of inclined two-dimensional compound parabolic concentrators (CPC). Both full and truncated CPC cavities (troughs) of different concentrations ($2<C<10$) with 1/3, 2/3 & full CPC heights are examined at different Rayleigh numbers ($1.0 \times 10^3 <Ra<2.5 \times 10^6$) & tilt angles ($\theta < r<60°$). Generalised charts and co-relations for estimating the average absorber plate Nusselt number as a function of Rayleigh number, concentration, and tilt angle are given. The results are useful for evaluating the convective loss coefficients from such collectors.

0219

Natural Convection in Compound Parabolic Concentrators - A Finite Element Solution : Abdel-Khalik S I, Li H W and Randall K R,"Journal of Heat Transfer (Trans.   ASME)", 100, 199, (1978).

Natural convection heat transfer coefficients between absorber surfaces and cover plates for vertically oriented two-dimensional compound parabolic concentrators (CPC) are evaluated using finite element technique. Values of critical Rayleigh number for different concentrators ($2<C<10$) with 1/3, 2/3 & full CPC heights are determined. Generalised charts for estimating the average absorber plate Nusselt number as a function of Rayleigh number and concentration for both full and truncated CPC cavities are given. The results are useful for evaluating the convective heat loss coefficients from such collectors.

0220

Development of a 3 x CPC with Evacuated Receiver : Ballheim R W, " Proceedings ISES Silver Jubilee, International Congress ", Georgia (USA), (1979).

This paper presents some of the considerations involved in the design and development of a 2.6x compound parabolic concentrator (CPC) with an evacuated receiver. A truncated version of a C P C trough reflector system and the General Electric Company tubular evacuated receiver have been integrated with a mass producible collector design suitable for operation at 250 to 450 deg F. A ray tracing program was used in conjunction with a heat gain mathematical model to compose the effect of collector parameters on the annual performance of the collector. The parameters studied include CPC acceptance angle, truncation height, reflector error, receiver placement error, glazing transmitivity, receiver tube transmitivity, reflector material reflectivity and insolation diffuse/beam ratio. An optimum design is selected and performance predictions on an annual basis are presented for specific design conditions.

0221

Parabolocylindric Reflecting Unit and its Properties: Baranov V K,"Applied Solar Energy",11,(3-4), 34, (1975).

An account is given of the properties of a solar energy concentrator in the form of a set of parabolocylindric elements. This type of concentrator is named as the "Foklin" and is produced by the translational displacement of a segment of a parabola and its mirror image in the plane of symmetry.The focus of the displaced segment of the parabola is located at the beginning of the mirror reflection, & the angle between the axis of the parabola and the plane of symmetry is equal to the given maximum angle of incidence of rays through the large slit of the concentrator. This angle is called the parametric angle of the Foklin. The properties of the Foklin are such that it can be used as a solar energy concentrator for a particular length of time and with a given concentration ratio without adjustment of its position.

0222

Focons and Foclines as Concentrators of the Radiation of Extended Objects : Baranov V K, "Soviet Journal of Optical Technology ", 44(2), 66, (1976).

Analytical expressions are given that make it possible to determine the illumination in the plane of the exit apertures of a parabolic toroidal focon and a parabolic-cylindrical focline in those c a s e s when the field of view of these elements is completely filled by an extended radiation source with a specified luminance. The possibilities of using focons for concentrating the radiation of extended sources are shown. Focons and concentrating lens systems are compared.

0223

Concentration of Diffuse Radiation : Baranov V K, " Applied Solar Energy ", 13(2), 21, (1977).

Results are reported for investigations that make it possible to determine the fundamental laws of focon and focline operation in concentrating the radiation of extended sources. Analytic relations are given that make it possible to determine the average flux density in the minor aperture of a focon or focline as a function of the flux density set up in the plane of the major aperture by objects that fall into the field of view of such concentrators. Focons and lens systems are compared in terms of their ability to concentrate the radiation of extended sources. It is shown that when a focon i s used to concentrate the diffuse radiation of extended sources it is in many cases possible to i n - crease the density of the concentrated flux.

0224

Phase Space Conservation in a Compound Parabolic Optical Concentrator : Barnett M E,"Optik", 55, 343, (1980).

Incoherent optical propagation through a two-dimensional compound parabolic concentrator is studied with particular reference to the development of the phase space profile. It is demonstrated analytically that conservation of phase space area is satisfied for the beam within the concentrator.

0225

Optical Flow in an Ideal Light Collector/The Q_i /Q_o Concentrator : Barnett M E, " Optik ", 57, 391, (1980).

Propagation in loss less two-dimensional optical collectors can be characterized in terms of ray, phase space or vector flux representations. The relationship between these modes of analysis is discussed and illustrated by reference to a $30°/60°$ solar concentrator.

0226

The Collection of Diffuse Light on to an Extended Absorber : Bassett I M and Derrick G H, "Optical and Quantum Electronics ", 10, 61, (1978).

If light comes from a fixed direction and is to be collected on to an absorbing body which is approximately a point or a straight line, it is well known that a mirror of parabolic profile serves t h e purpose efficiently. If the light is diffused , or its direction of incidence variable, or if the absorber is not approximately a point or a straight line, the parabola is less helpful. This paper describes various particular collecting problems of this more general sort, together with efficient solutions for them, mainly by means of mirrors. The subject represents a corner of geometric o p t i c s little explored until recently, in which the notion of image formation scarcely appears. A unifying principle is the invarious of the 'Lagrange-Helmhollz' invariant, which is related to Louiville's theorem of Hamiltonian mechanics and there is a unifying method of construction of the mirror profiles by means of string.

0227

Diffuse Reflectors in Non Imaging Optics : Bassett I M & Derrick G H, " Proceedings ISES Silver Jubilee International Congress",Georgia (USA), p.557, (1979).

Let a cylindrical body of convex cross section be partly surrounded by a cylindrical bowl whose inner surface is a perfect diffuse reflector. The entrance aperture of the system, determined by the positions of the rims of the bowl, is assumed to be illuminated over the whole inward hemisphere bv the light o f uniform brightness. A method is presented for determining the reflector profiles which deliver the greatest possible power to the absorbing bodies, and a simple expression is found for the power delivered. Extensions of the method to wider problems are considered briefly. The same reflector profiles solve the complementary problems of light emission, when the absorbing body is replaced by a uniformly bright diffuse emitter.

0228

Design of Diffuse Reflectors for Collection of Diffuse Light onto an Absorbing Body : Bassett I M and Derrick G H, "Proceedings ISES Silver Jubilee International Congress",Georgia (USA), (1979).

A closed expression is given for the power delivered from a given diffuse incident radiation f i e l d to a given body, when a given diffuse reflector is employed. The condition that this closed expression has a maximum value (for fixed aperture of the reflector bowl) leads to a prescription for the reflector shape; it also leads, if the reflection coefficient is assumed to be unity, to a simple prescription for the power delivered by the "optimum" reflectors, in terms of the geometry of the absorb - ing body and the aperture, not explicitly involving the reflector shape. Examples are given exhibiting power delivered and reflector shape, for cylindrical geometry, for fully diffused incident light (i.e the whole forward hemisphere uniformly illuminated) and assuming a reflection coefficient of unity. Examples include both single absorber and arrays of parallel absorbers.

0229

The Collection and Dissemination of Light with the Aid of Diffuse Reflectors : Bassett I M and Der - rick G H, " Optica Acta ", 27, 215, (1980).

Let a cylindrical absorbing body of convex cross section be partly surrounded by a cylindrical bowl whose inner surface is a perfect diffuse reflector. The entrance aperture of the system, determined by the position of the rims of the bowl is assumed to be illuminated over the whole inward hemisphere by light of uniform brightness. A method is presented for determining the reflector profiles which deliver the greatest possible power to the absorbing body, and a simple expresssion is found for the power delivered. The same reflector profiles solve the complementary problem of light emission when the absorbing body is replaced by a uniformly bright diffuse emitter.

0230

A Concentrating Flat Plate Solar Collector : Boyd D A, Gajewski R and Winston R, " Proceedings ISES Conference", Los Angeles (USA), (1975).

A concentrating flat plate solar collector (CFP), presently under construction at American Science and Engineering, combines the concept of a compound parabolic concentrator (CPC) with that of a cylindrical black body (CBB) heat receiver. It consists of a row of compound parabolic concentrators, and a corresponding row of CBB receivers each concentrator illuminating the aperture of its respective receiver. The concentrators are positioned with their entrance apertures adjacent to each other so that essentially all radiant energy incident within the acceptance angle of the CPC is concentrated on to the CBB receivers. In order to minimise the heat loss, the receivers are embedded in a thermal insulator. The working fluid circulating in the receivers is water glycol and the target operating temperature is 120 deg C. A performance analysis of the CFP will be presented, based on the calculation of heat losses from the concentrator and the receiver. The approach used is thermal modelling of the collector in which account is taken of all heat loss mechanisms. Experimental data on the collector performance will be presented if available at the time of the meeting.

0231

Truncated "Focones" & "Foclines" : Braslavskaya M V & Baranov V K, "Applied Solar Energy", 13(3), 19, (1977).

Truncation of focones and foclines at the large aperture end leads to a relatively small reduction in the concentration coefficient but substantially reduces the depth of the concentrator. Thus, the truncation of a focone (with parameter angle of $20°$) by one third and the truncation of the analogous focline by one half results in a reduction in the concentration factor by only 10%. The formulas, graphs and tables allow us to determine readily and simply the parameters of truncated focones and foclines for a given depth and acceptable reduction in the concentration coefficient.

0232

Principles of Solar Concentrator Design : Burkhard D G, " Proceedings 1st South-Eastern Conference on Solar Energy", Alabama (USA), (1975).

A differential equation is obtained for the shape of a reflecting surface which will distribute axially symmetric light intensity into a specific irradiance over a receiver which is symmetric about the direction of the incident light. Results are applied to the design of rotationally symmetric solar reflectors and also to a two-dimensional geometry that is, one in which the reflector is a cylinder with its axis perpendicular to the incident beam.

0233

Design of Reflectors which will Distribute Sunlight in a Specified Manner : Burkhard D G & Shealy D G, " Solar Energy ", 17, 221, (1975).

A differential equation is obtained for the shape of a reflecting surface which will distribute axially symmetric light instensity into a specified irradiance over a receiver which is symmetric about the direction of the incident light. Results are applied to the design of rotationally symmetric solar reflectors and also to a two dimensional geometry, that is one in which the reflector is a cylinder with its axis perpendicular to the incident beam . The procedure is used to calculate numerically the shape of reflectors which will uniformly concentrate collimated light and also light from a point source over a flat receiver . Results are also applied to determine the shape of a reflector which will distribute collimated light uniformly over the surface of a cylinder and also over a sphere .

0234

The Winston Solar Concentrator Described as an Ellipse : Canning J S, "Solar Energy", 18, 155, (1976).

It is shown that a compound parabolic concentrator may be described as a compound elliptical concentrator.

0235

Long Term Average Performance Predictions for Compound Parabolic Concentrator Solar Collectors : Cole R, " Proceedings ISES Conference (American Section)", Orlando (USA), (1977).

This paper describes how the method of B Y H Liu & R C Jordan can be extended to calculate the performance of collectors such as the compound parabolic concentrating (CPC) collector. The method allows calculations of the monthly average of daily heat collection given only the monthly average of t h e measured daily insolation on a horizontal plane and the monthly average day time ambient temperature. The calculations were made for two plane flat plate collectros at three locations and were found to be in good agreement.

0236

Conceptual Design of a 5x CPC for Solar Total Energy Systems : Cole R, Schertz W W and Teagan W P, " Proceedings ERDA Conference on Concentrating Solar Collectors", Georgia(USA), p.3-17, (1977).

This paper describes the results of a conceptual design of a non tracking collector for a solar total energy system. Sandia Laboratories has responsibility for the evaluation of concentrating collectors in a total energy test bed. A Rankine cycle-turbine, generator, controls, thermal storage & air-conditioning equipment has been installed and checked out. The thermal energy for the facility is to be provided by a large (~800 sqm) concentrating collector field. At present a portion of the area is installed as E-W oriented linear parabolic troughs. Three additional concepts for the remaining area have been selected; a fixed mirror moving receiver system, fixed receiver moving reflector slats, and a two-axis tracking parabolic dish. All four systems use diurnal tracking and have the reflecting surfaces exposed to the elements.

0237

Performance and Testing of a Stationary Concentrating Collector : Cole R L, Allen J W, Levitz N M, McIntire W R and Schertz W W, "Proceedings ERDA Conference on Concentrating Solar Collectors ", Georgia (USA), p.3-31, (1977).

This paper reports on the development of non imaging solar collectors for heating and cooling applications. A totally stationary concentrating collector has been designed, built and tested. Performance of the collector is substantially better than flat plate collectors, and the collectors are suitable for powering mechanically driven air conditioning systems as well as conventional absorption cycle machines.

0238

Non Evacuated Solar Collectors with Compound Parabolic Concentrators : Collares Pereira M, Goodman N B, Greenman P, O'Gallagher J, Rabl A, Wharton L and Winston R, " Proceedings ISES (American Section) Conference ", Orlando (USA), (1977).

Properly designed solar collector panels using compound parabolic concentrator (CPC) troughs with concentration ratio in the range 3 to 10 are suitable for operation (assuming at least 40% efficiency)at 100 to 150 deg C above ambient even if not surrounded by vacuum. The first generation of CPC collector fell short of the expected performance because of various parasitic heat losses through the ref-lector and through the insulation. Analysis of the data indicated, however, that optical performance and frontal heat losses agreed with the predictions and that CPC collectors with a reasonable perf-ormance could be built. This paper describes the design of CPC collectors with concentrations 3 and 6.5 and reports the test results.

0239

Compound Parabolic Concentrators with Non-Evacuated Receivers - Prototype Performance & a Large Scale Demonstration in a School Heating System : Collares Pereira M, Goodman N B, Greenman P, O'Gallagher J, Rabl A, Simmons H, Wharton L and Winston R, " Proceedings ISES Congress & Exposition", New Delhi (India), p. 1227, (1978).

The design and fabrication of two different prototype CPC concentrating collectors with non-evacuated receivers is described and the results of performance tests are reported. In addition, the installation of an array of CPC concentrating collectors conforming to one of these prototype designs in an operating space heating application is also described.

0240

Approximations to the CPC - A Comment on Recent Papers by Canning and Shapiro : Collares Pereira M, O'Gallagher J and Rabl A, " Solar Energy ", 21 (3), 245, (1978).

This note clarifies some of the issues involved in approximating a compound parabolic concentrator surface by reflector surfaces which can be manufactured easily. It is made clear that in any case the maximum possible concentration ratio which can be achieved by using such approximated surfaces will not be higher than the ideal non imaging solar concentrators.

0241

Preliminary Results from a Test Array of 3 x CPC Collectors in a School Heating Application : Collares Pereira M, O'Gallagher J J, Rabl A, Simmons H, Stein C, Wharton L, Winston R and Ziteck W, " Proceedings ISES (American Section) Conference ", Denver, Colorado (USA), 2.1, 347, (1978).

An array of 56 trough compound parabolic concentrator (CPC) solar collector modules has been installed on the roof of an elementary school on the Navajo Indian Reservation near Gallup, New Mexico. The array represents an effective net collective area of 72.9 sqm and has been fully instrumented to provide a data base for analysing CPC collection characteristics. This is the first quantitative field test of these collectors conducted elsewhere.

0242

Design & Performance Characteristics of Compound Parabolic Concentrators with Evacuated & with Non-Evacuated Receivers : Collares Pereira M, O'Gallagher J, Rabl A, Winston R, Cole R, McIntire W, Reed K & Schertz W, "Proceedings ISES Silver Jubilee International Congress",Georgia (USA), (1979).

An experimental higher concentration 5 x CPC requiring approximately 12 tilt adjustments annually when used with similar available evacuated absorber has been built having a measured efficiency of 60 % at 220 deg C above ambient and is capable of efficiencies near 50 % at 300 deg C. With such thermally ef-ficient absorbers higher concentrations are not necessary or desirable. The temperature capabilities for CPC's with non evacuated absorbers are somewhat more limited. However, with proper design, taking care to reduce parasitic thermal losses through the reflectors, non evacuated CPC's will out-per-form the best available flat plate collectors at temperatures above 50 deg C to 70 deg C.

0243

Applications of CPC's in Solar Energy - An Overview : Collares Pereira M, O'Gallagher J, Cole R, Gorski A,

Rabl A, Winston R, McIntire W, Reed K and Schertz W, "Proceedings ISES Silver Jubilee International Congress ", Georgia (USA), (1979).

The CPC is not a specific collector, but a general design principle for maximizing the geometric concentration, C, for radiation within a given acceptance half angle $\pm\,\theta_c$. The maximum limit exceeds by a factor of 2.4 that attainable by systems based on focussing optics. The wide acceptance angles permitted using these techniques have several unique advantages for solar concentrators, including the elimination of the diurnal tracking requirement at intermediate concentrations (upto $\cong$10X),collection of circumsolar and some diffuse radiation and relaxed tolerances. Because of these advantages, C P C type concentrators have applications in solar energy whenever concentration is desired e.g. for a wide variety of both thermal and photovoltaic uses.

0244

Design of Fixed Evacuated Concentrating Collectors for Operation at 80 deg and 200 deg C : Garrison J D, " Proceedings ERDA Conference on Concentrating Solar Collectors", Georgia (USA),p.3.45, (1977).

The design of a low cost, fixed evacuated concentrating collector is presented in this paper. An east-west alligned all glass Winston collector geometry with a selective absorber is found suitable.   The collector length and width is optimized by heat transfer calculation while radiation collection calculated determine the acceptance angle and the other parameters.

0245

Evaluation of an Optimized Solar Thermal Collector by a New Method : Garrison J D, "Proceedings   ISES (American Section) Conference ", Orlando (USA), (1977).

A model for calculating, on the average, the angular distribution and intensity of solar radiation at the site of a solar collector is presented. The irradiance on a horizontal surface and the time o f day and year are required as input data.   This model is used to complete the design of an optimum fixed solar thermal energy collector.   It is then also used to examine the performance of this collector under various operating conditions, and to compare this performance with other fixed collectors.

0246

Optimisation of a Fixed Solar Thermal Collector : Garrison J D, " Solar Energy ", 23, 93, (1979).

Criteria are presented for optimizing solar thermal energy collection.   These criteria are used i n setting the design of a fixed solar thermal energy collector.   This design is obtained by proceeding carefully through a series of optimization steps. Initial optimisation steps lead to an all glass vacuum collector tube whose side and lower walls are internally silvered to provide optimal Winston concentration on an interior glass coated with a selective absorber. Heat transfer calculations, performed for an array module of these collector tubes, produce values for the radiation, heat conduction & pumping losses and indicate operating conditions which minimize these losses.

0247

Radiation Collection by an Optimized Fixed Collector : Garrison J D," Solar Energy ", 23, 103, (1979).

A model for determining, on the average, the angular distribution and intensity of solar radiation at the site of a solar collector as a function of the irradiance on a horizontal surface and the time of day and year is presented. The model is used to study radiation collection by a collector whose design has been approximately optimized.   The variation of radiation collection by this collector with changes in the clearness index, operating temperature, ambient temperature, acceptance angle, tilt angle, selective absorber properties and window reflection is studied.   This indicates how to further optimize the collector design.   The performance of this optimized collector is compared with that of other existing fixed collector designs.

0248

Study of Solar Thermal Energy Collection Using Fixed and Tracking Collectors : Garrison J D, Craig G T and Morgan C, " Proceedings ISES (American Section) Conference ", Denver, Colorado(USA),2.1, 919, (1978).

Energy collection at several sites is calculated for air and vacuum flat plate collectors and a vacuum collector using cylindrical Winston concentrator.   The hourly intensity and angular distribution of solar radiation is predicted using insolation measurement and an insolation model. Tracking configurations include two axis, vertical axis, east - west axis tracking.   It is the purpose of this study to calculate estimates of energy collection by a variety of fixed and tracking collectors operat i n g at temperatures from ambient to 30 deg C.

0249

Compound Parabolic Concentrator : Giugler R, Rabl A, Sevick V J and Winston R, "Proceedings ISES Conference", Los Angeles (USA), (1975).

Several different versions of CPC solar collectors with concentration ratios 3 ,5 and 10 have b e e n constructed and tested.   They were mounted as a tilted platform approximately normal to solar n o o n. Insolation, as well as ambient air temperature and wind velocity, were monitored.   The heat output of the collectors was determined from flow rate and temperature rise of the fluid (50% water and 50% ethylene glycol). For diagnostic purposes, temperatures of various points of the collectors were monitored and heat losses were measured at night.   In order to compare the findings with theory, one has to know the loss through the back and sides of the collector, and this too has been measured. The experimental results are summarised in terms of effective U values, and in graphs of collection effici - ency versus temperature.

0250

On the Use of Solid-Dielectric Compound Parabolic Concentrators with Photovoltaic Devices : Goodman N B, Ignatins R, Wharton L and Winston R, "Proceedings ISES Conference - Sharing the Sun", Winnipeg (Canada), 6, 325, (1976).

Prototype solid dielectric compound parabolic concentrators are tested. By means of the geometry and refractive properties of a transparent solid they provide a technique for increasing the power output of silicon solar cells exposed to the sun by an amount equal to the increase in effective collecting area. The response is uniform over a large angle which eliminates the necessity of diurnal tracking of the sun. The technique can be applied to the construction of thin panels and has the potential for significantly reducing their cost per unit area.

0251

Optical and Thermal Design Considerations for Ideal Light Collectors : Goodman N B, Winston R & Rabl A, " Proceedings ISES Conference - Sharing the Sun ", Winnipeg (Canada), 2, 336, (1976).

The design of practical CPC collectors may require a gap between reflector and absorber to avoid heat leaks. Generic solutions are given which minimize the total losses. The merits of second stage concentrators are described and analysed. Prescriptions are given for maximizing the concentration o f lens - mirror combinations.

0252

Relation Between Concentration and Acceptance Angle in Solar Collectors : Grasso F, Musumoci F a n d Triglia A, " Solar Energy ", 22, 521, (1979).

The maximum concentration ratio has been derived by Winston for optical systems having an angular acceptance described by a step function. In this paper a more comprehensive relation between maximum concentration ratio and angular acceptance is obtained. This relation is valid for any acceptance function ; the proof is based on thermodynamics and is valid for large angles as well. The total energies collected yearly by concentrators having different acceptance functions are compared.

0253

Reduction of Intensity Variations on the Absorbers of Ideal Flux Concentrators : Greenman P, "Applied Optics", 19, 2812, (1980).

Although ideal concentrators such as the compound parabolic concentrator have the property that concentrated flux is uniformly distributed on the absorber when the angular acceptance is filled by the incident flux, the instantaneous flux distribution may be highly non-uniform when illuminated by a point source. These non-uniformities may be reduced by texturing the surface with small distortions. The necessary reduction of concentrator throughput is small enough to allow such textured concentrators to accommodate a wide range of tolerances in concentrator efficiency and uniformity of the flux distribution. In particular the suitability of such concentrators for some space application is discussed. Results of measurements on a test model are presented that demonstrate the effective reduction of flux non-uniformities by reflectors that have been textured by a simple process. Two - stage concentrators for attaining high concentration ratios are shown to distribute the concentrated flux fairly uniformly across the absorber when illuminated by point sources within the field of view.Again the remaining non-uniformity may be removed by texturing the reflecting surfaces (entailing some reduction in concentrator efficiency).

0254

Maximally Concentrating Collectors for Solar Energy Applications : Guay E J, " Solar Energy ", 24, 265, (1980).

The performance characteristics of a solar energy collector are described by a function relating the concentration to the direction of incident radiation. A strict bound on the integral of this function is established. This is used to define a realistic class of collectors with maximal concentration which can serve as a basis for design and optimal studies.

0255

Design of Non Tracking Concentrator Capable of Avoiding Shadow on the Absorber Plate : Gupta A a n d Kumar S, " Proceedings Indian National Solar Energy Convention ", Bhavnagar (India), p.127, (1978).

In this paper the design of a non tracking concentrator in which the absorber plate is completely illuminated lengthwise as well widthwise throughout the operating period is presented. The calculations have been done using an averaged out 6 hour operating period. This requires taking half-acceptance angle $\phi = 60^\circ$ in the north - south vertical plane and a half-acceptance angle $\theta = 45^\circ$ in the east - west plane containing the symmetry axis of the concentrator. It is shown that the length to width ratio of the absorber area depends on the two half-acceptance angles ϕ and θ and on the two design parameters C and M for the concentrator, which are defined in the paper. This ratio comes out to be 48 : 1 for the case where $\phi = 6^\circ$, $\theta = 45^\circ$, C = 10 and M = 1.4 and reduces to 42 : 1 as C and M become infinitely large, ϕ , θ remaining same.

0256

Comparison of Elliptical and Parabolic Non-Imaging Concentrators : Gurnee E F, " Solar Energy ", 19, 323, (1977).

In this note a comparison is made between two non-imaging concentrators with elliptical and parabolic cross sections respectively and it is shown that the elliptical cross section does not fulfill the requirements for an ideal cylindrical concentrator.

0257

Hyperbolic Cone Channel Condenser : Gush H P, " Optics Letters ", 2, 22, (1978).

An optical condensing system is described that consists of a lens and the polished interior of a cone generated by rotating a hyperbola about an appropriate axis. An optimum concentration of light is achieved in a relatively compact system. In certain circumstances the hyperbola reduces to a straight line simplifying the construction of the cone.

0258

Investigations on the Prediction of Thermal Performances of Compound Parabolic Concentrators : Hariprasad C R, Natarajan R and Gupta M C, " Proceedings 2nd International Conference on Alternative Energy sources", Miami (USA), (1977),

The present work proposes an iterative scheme to predict the temperature of the absorber, the glazing, and the outlet fluid, and for the evaluation of convective and radiative losses and the collector efficiency, for given values of mass flow rate and insolation, under steady state conditions for a CPC. The computations are performed for different concentration ratios for a simple glazed CPC collector.

0259

Practical Concentrators Attaining Maximal Concentration : Harting E, Mills D R and Giutronich J E, " Optics Letters ", 5(1), 32, (1980).

A new class of radiation concentrators is described that achieve maximal concentration of radiation from a uniform source. Unlike ideal concentrators, which accept all radiation within a given acceptance angle and none outside, the new maximal concentration collectors may reject some radiation from within the nominal acceptance angle. However, the new concentrators offer small mirror or refractor area, high practical concentration levels (unlike ideal designs, which must be truncated), and an adaptable concentration response versus radiation incident angle. The new concentrators are exceptionally well suited to solar energy applications and should also prove useful for radiation detection or distribution.

0260

Efficient Light Coupler for Threshold Cerenkov Counters : Hinterberger H and Winston R, " Review of Scientific Instruments ", 37, 1094, (1966).

The design of a light funnel which performs a maximum reduction in spot size consistent with high efficiency is discussed. It is shown that the funnel is efficient over its entire aperture. The average number of reflections is 1.4 and the mean transmission is 78 %. Such a funnel or a cluster of them hexagonally close packed are found useful for collecting Cerenkov light with high efficiency.

0261

Efficient Design of Lucite Light Pipes Coupled to Photomultipliers : Hinterberger H and Winston R, " Review of Scientic Instruments", 39, 419, (1968).

An efficient design for Lucite light pipes coupled to photomultiplier tubes is presented. It is shown that the ratio of the entrance aperture to the exit aperture of the pipe can be increased by a substantial factor $n^2/(n^2-1)$ where 'n' is the index of refraction of the light pipe material, while maintaining high efficiency. For Lucite ($n \cong 1.5$) this factor equals 1.8 and represents an effective increase of photocathode area by this amount. The light funnel has aluminized walls to reflect light incident at angles less than the acceptance angle and should be joined to both light pipe and photocathode with some optical coupling material such as silicone grease.

0262

Use of a Solid Light Funnel to Increase Phototube Aperture Without Restricting Angular Acceptance : Hinterberger H and Winston R, "Review of Scientific Instruments", 39, 1217, (1968).

The possibility of using a solid light funnel increasing phototube aperture without restricting angular acceptance is explored. Such a funnel may also be used with Cerenkov counters. It is estimated that the cost of such a funnel will be considerably less than the additional phototubes and associated electronics otherwise required.

0263

Solar Collector with no Convection Losses : Hinterberger H and O'Meara J, "Proceedings ISES Conference - Sharing the Sun ", Winnipeg (Canada), 2, 138, (1976).

A geometry inherently free of convective circulation by orienting the panel horizontally facing downward is presented. A parabolic mirror below the panel concentrates the sun light on to the panel. A second advantage of this geometry is that it is less expansive than a conventional flat panel. Since the mirror concentrates the sun light, the panel area is a small fraction of the area of the intercepted sunshine. A model of this collector reaches a stagnation temperature of 370 deg C.

0264

Design of a Specular Aspheric Surface to Uniformly Radiate a Flat Surface Using a Non Uniform Collimated Radiation Source : Horton T E & McDermit J H ,"Journal of Heat Transfer (Trans. ASME)", 94 , 453, (1972).

When uniform radiative heating is desired on a surface the radiative heating system normally suffers from either (i) low efficiency or (ii) the heating not being as uniform as desired. Using an axisymmetric collimated but non uniform source of radiation and a properly designed specular surface, uniform heating can be obtained at 100 % efficiency. The differential equation which describes such a surface is derived. The derivation is facilitated by the use of a vectorial ray trace formulation. The differential equation has been solved and results are presented for the case of an axisymmetric non uniform collimated source uniformly heating a flat surface.

0265

Optical Properties of Cylindrical Elliptic Concentrator : Jones Jr. R E, " Proceedings ISES (American Section) Conference ", Orlando (USA), (1977).

The optical properties of elliptic concentrators are investigated using numerical ray tracing. Elliptic concentrators are shown to be non-ideal concentrators and the angular acceptance is studied for two different designs. The relationship between angular acceptance and concentration for a general cylindrical concentrator is derived using the second law of thermodynamics.

0266

Collection Properties of Generalized Light Concentrators : Jones Jr. R E, "Journal Optical Society of America ", 67, 1594, (1977).

The collection properties of generalised non imaging radiation concentrators are studied for the case of geometrical optics. The second law of thermodynamics is applied to a concentrator located at the center of a cavity radiation source. Generalised relationships are obtained between concentration & angular acceptance for both two dimensional and three dimensional geometries. These relationships are shown to reduce to those previously known for the special case of ideal concentrators.

0267

The Winston Solar Concentrator is not an Elliptical Concentrator : Jones Jr. R E, "Solar Energy ", 20 (2), 179, (1978).

It is pointed out that a cylindrical elliptical concentrator is not an ideal Winston concentrator. The position and slopes of the end points of their respective curves are the same but between the end points the two curves are not identical and this results in different optical performances.

0268

Optical Properties of Compound Circular Arc Concentrators : Jones Jr. R E and Anderson G C , " Solar Energy ", 21, 149, (1978).

A complete analysis of the optical properties of compound circular Arc Concentrators is presented and compared with those of a compound parabolic concentrator.

0269

Optical and Thermal Properties of Truncated Compound Circular Arc Concentrators : Jones Jr. R E and Anderson G C, " Proceedings ISES Silver Jubilee International Congress ", Georgia (USA), (1979).

An economical way to form non focussing concentrators would be to support aluminized plastic films at the edges, and inflate them into circular cylinders. Here, the optical and thermal properties of circular arc approximations to truncated compound parabolic concentrators are investigated. Annual solar fractions for residential heating loads are found to be nearly the same for low and medium concentrations.

0270

Circular Arc Approximations of Truncated CPC Collectors : Jones Jr. R E and Anderson G C, "Solar Energy ", 25, 139, (1980).

The optical and thermal properties of circular arc approximations to truncated compound parabolic concentrators are investigated. It is shown that the resulting compound circular arc concentrators can be economically formed by inflating aluminized plastic films which are held at the glaze and absorber. The annual performance of systems incorporating such collectors is studied for residential and process heat loads.

0271

An Investigation of Experimental Performance of a Compound Parabolic Concentrator : Krishna Rao K, Ahmed S B, Natarajan R and Gupta M C, " Proceedings Indian National Solar Energy Convention ", Bhavnagar (India), p. 137, (1978).

The experimental performance of a header type Compound Parabolic Concentrator and a Flat Plate Collector (FPC) has been investigated here. Two configurations of collectors have been considered : one in which the absorber tubes were all oriented in the east - west direction, and another in which the absorber tubes of the FPC were oriented in the north - south direction. The effects of fluid flow rates (5, 10 and 15 l/h) and angle of tilt (latitude $\sim$13; declination of the day and horizontal) on hourly efficiency and useful thermal energy gain have been determined as a function of the hour of day. In the present experimental configuration considered, the CPC was more efficient than FPC for about two hours around noon when the tilt was equal to declination, but when tilted to an angle equal to latitude the FPC was found to have better overall performance.

0272

Distribution of Beam Radiation on the Receiver Plane of a CPC Solar Concentrator : Kreider J F,"Proceedings ISES Conference ", New Delhi (India), p. 1299, (1978).

The effect of a non uniform flux distribution over the absorber plane of a CPC on its optical & thermal performance is discussed. It is found that the thermal effects of non-uniform absorber flux are of no importance in engineering designs of CPC's, whereas a variation of upto 9% may be observed in optical efficiency with incidence angle.

0273

Influence of Circumsolar Radiation on the Performance of Nonfocussing - Concentrating Collectors : Kreith F, Oswald R and Winston R, " Proceedings ERDA Conference on Concentrating Solar Collectors ", Georgia (USA), (1977).

This paper compares the thermal performance predicted on the basis of the idealized model which assumes isotropic diffuse radiation with a more realistic model which considers the circumsolar radiation distribution. A preliminary distribution of circumsolar radiation applicable to various weather conditions is presented.

0274

The Corneal Cones of Limulus as Optimized Light Concentrators : Levi Setti R, Park D A and Winston R, " Nature ", 253, 115, (1976).

In this paper the qualitative similarity between the profile shape of an ideal light concentrator and that of the crystaline cones of Limulus is discussed.

0275

Sunshine on Tap : The Focon : Litynski Z, " Foreign Science Bulletin "(Aerospace Technology Division Library of Congress, Washington D.C. USA), 2 (9), 1, (1966).

A brief account is given of recent Soviet research on light conductors, particularly the focon. It is shown that, even though the Soviet studies are mainly based on the work of Kapany in the United

040

States,the Soviet is also developing original ideas aiming at the application of conic lightguides to
solar concentrators.

0276

Quasi – Optimum Pseudo – Lambertian Reflecting Concentrators : An Analysis : Luque A, " Applied Optics",
19, 2398, (1980).

A two stage concentrator in which the first stage is reflective and the second stage considers t h e
first one as a Lambertian source in order to obtain the highest possible gain is analysed.The position
and profile of the first stage is determined, which happens to be a parabola and the value of the gain
as a function of the acceptance angle and the focal length of the first stage is calculated.

0277

Performance of an Evacuated Tubular Collector Using Non – Imaging Reflectors : Mather Jr. G R & Beekley
D C, " Proceedings ISES Conference – Sharing the Sun ", Winnipeg (Canada), 2, 64, (1976).

Specular reflectors have been found to improve significantly the performance of the Owens- Illinois
evacuated tube collector. Two types of specular reflectors, both of which are non imaging and result
in moderate concentration ratios, have been investigated in detail. Both collect diffuse radiation efficiently, and neither requires tracking of the sun. The results of optical analysis and experimental tests of the two reflectors are described and compared,with similar results for the diffuse reflector used with present Owens – Illinois tubular collectors. Energy outputs of the collector are found
to increase by 25 – 35% when the diffuse reflector is replaced with either of the specular reflectors.

0278

Design Considerations for Solar Radiation Concentrators : Mather Jr. G R & Beekley D C, " Proceedings
ERDA Conference on Concentrating Solar Collectors ", Georgia (USA), p. 3 – 87, (1977).

The character of insolation as to beam and diffuse components must enter into the design considera –
tions for proper evaluation of a concentrating surface. Various aspects of this problem are discused in terms of tests and evaluation of several concentrating surfaces. The surfaces are shown to provide various degrees of collector performance enhancement both for the beam and diffuse components of
insolation. Modest concentration ratios are achieved b y t h e use of these nonimaging optical surfaces but collector performance is increased regardless of the character of the insolation. The problems concerning the manufacure and durability of the surface are discussed.

0279

A Compound Parabolic Concentrator's Array Optimized for Northern Climates : Stein M L, " Proceedings
ISES Conference – Sharing the Sun ", Winnipeg (Canada), 2, 375, (1976).

The radiative exchange between collector and source sets the limit for efficient collector operation.
When actual insolation conditions and performance requirements are considered for northern latitudes,
the compound parabolic concentrator is shown to perform better than t h e flat plate since it approaches the limit for optimum radiative exchange. The relationship between concentration and acceptance
angle is analyzed for its effect on the collector's overall thermal performance. The resulting collector is compared with an equivalent flat plate type.

0280

An Analysis for Designing Solar Concentrator Collector Configurations : McDermit J H, " Proceedings 1st
South – Eastern Conference ", Huntsville, Alabama (USA), p.405, (1975).

An analysis for predicting the heat flux distribution on collectors in the focal region of solar concentrators has been formulated. The analysis is rather general and can consider limb darkening e f-
fects, collector placement errors, concentrator errors and concentrator pointing errors. By coupling
finite differencing techniques and heating requirements to this analysis, it can also be used to design concentrators that give prescribed heating on the collector surface. Illustrative results a r e
presented. It is concluded that the concentrator and collector should be designed as a unit and that
concentrator configurations that are more desirable than the paraboloid and cylindrical parabola can
be designed.

0281

Reflective Optics for Obtaining Prescribed Irradiative Distributions from Collimated Sources :
McDermit J H and Horton T E, " Applied Optics ", 13, 1444, (1974).

The differential equations are derived for reflective optical surfaces in systems that give prescrib –
ed spatial irradiation from collimated sources. The equations are derived for axisymmetric configurations assuming specular reflective surfaces. The formulation considers general sources radiant emittance distributions, general receiver shapes and receiver irradiance distributions. The formulat i o n
of the differential equations is effected through he vector formulation of ray trace equations for the
systems, formulation of differential radiant energy balance equations, and appropriate differentiation
of the ray trace equations. The differential equations are derived for two configurations.

0282

Optical Design of Solar Concentrators : McDermit J H and Horton T E, " Proceedings 2nd A I A A / A S M E
Thermophysics & Heat Transfer Conference ", Palo Alto, California (USA), (1978).

A generalized technique for the optical design of solar concentrators has been developed. The design
technique considers limb darkening effects, collector placement errors, concentrator optical errors &
concentrator pointing errors. In addition to giving the designer the ability to explore the sensitivity of a design to the above parameters, the technique also allows the designer to prescribe the heating distribution on the collector surface. A detailed derivation of the design technique and illustrative results are presented.

0283

Truncation of Nonimaging Cusp Concentrators : McIntire W R, " Solar Energy ", 23, 351, (1979).

Reflector shapes for truncated nonimaging cusp concentrators having various acceptance angles a r e presented , as well as curves for height – aperture and mirror arc length – aperture ratios versus concentration ratio. In addition to their general utility in concentrator design, the latter curves have special significance for thermoformed plastic reflector substrates. The reflector height – aperture ratio is the 'draw' and the reflector arc length – aperture ratio is the 'stretch' to which the material is subjected during forming.

0284

Stationary Concentrators for Tubular Evacuated Receivers: Optimization and Comparison of Reflector Designs : McIntire W R, " Proceedings ISES (American Section) Conference", Delaware (USA),p.505,(1980).

In this paper, a comparison of the optical efficiencies of the different collector designs as a function of incident angle is presented and the factors which determine the optical performance of a given concentrator design are discussed.

0285

Elimination of the Optical Losses Due to Gaps Between Absorbers and Their Reflectors : McIntire W R, " Proceedings ISES (American Section) Conference ", Delaware (USA), p. 600, (1980).

In this paper a new type of a nonimaging concentrator design for tubular absorbers is presented.These reflector designs have higher optical efficiencies than the original nonimaging optics designs, because they eliminate the gap losses, and because they allow some of the light reflected from the absorber tube and its cover shroud to be reflected back onto the absorber. The new reflector designs have potential for stationary concentrators for tubular evacuated receivers and for secondary concentrators for linear focussing systems.

0286

Optimization of Stationary Nonimaging Reflectors for Tubular Evacuated Receivers Aligned North – South: McIntire W R, " Solar Energy " , 24, 169, (1980).

This detailed study optimizes the reflector design of stationary nonimaging reflectors for tubular evacuated receivers aligned north-south for maximum daily energy collection. It includes the effects of reflection losses, reflector-receiver alignment errors, variation of selective surface absorptance with angle of incidence on the receiver, and losses through the gap between the receiver and reflector.

0287

New Reflector Design which Avoids Losses Through Gaps Between Tubular Absorbers and R e f l e c t o r s : McIntire W R, " Solar Energy ", 25, 215 (1980).

A reflector design has been developed which eliminates the loss of solar radiation through the g a p between the tubular absorber and the reflector. The new design gives higher optical efficiencies by eliminating the gap losses and enhancing the net absorptance of the receiver tubes.

0288

Development of a Stationary Concentrator : McVeigh J C, " Proceedings 2nd International Solar Forum ", Hamburg (West Germany), p 141, (1978).

A stationary collector with a compound parabolic concentrator has been developed for countries with a relatively high proportion of diffused radiation. These collectors have several advantages over the traditional flat plate collector, for example the area of the absorber surface relative to the collector panel area is small. This enhances the performance under moderate radiation conditions and leads to economies in production.

0289

The Place of Extreme Asymmetrical Non-Focussing Concentrators in Solar Energy Utilisation : Mills D R, " Solar Energy ", 21, 431, (1978).

The extreme asymmetrical concentrator (EAC) is discussed with reference to solar energy application, EAC's in E-W trough systems are found to offer advantages compared with conventional symmetrical [CPC] troughs in, stationary, occasionally tilting, and frequently adjusted versions. For the first time, the matching of yearly output to a yearly demand cycle is shown to be a possibility in completely stationary collector systems.

0290

Two Stage Tilting Solar Concentrators : Mills D R, "Solar Energy", 25, 505, (1980).

In this paper a novel type of two-stage non-focussing concentrator is described which makes economic photovoltaic and thermal solar conversion possible at current prices in many locations. The new concentrators use exceptionally little mirror (1.3 - 1.8 times full aperture), yet achieve concentrations upto 12X without diurnal tracking. Optical efficiency is high, and no moving parts are required except for a simple pivot and securing device.

0291

Developments in Asymmetrical Nonimaging Solar Concentrators / Wedge-shaped Totally Internally Reflecting Concentrators : Mills D R & Giutronich J E, " Proceedings ISES Conference ", New Delhi (India),(1978)

The design principle and performance of asymmetrical nonimaging solar concentrators is described. The development of a new family of refracting nonimaging concentrators, called the wedge concentrators , which make use of the phenomenon of total internal reflection is also reported.

0292

Asymmetrical Non-Imaging Cylindrical Solar Concentrators : Mills D R & Giutronich J E, " Solar Energy ", 20, 45, (1978).

In this paper both Parabolic and Non-Parabolic Asymmetrical Concentrators are examined and compared with symmetrical designs. Among the advantages of asymmetrical systems are (i) a concentration versus

042

time of day relationship which can be designed to compensate for projected solar area fall-off in early morning and late afternoon, allowing more uniform output where this is desirable, (ii) greater operational flexibility for accommodating unexpected fluctuations in demand, (iii) easier adaption to vacuum-insulated receivers in one configuration and (iv) the possibility of substantially increased concentration and energy collection per unit of mirror area for systems with receivers which can make use of large daily variations in energy input.

0293

Ideal Prism Solar Concentrators : Mills D R and Giutronich J E, " Solar Energy ", 21, 423, (1978).

A completely seperate but parallel family of nonimaging concentrators is proposed which utilizes the phenomenon of total internal reflection within materials of high refractive index to achieve concentration. In both symmetrical and asymmetrical forms, the new concentrators satisfy the maximum concentration limits for ideal radiation transformers for a given acceptance angle. Light radiating from the exit aperture is restricted in angle, but concentration performance is good and the irradiation of the exit aperture is much more uniform for a distant point source than in any other design.

0294

A Stationary Prism Concentrator for Solar Cells : Mills D R, Giutronich J E and Harting E, " Proceed - ings ISES Silver Jubilee International Congress ", Georgia (USA), (1979).

Prism systems which are allowed to tilt occasionally throughout the year have been formulated & tested. A possible future development is a hybrid system which would extract low grade heat at 60-70 deg C as well as the present electrical output. Progress in this area is outlined. Extensive computer calculations have been carried out and these are presented and compared with experimental results. Anticipated future prism development projects in Australia are described.

0295

Symmetrical and Asymmetrical Ideal Cylindrical Radiation Transformers and Concentrators : Mills D R & Giutronich J E, " Journal of Optical Society of America ", 69, 325, (1979).

Ideal cylindrical radiation transformers and concentrators are examined in both, asymmetrical and symmetrical forms. Symmetrical ideal transformers are found to give the lowest peak concentration for a given angle of acceptance. Average concentration for uniform diffused radiation, or a distant point source of constant angular velocity, is found to be independent of the symmetry of the transformer; for an apparently accelerating source such as the sun, however, an asymmetrical transformer can give higher performance than a symmetrical unit when averaged over time.

0296

Performance of a Seasonally Adjusted Concentrator with Modified Absorber : Mullick S C and Nanda S K, " Proceedings Indian National Solar Energy Convention ", Bombay (India), p. 104, (1979).

The heat losses from concentrators can be reduced by thermally isolating the absorber with optimal air gap between the absorber and the glass cover. However, some of the reflected rays will escape through this gap resulting in increased optical losses. The present paper describes the design procedures for modifying the reflector and the target shapes to minimise the optical losses through the air gap. The concentrator has been fabricated and tested. The results for the modified absorber are compared with the plain tubular one. The intercept factor is improved from 0.775 to 0.858.

0297

Collimated Radiation in Conical Light Guides : Myer J H, "Applied Optics ", 19, 3121, (1980).

A simple method is presented for finding the dimensional proportions of conical light guides for collimated radiation.

0298

Design and Performance of a Seasonably Adjusted Solar Concentrator : Nanda S K, Mullick S C & Chawla O P, " Regional Journal of Energy Heat & Mass Transfer ", 2 (1), 75, (1980).

A concentrating collector has been designed and fabricated which is inexpensive, non-tracking and reasonably efficient at temperatures of 100 deg C above the ambient. Problems involving materials & construction, the optical efficiency, and heat losses are analysed in detail. The concentrator is made of commercially available glass mirror strips with a thermally isolated absorber. Its optical and thermal performance has been studied by experiments performed over a large number of days. The heat loss factor is obtained as a function of operating temperature.

0299

Compound Parabolic Concentrator for Operation at 300 C : O'Gallagher J O, Collares Pereira M, Rabl A & Winston R, " Proceedings ISES (American Section) Conference ", Denver, Colorado (USA), 2.1, 885, (1978).

Because nonimaging concentrators of the compound parabolic concentrator (CPC) type are limited to concentration factors less than 10, in nontracking applications, it is sometimes not recognised that they are capable of competitive performance at the high temperature required for power generation. In particular, when an intermediate concentration is combined with other recent high technology improvements in absorber design i.e. selectively coated vacuum enclosed receivers, collector heat losses can be reduced to values per unit aperture area, usually associated with much higher concentration (≥ 15 X) linear focussing tracking collectors.

0300

Absorption Enhancement in Solar Collectors by Multiple Reflections : O'Gallagher J J, Rabl A, Winston R and McIntire W, " Solar Energy ", 24, 323, (1980).

Evacuated dewar type absorbers currently available reflect away a significant fraction (20 - 30%) of the incident radiation so that in such cases, increasing absorption may be more important than maximizing concentration, at least upto moderate operating temperatures. A fraction of this reflected energy may be recaptured by increasing the ratio of the receiver to aperture area to take advantages of cavity effects. This note examines the problem of the corresponding enhancement from the point of view of the second law of thermodynamics. A limit for the enhancement factor is derived and it is shown that the limit can indeed be reached in cases of practical interest if the receiver is coupled to an involute reflector with concentration less than unity. A simple formula is obtained for enhancement factor which agrees well with ray trace results.

0301

Process Steam Generation by a Facetted Non-Imaging Solar Concentrator : Ortabasi U, Soderstrom K G , Lopez A M, Atienza J A and Winston R, " Proceedings ISES Silver Jubilee International Congress ",Georgia (USA), (1979).

A modular non-imaging trough collector for the generation of process steam in a tropical environment has been designed and built. The concentrator design has the desirable features of low cost and light weight. Transporting and installation costs are reduced by its modular aspect. The most innovative feature of this concentrator is that the mirror surface consists of long and narrow planer facets sealed inside low cost glass tubes. The absorber for this collector is also a new design.

0302

Cusp Mirror – Heat Pipe Evacuated Tubular Solar Thermal Collector : Ortabasi U and Fehlner F P,"Solar Energy ", 24, 477, (1980).

A solar thermal collector w a s constructed based on an internal 1.15 x cusp concentrator, thermal inulation involving a vacuum and selective absorber, and thermal transfer to a manifold via heat p i p e action. Performance of the collector was compared with that of an evacuated, selectively coated, flat plate absorber equipped with flow – through heat transfer. It was shown that with single collector , collector tubes, mirror losses lowered the optical efficiency of the cusp heat pipe collector below that of the flat plate, while the smaller absorber area of the heat pipe reduced thermal losses at absorber temperatures above ambient. Thus, a crossover in efficiency occurred such that the flat plate was more efficient at low $\Delta T/H$ while the cusp heat pipe was more efficient at high $\Delta T/H$. Testing of modules showed that manifold losses and gains could dominate these collector effects when the collector area approximately equalled the manifold area.

0303

An Internal Cusp Reflector for an Evacuated Tubular Heat Pipe Solar Thermal Collector : Ortabasi U and Buehl W H, " Solar Energy ", 25, 67, (1980).

This study involves the optical analysis of a slightly concentrating, symmetric cusp reflector inside a tubular glass envelope with a cylindrical heat pipe as the solar absorber. Differential equations of the cusp reflector optics, given the geometrical restriction, are derived and solutions for the largest possible aperture inside a given diameter envelope and acceptance angle are presented. As an extension of the same study, the optical efficiency of a single collector tube has been simulated by means of a Monte Carlo Ray Tracing Program. For a concentration ratio of 1.15, the flux distribution around the heat pipe is computed as a function of incidence angle. In addition, the impact of mirror defects and absorber misalignment on the optical performance is analyzed.

0304

Irradiance on the Receiver of a General Optical Concentrator : Patera R P, "Journal of Optical Society of America ", 70, 986, (1980).

A general expression is obtained for the maximum radiant power density at the receiver of a general optical concentrator in terms of the acceptance function and input distribution of radiation.   As an example of the result, the radiant power density for two and three dimensional symmetric and asymmetric ideal concentrators is found without reference to any particular concentrator design. For parti – cular input distributions both two and three dimensional ideal asymmetric concentrators have greater power density than their symmetric counterparts.

0305

Information Theory and Solar Energy Collection : Patera R S & Robertson H S,"Applied Optics", 19, 2403, (1980).

Information theory is applied to the problem of solar radiation collection. We find that the optimum solar concentrator corresponds to a perfect imaging system i.e. one that images the entire sky on the absorber with no aberrations. For a non isotropic distribution of radiation at the collector aperture, many thermally separated absorber segments are necessary at the absorber for optimum performance The heat transfer fluid is first passed through the warm segments and then passed sequentially through the progressively hotter segments.

0306

Design Considerations for a Stationary Concentrating Collector : Patton R, " Proceedings ERDA Conference on Concentrating Solar Collectors ", Georgia (USA), p. 3 – 37, (1977).

Several configurations of compound parabolic concentrators (CPC) have been evaluated for non evacuated receiver, intermediate temperature use. Collector models were constructed and tested for overall heat loss coefficients. The model variables included concentration ratios and absorber sizes.A mathematical model of the various reflector absorber configurations was integrated with a ray trace program to evaluate the average number of reflections and the actual percentage of the incident insolation that intercepts the absorber for incidence angles from $0°$ to $34°$. These results were incorporated into a computer model with hourly insolation data to predict monthly and annual gains for each configuration. The most effective design was chosen and a prototype solar collector has been designed. The major components of the optimum design have been fabricated and a prototype collector is being built to verify the performance prediction.

0307

Lichtführungseinrichtungen mit Starker Konzentrationwirkung / (Funnel Shaped Solar Concentrator with High Concentration Results : Ploke V M , " Optik ", 25, 31, (1967).

Funnel shaped mirror chambers and optical elements are described, which concentrate light p a s s i n g
through a diaphragm at the primary focus on a second stop. If the second stop coincides with the exit
opening of the light conductor, half space radiation can be obtained on the exit side.

0308

Comparison of Solar Concentrators : Rabl A, " Solar Energy ", 18, 93, (1976).

The connection between concentration, acceptance angle and operating temperature of a solar collect-
or is analysed in simple intuitive terms, leading to a straightforward recipe for designing collectors with maximal concentration (no radiation emitted by the absorber must be allowed to leave the concentrator outside its acceptance angle). Some new concentrators, including the use of compound parabolic concentrators as second stage concentrators for conventional parabolic or Fresnel mirrors a r e
proposed. Such a combination approaches the performance of an ideal concentrator without deman d i n g
large reflectors. It may offer significant advantages for high temperature solar systems.

0309

Optical and Thermal Properties of Compound Parabolic Concentrators : Rabl A," Solar Energy ", 18, 497,
(1976).

Compound parabolic concentrators (CPC) are relevant for solar energy collection because they achieve
the highest possible concentration for any acceptance angle. The convective and radia t i v e h e a t
transfers through a CPC are calculated, and formulas for evaluating the performance of solar collectors based on the CPC principle are presented. A simple analytic technique for calculating the average
number of reflections for radiation passing through a CPC is developed ; this is useful for computing
optical losses. In most practical applications, a CPC will be truncated because a large portion of the
reflector area can be eliminated without seriously reducing the concentrator. The effects of t h i s
truncation are described explicitly. The paper includes many numerical examples, displayed in tables
and graphs, which should be helpful in designing CPC solar collectors.

0310

Solar Concentrators with Maximal Concentration for Cylindrical Absorbers : Rabl A, " Applied Optics",
15, 1871, (1976).

The differential equation is derived that describes the reflector of an ideal two dimensional radia -
tion concentrator with an absorber of arbitrary convex shape. For the special case of an absorber with
circular cross section, the equation can be solved in closed form if suitable coordinates are u s e d.
The effect of absorption at the reflector is considered, and formulas are presented f o r determining
the attenuation of radiation on its passage from aperture to absorber.

0311

Ideal Concentrators for Finite Sources and Restricted Exit Angles : Rabl A and Winston R, " Applied
Optics ", 15, 2880, (1976).

Design procedures for ideal radiation concentrators are described which are applicable to f i n i t e
sources and or restricted exit angles. Finite sources are relevant for second stage concentrators which
collect and further concentrate radiation from a primary focussing element (mirror or lens) in a manner similar to the field optic element in a telescope. Restricting the exit angle is useful for improving the optical efficiency of solar collectors by eliminating grazing angles of incidence of the absorber. It also serves to extend the useful range of angular acceptance values available from solid d i -
electric concentrators that function by total internal reflection. Concentrators of this type can be
used to construct highly efficient radiation traps (spectrally selective filters).

0312

Prisms with Total Internal Reflection as Solar Reflectors : Rabl A, " Solar Energy", 19, 555, (1977).

In this paper the suitability of Total Internal Reflection (TIR) prismatic reflectors for solar energy
collection is investigated systematically. Several promising applications of TIR prismatic reflect -
ors have been found, in particular, heliostats for the central receiver, point focus parabolas, line
focus concentrators with polar mount and seasonal tilt adjustments and, in some circumstances, compound
parabolic concentrators.

0313

Solar Concentrators with Restricted Exit Angles : Rabl A & Winston R, US Patent 4, 130, 107, Dec. 19, 1978.

A device is provided for the collection and concentration of radiant energy and includes at least one
reflective side wall. The wall directs incident radiant energy to the exit aperture thereof or onto
the surface of energy absorber positioned at the exit aperture so that the angle of incidence of radiant energy at the exit aperture or on the surface of the energy absorber is restricted to desired values.

0314

Design and Test of Non-Evacuated Solar Collectors with Compound Parabolic Concentrators : Rabl A,
O'Gallagher J and Winston R, "Enrico Fermi Institute (USA) Report EFI- 79 - 42", (1979)

The present paper summarises more than 3 years of research on non-evacuated CPC's and reviews measured performance data and critical design considerations. Concentrations in the upper portions of the
practical range (e.g. 6 X) in the 100 to 160 deg C temperature range with relatively frequent tilt adjustments (12 to 20 times per year). At lower concentrations (e.g. 3 X), performance will still be
substantially better than that for a double glazed flat plate collector above 70 deg C and competitive
below, while requiring only semi annual adjustments for year round operation. In both cases the cost
savings associated with inexpensive reflectors, and the optimal coupling to smaller, simple inexpensive absorbers (e.g. tubes, fins, etc.) can be as important an advantage as the improved thermal p e r -
formance.

0315

Practical Design Considerations for CPC Solar Collectors : Rabl A, Goodman N B and Winston R, " Solar
Energy ", 22, 373, (1979).

Several practical problems are addressed which arise in the design of solar collectors with compound parabolic concentrators (CPC's). They deal with the selection of a receiver type, the optimum method for introducing a gap between receiver and reflector to minimize optical and thermal losses and the effect of a glass envelope around the receiver. This paper also deals with the effect of mirror errors and receiver misalignment and the effect of the temperature difference between fluid and absorber plate. The merits of the CPC as a second stage concentrator are analyzed.

0316

Use of Compound Parabolic Concentrator for Solar Energy Collection : Rabl A, Sevick V J, Guigler R M and Winston R, " Argonne National Laboratory (USA) Report ANL – 75 – 42 ", (1974).

Results of a proof of concept investigation of the compound parabolic concentrator (CPC) for solar energy collection are reported. A 3 X concentrating flat plate collector is fabricated and tested in a trailer laboratory facility built at Argonne National Laboratory. The optical and thermal performance of this collector was in good agreement with theory. A detailed theoretical study of the optical and thermal characteristics of the CPC design has been performed.

0317

Design and Test of Non – Evacuated Solar Collectors with Compound Parabolic Concentrators ; Rabl A , O'Gallagher J and Winston R, " Solar Energy ", 25, 336, (1980).

This paper summarizes more than 3 years of research on nonevacuated CPC's and reviews measured performance data & critical design considerations. Concentrations in the upper portion of the practical range 1.5 X – 10 X (e.g. 6 X) can provide good efficiency (40 to 50%) in the 100 to 160 deg C temperature range with relatively frequent tilt adjustments (12 – 20 times per year). At lower concentrations (e.g. 3 X) performance will still be substantially better than that for a double glazed flat plate collector above about 70 deg C and competitive below, while requiring only semi– annual adjustments for year round operation. In both cases the cost savings associated with inexpensive reflect – ors, and the optimal coupling to smaller simple inexpensive absorbers (e.g. tubes, finns, etc.) can be as important an advantage as the improved thermal performance.

0318

CPC Thermal Collector Test Plan : Reed K A, " Proceedings ERDA Conference on Concentrating Solar Collectors ", Georgia (USA), p.6 – 15, (1977).

A comprehensive set of test procedures has been evolved at ANL for establishing the performance of compound parabolic and related concentrating thermal collectors with large angular fields of view. The procedures range from separate thermal and optical tests, to overall performance tests. A calorimetric ratio technique has been developed to determine the heat output of a collector without knowledge of the heat transfer fluid's mass flow rate and heat capacity. Special attention is paid to the problem of defining and measuring the incident solar flux with respect to which the collector efficiency is to be calculated.

0319

Performance Enhancement of Compound Parabolic Concentrators Using Air and a Liquid Simultaneously as Heat Transfer Mediums : Scheier LM & Kuehn T H, "Proceedings AIChE Symposium", San Francisco (USA), 76, (198), 75, (1980).

The thermal efficiency of a Compound Parabolic Concentrator (CPC) solar collector can be enhanced by extracting the solar energy absorbed not only by the receiver but also by the reflectors and aperture cover. Much of the energy that would normally be lost from the front of the collector is used to heat air blown through the collector, which can then be utilised for space or process heating. The present analysis uses a thermal network analogy to predict the radiation, conduction and convection heat transfer between the CPC collector and its surroundings. Results are presented for concentration ratios of 1 . 5 and for selective and non–selective absorbers and various air and liquid flow rates.

0320

Non – Imaging Concentrators Deliver Higher Temperatures for Industry : Schertz W W, " Solar Engineering Magazine ", [Pub. from 8435 N. Stemmons Freeway, Suite 880, Dallas, Texas (USA)].

This article describes the characteristics and advantages of non-imaging concentrators for high temperature applications.

0321

Performance of a Stationary Concentrating Collector for Heating,Cooling and Process Heat Applications: Schertz W W, Peters T, Levitz N, Allen J, Rabl A and Cole R, "Proceedings Flat Plate Solar Collector Conference ", Orlando, Florida (USA), p. 75, (1977).

The design and operation of a prototype collector suitable for operation in the intermediate temperature range (100 – 170°) is described. The collector is a non-imaging concentrator with an evacuated glass tube receiver for reducing the heat loss. A concentration ratio of 1.5 based on the circumference of the absorber tube, allows efficient operation at temperatures in the intermediate tempera – ture range (100 – 170 deg C) while allowing the collector to remain totally stationary throughout the year .

0322

Compound Parabolic Concentrators : Sevick V J and Winston R, " Proceedings Workshop on Solar Collectors for Heating and Cooling of Buildings ", New York (USA), p. 143, (1974).

The objective of the program discussed is to explore the performance of several versions of the compound parabolic concentrator (CPC) collector and to assess its applicability to economic utilization of solar energy for heating and cooling of buildings,central power, and photovoltaic power generation. With the CPC, it is possible to concentrate solar energy by a factor of 10 without diurnal tracking. The CPC is a nontracking solar collector consisting of two sections of parabola located symmetrically about the collector midplane. The two sections form a single curvature, or trough like solar concentrator with an angular acceptance of 2θ max.

0323

<u>Non – Focussing Solar Concentrators of Easy Manufacture</u> : Shapiro M M, " Solar Energy ", 19, 211, (1977).

The possibility of approximating a perfect ideal parabolic trough with mirrors of circular and polygonal cross section is explored. Two particular designs are presented here which give more than 90%
of the useful heat output of a system based on Winston's trough, when coated aluminium surfaces are
used, but have mirror shapes easier to produce than parabolic cylinders.

0324

<u>A Note on the Economics of Deep Cylindrical Mirror Concentrating Collectors</u> : Tabor H, " Solar Energy ",
19, 573, (1977).

A cylindrical concentrating system is analysed in a view to determine optimum depth. It is shown that
practical thermal systems will optimize at very much shorter length mirrors than those giving maximum
theoretical concentration ratio.

0325

<u>Predicted Heat Transfer Performance of an Evacuated Glass Jacketed CPC Receiver Counter-current Flow
Design</u> : Thodos G, " Argonne National Laboratory (USA) Report 76 – 67 ", (1976).

The heat transfer performance of an evacuated glass jacketed CPC receiver facility, free on one end
and fixed onto the glass jacket at the other, was carried out using heat transfer relationships & the
least information available in the literature. Specifically the collector examined was a 3 X — C P C
facility, 8 ft long with an entrance aperture 4.5 in wide covered with a single glass cover and provided with an aluminium reflecting surface $\rho = 0.88$ to maximise heat retention, a selectively treated receiver surface, $\alpha = 0.11$, was used. The optical efficiency of this CPC collector facility
was calculated t o b e $\eta_o = 0.536$.

0326

<u>On the Problem of Ideal Flux Concentrators</u> : Welford W T, & Winston R, " Journal Optical Society of
America ", 68, 531, (1978).

It has been shown that ideal flux collectors in the form of systems of axially symmetrical s y s t e m s
which concentrate flux from a given angular extent with no loss in etendue are impossible; on   the
other hand, systems approaching this limit very closely can be designed.

0327

<u>On the Problem of Ideal Flux Concentrators Addendum</u> : Welford W T and Winston R, " Journal Optical
Society of America ", 69, 367, (1979).

The argument that an ideal imaging forming concentrator cannot be designed are strengthened in this
note and also extended to general image forming systems with an axis of symmetry.

0328

<u>Two Dimensional Non Imaging Concentrators with Refracting Optics</u> : Welford W T and Winston R, " Journal
Optical Society of America ", 69, 917, (1979).

The possibility of designing a concentrator with refracting optics having the maximum theoretical concentration ratio has been explored. The chromatic aberration is neglected and skew rays are ig n o r e d
in this treatment.

0329

<u>Light Collection within the Framework of Geometrical Optics</u> : Winston R, " Journal Optical   Society
of America ", 60(2), 245, (1970).

The problem of light collection is examined from first principles within the framework of geometrical
optics. From the outset a distinction is made between light collection and the usual theory of image
formation. From phase space considerations, the sine inequality, a generalization of the Abbe sine law
appropriate to non imaging systems is derived. Two and three dimensional non imaging systems   that
reduce the f number to the least allowed by the sine inequality are constructed. Such systems g i v e
substantially improved light collection as compared with conventional systems.

0330

<u>Principles of Solar Concentrators of a Novel design</u> : Winston R,   " Solar Energy ", 16, 89, (1974).

A new principle for collecting and concentrating solar energy, the ideal cylindrical light collector
has been invented. This development has its origin in detecting Cerenkov radiation in high e n e rgy
physics experiments. In its present form, the collector is a trough like reflecting wall light channel of a specific shape which concentrates radiant energy by the maximum amount allowed by phase space
conservation. The ideal cylindrical light collector is capable of accepting solar radiation over an
~ 8 hr/day and concentrating it by a factor of 10 without diurnal tracking of the sun. This i s
not possible by conventional imaging techniques. The ideal collector is non imaging and possess e s
an effective relative aperture (f number) = 0.5. This collector has a larger acceptance for diffuse
light than concentrating collectors using imaging optics. In fact, the efficiency for collect i n g
and concentrating isotropic radiation, in comparison with a flat plate collector, is just the reci-
procal of the concentration factor.

0331

<u>Development of the Compound Parabolic Collector for Photo – Thermal and Photovoltaic Applications</u> :
Winston R, " Proceedings SPIE Conference ", 168, 136, (1975).

Principles of design and experimental tests of the compound parabolic collector (CPC) concept for concentrating solar energy onto a receiver are described. Because of its wide angle of acceptance, t h e
CPC can concentrate solar energy upto a factor of ten without diurnal tracking of the sun, making it
well suited for use in applications as a source of relatively high temperatures (150 – 400 deg F) heat

or an optical concentrator to reduce surface area requirements of photovoltaic cells.

0332

<u>Dielectric Compound Parabolic Concentrators</u> : Winston R, " Applied Optics ", 15 (2), 291, (1976).

A dielectric compound parabolic concentrator (CPC) which permits no leakage of radiation is proposed. The dielectric CPC profile curve is generated by allowing the maximum possible slope consistent with totally internally reflecting the extreme rays onto the receiver. This guiding principle specifies the design for quite general combination of angular acceptance, indices of refraction a n d receiver shapes.

0333

<u>Cylindrical Radiant Energy Direction Device with Refractive Medium</u> : Winston R, " U S Patent Application ", 714, 863, 16th August 1976.

A device is described for directing radiant energy and includes a refractive element and a reflective boundary. The reflective boundary is so contoured that incident energy directed by the refractive element is directed to the exit surface onto an energy absorber positioned at the exit surface.

0334

<u>Non— Imaging Concentrators for Wide Angle Collection of Solar Energy</u> : Winston R, "Progress R e p o r t 11, July 1, 1976 – April 30, 1977, No: COO – 2446 – 8, University of Chicago (USA), (1977).

Research at the University of Chicago on non-imaging concentrating collectors is briefly outlined. Activities in the design, construction, optical studies, and performance testing of concentrating collectors is reviewed.

0335

<u>Ideal Light Concentrators with Reflector Gaps</u> : Winston R, "U S Patent Application ", 909, 864, (1978).

A cylindrical or trough like radiant energy concentrating collector is described. The device includes an energy absorber, a glazing enveloping the absorber and a reflective wall. The ideal contour of this wall is determined with reference to a virtual absorber and not the actual absorber cross section.

0336

<u>Cone Collectors for Finite Sources</u> : Winston R, "Applied Optics ", 17, 688, (1978).

A brief description of the cone collectors which concentrate radiation but do not form an i m a g e is given in this paper.

0337

<u>Retinal Cone Receptor as an Ideal Light Collector</u> : Winston R and Enoch J M, Journal Optical Society of America ", 61, 1120, (1971).

In this note a striking similarity between the ellipsoid portion of retinal cone receptors and t h e design of an ideal light collector is pointed out. This may be helpful in understanding the mechanism contributing to the directional sensitivity of the retina.

0338

<u>Principles of Cylindrical Concentrators for Solar Energy</u> ; Winston R and Hinterberger H, "Solar Energy," 17, 255, (1975).

Ideal cylindrical light collectors are trough like reflecting wall light channels of a specific shape which concentrate radiant energy by the maximum amount allowed by phase space conservation. A principle for maximally concentrating radiation onto a tube receiver of general shape is proposed. Employing this principle a general prescription for designing concentrators appropriate to such tube receivers is given. This design may have advantages for solar thermal and photovoltaic applications.

0339

<u>Two Dimensional Concentrators for Inhomogeneous Media</u> : Winston R and Welford W T, "Journal   Optical Society of America ", 68, 289, (1978).

A two dimensional concentrator for collecting light over a specified range of angles and concentrating it at a specified exit surface without loss of etendue can be postulated. A design procedure for such systems is developed and the conditions under which the procedure is applicable are derived.

0340

<u>The Geometric Vector Flux : A New Description of Light Rays</u>: Winston R and Welford W T, " Proceedings ISES Silver Jubilee International Congress ", Georgia (USA), (1979).

Light rays are described by a vector flux field. In the new picture, ray tracing is replaced by radiative transfer concepts. The lines of flow of the vector flux form surfaces along which mirrors can be placed without disturbing the flux field. Refracting surfaces perpendicular to the lines of flow also leave the flux field undisturbed. However, the law of refraction of flow lines is different from Snell's Law. The new picture is very powerful in generating designs of solar concentrators. The known nonimaging designs follow in a natural way. More importantly, new designs emerge which were not anticipated by previous work. For example, the lines of flow from a Lambertian disc radiator are confocal hyperbolas. The corresponding concentrator in the form of a hyperboloid of revolution offers significant advantages for final stage solar concentration in a point focal parabolic dish or Fresnel lens collector.

0341

<u>Geometrical Vector Flux and Some New Non— Imaging Concentrators</u> : Winston R and Welford W T, " Journal Optical Society of America ", 69, 532, (1979).

The geometrical vector flux, a quantity related to measures of illumination, is defined, some of its properties are explained and used to develop new forms of non-imaging concentrators.

0342

Ideal Flux Concentrators as Shapes that do not Disturb the Geometrical Vector Flux Field : A New Derivation of the Compound Parabolic Concentrator : Winston R and Welford W T, " Journal Optical Society of America ", 69, 536, (1979).

The lines of flow of the geometrical vector flux form surfaces along which mirrors can be placed without disturbing the flux field. This result is used to give new insights into the design of non-imaging concentrators.

0343

Preliminary Results from a Test Array of 3 X CPC Collectors in a School Heating : Ziteck W," Proceedings ISES (American Section) Conference ", Denver, Colorado (USA), (1978).

An array of 56 two trough compound parabolic concentrators (CPC) solar collector modules has been installed on the roof of an elementary school on the Navajo Indian Reservation near Gallup, New Mexico, USA. The array represents an effective net collecting area of 72.9 sqm and has been fully instrumental to provide a data base for analysing CPC collection characteristics. This is the first quantitative field test of these collectors conducted anywhere. The basic collector design is a 3 X CPC matched to a finned tubular absorber with an acceptance half angle of $\pm18^{\circ}$, allowing collection for half a year without adjustments. Very preliminary results from late spring are reported.

0344

Solar Energy Concentration : Argonne National Laboratory, Illinois (USA), " Annual Progress Report - July 1 - December 31 ", (1975).

A concentrating solar heater based on the compound parabolic design was constructed and tested. The principal characteristics of the collector are that it has a concentration factor = 3 , and an angular acceptance (full angle = 38°). The large angular acceptance implies that bi-annual adjustment of the collector orientation is sufficient to accept direct solar radiation. Experimental results from outdoor tests show good agreement with theory and confirm the expected improvement in performance resulting from concentration. Such performance would be useful for space conditioning applications in a temperature range where flat plate collectors are marginal (130 deg F above ambient).

0345

Goal Studies for the Technical and Economic Evaluation of the Compound Parabolic Concentrator (CPC) Concept Applied to Solar Thermal and Photovoltaic Collectors : Argonne National Laboratory, Illinois (USA), " Final Report ANL - K - 75 - 3192 - 1 ", (1975).

This report presents the results of a quick six weeks technical and economic evaluation of the compound parabolic concentrator (CPC) solar collector. The purpose of this effort was to provide an initial phase of a goals study that is directed towards recommending relative priorities for development of the compound parabolic concentrator concept. The findings of this study are of a very preliminary nature. Conclusions based on study finding at this depth should be considered preliminary and subject to revision and review in later phases.

Paraboloid Of Revolution / Solar Furnaces

0346

High Vacuum Treatment in the Solar Furnace Experimental Set up, Preparation and Study of New Rare - Earth Compounds : Achard J C, " Proceedings UN Conference on New Sources of Energy ", Rome (Italy), 6, 327, (1964).

The behaviour of europium sesquiouide at high temperature in vacuo, and the reduction of this compound in the solid phase by reducing agents such as carbon and boron, have been studied in the solar furnace. The power of the solar furnace used was about 2 Kilowatts. It had a fixed parabolic mirror, 2 meters in diameter with a focal length of 85 cm, receiving the solar radiation reflected vertically onto it by an automatically oriented plane mirror. The vacuum chamber was of steel, except for the portion subjected to the convergent radiation, which was of Pyrex glass. It was rigidly connected to the secondary pump, located outside the path of the solar rays. The secondary pump and vacuum chamber were both movable, permitting adjustment of the relative positions of the specimen and focus, and the positioning of the product in the chamber. The group of primary pumps was fixed and the mutual displacement of the two parts — secondary vacuum and primary vacuum, was accomplished by means of special joints located in the primary vacuum ducts.

0347

Measurement of Electrical Conductivity of Magnesium Oxide Single Crystal at High Temperature Using a Solar Furnace : Afzal F A and Giutronich J E, " Solar Energy ", 15, 125, (1973).

This paper reports the technique developed for the measurements of electrical conductivity of magnesium oxide crystals at temperatures upto 1700 deg C using a solar furnace and also the results obtained.

0348

Efficiency of a Thermogenerator with a Cavity Solar Energy Collector : Akramov Kh. T and Makov N Y , " Applied Solar Energy ", 6 (2), 75, (1970).

The cylindrical cavity efficiency of a helio collector, obtained by collection and by calorimetric measurement is given. The difference between the calculated and experimental values does not exceed 10%. The article considers the dependence of cavity and flat radiant solar energy collector efficiency, and geometric concentration factor on the light absorbing surface at three temperatures (600 deg K, 800 deg K and 1000 deg K).

0349

Investigation of Solar Concentrators Made up from Facets with Double Curvature : Alavutdinov D N, Alimov A K and Umarov G Ya., " Applied Solar Energy ", 3 (3), (1967).

A parabolic solar concentrator manufactured from round or hexahydral facets is described. Characteristics for concentrators that are made up from 60 round and 60 hexahydral facets are given.

0350

Properties of a Solar Concentrator with Hexagonal Glass Facets : Alimov A K , Alavutdinov D N and Abduazizov A, " Applied Solar Energy ", 11 (3-4), 14, (1975).

The enrgy parameters of a faceted concentrator 5 m in diameter have been investigated. The facets were in the form of hexagonal facets mounted on a paraboloidal base. This gives an effective working area that is 10% greater as compared with a concentrator using circular facets. The power generated in the focal plane of the concentrator was 9 Kw for an incident radiation of 784 W/m^2 .

0351

Solar Furnaces with Inclined Collector Axis : Annayev A, " Applied Solar Energy ", 1(13), (1965).

A solar furnace whose collector axis is inclined towards the north is described. This arrangement allows the height of the collector to be reduced as well as the distance from the collector to the heliostat, which directs the energy into the furnace. It is pointed out, however, that this arrangement necessitates a larger heliostat, and one that is ellipsoidal in shape rather than spherical. The calculations are presented giving the dimensions of an inclined axis system which has a collector with a 2 m diameter and an angle of inclination β .

0352

Design of Heliostats for Solar High Temperature Furnaces : Annayev A A and Roziev Kh., " Applied Solar Energy ", 6 (6), 56, (1970).

The shapes and dimensions of heliostats with the optical axis of the paraboloid concentrator taking on an arbitrary angle with the horizon are determined. Equations for the heliostat plane are given.

0353

Energy Trapping Characteristics of Non Circular Receivers in Paraboloidal Solar Devices : Aparisi R R, Kolos Ya. G and Teplyakov D I, " Applied Solar Energy ", 4 (2), 58, (1968).

A method for calculating the distribution of energy passing through a non-round shaped orifice situated in the focal image is described. In this calculation, the Gauss probability distribution formula is used. To a large extent it simplifies the procedure.

0354

Simplified Analytic Treatment of Reflected Energy Field in Paraboloidal Solar Concentrators : Aparisi R R and Kolos Ya. G, " Applied Solar Energy ", 4 (6), 53, (1968).

The proposed solution permits simple field calculation of the quantitative value of energy density in focal and other planes oriented normal to the optical paraboloid axis. The results are coordinated with the experimental measurements, thus validating the use of the simplified formulas for calculations.

0355

Concentration Capacity of Paraboloidal Mirror with Angular Defocussing : Aparisi R R and Kolos Ya.G, " Applied Solar Energy ", 6(5), 9, (1970).

The problems connected with determining the direction of sunlight which has been reflected from a parabolic mirror are considered. The light drops on the mirror surface with a certain angle of declination on the mirror axis.

0356

SHADE - A Computer Model for Evaluating the Optical Performance of Two Axis Tracking Parabolic Concentrators : Apley W J, " ASME Paper No. 79 WA/Sol - 13 ", (1979)

A computer model SHADE (Selection of Heliostat Arrangement for Distributed Engines) has been developed at the Pacific Northwest Laboratory to aid in determining the optical performance of two-axis tracking parabolic concentrators. The shading of individual mirror assemblies in a field of parabolic dishes determines the optimal field arrangement and the most efficient of plant operations. SHADE provides a simple and inexpensive analytical tool for examining certain design aspects of solar thermal power systems using a network of point focussing parabolic concentrators.

0357

A Method for Manufacturing Parabolic Mirrors : Archibald P B, " Solar Energy ", 1(2-3), 102, (1957).

A simple method is described for producing parabolic mirrors by revolving a horizontal pan of liquid plastic and allowing it to harden into the parabolic shape which has been assumed by the plastic . This surface can be plated without further polishing into a mirror of a quality suited to various laboratory and industrial applications.

0358

Experimental Investigation of Temperature Field in the Cylindrical Cavity Receiver of a Solar Thermo-electric Converter : Arifov U A, Akramov H T, Djalilov B N, Kulagin A I & Makov N Y,   " Applied Solar Energy ", 4 (4), 49, (1968).

Results are given of an experimental investigation of a temperature field of solar receiver with cylindrical cavity. Analysis is given of non uniformity of temperature field and optimal geometry is recommended for low temperature solar thermogenerators.

0359

Chemical Vapour Deposition of Molybdenum and Tungsten Borides by Thermal Decomposition of Gaseous Mixtures of Halides on a Solar ' Front Chaud ' : Armas B and Trombe F, " Solar Energy ", 15, 67, (1973).

The authors describe an experimental device in which the thermal decomposition of mixtures of a metal chloride and boron tribromide takes place by means of solar furnace heating.

0360

Elaboration of Fe – B Alloys with High Boron Concentration : Armas B, Combesure C and Morales M, " Proceedings ISES Silver Jubilee International Congress", Georgia (USA), (1979).

The preparation of Fe – B alloys with high boron concentration in a solar furnace by direct melting in order to obtain thermoelectrical materials is reported.

0361

Asymmetric Temperature Distribution on the Surface of a Paraboloidal Reflector : Bairamov R, Toiliev K and Galkanov A, " Applied Solar Energy ", 10 (3), 38, (1974).

An aim of this paper is to find the temperature distribution on the surface of a paraboloidal reflector as a special case in the theory of heat conduction. A differential equation is established for the time dependent distribution of temperature on the surface of the body. The equation is then used to find the approximate solution of a boundary value problem using boundary condition of the third kind.

0362

Method of Calculating Radiant Flux Distribution on Walls of a Cavity Solar Receiver : Baranov V I and Muchnik G F, " Applied Solar Energy ", 2(5), 1, (1966).

A mathematical method of computing the radiation flux distribution over the walls of a cylindrical solar receiver is described. Expressions for the energy reflected by the annular mirror surface of the receiver; and for the flux density changes in three different regions of the receiver (cylindrical region, bottom part and angular region) are given. A geometrical diagram of the problem is depicted.Also calculated were the energy distribution in the solar ray beam itself, and the flux density around the focal point of radiation collector (uniform Gaussian distribution of the ray was assumed).   The computed results for flux densities are shown graphically along with curves showing the change in flux density at a constant cavity length to diameter ratio with variable cavity diameters.

0363

Optical Furnace with an Artificial Source : Baum I V, " Applied Solar Energy ", 10(4), 111, (1974).

The theoretical principles that can be used in the design of optical furnaces are reviewed. The possibility of producing a uniformly illuminated zone with relatively low flux density is est i m a t e d. Special formulas are given for the parameters of optical systems.

0364

Calculations on an Optimized Faceted Solar Concentrator : Baum I V, Bairiev A Ch. and Saiylov N," Applied Solar Energy ", 12(1), 27, (1976).

A procedure is proposed for the design of an optimum variable curvature faceted concentrator in which each of the facets is a trapezium in plan. The degree of concentration is specified and the corres – ponding number of facets and angular aperture of the concentrator are determined. The next step is to calculate the area of the paraboloid and the remaining parameters, and these are used to improve the final figure for the number of facets. Computation at formulas and tabulation of functions necessary for the calculation of the concentrator parameters are reproduced.

0365

High Power Solar Installations : Baum V A, Aparisi R R and Garf B A, " Solar Energy ", 1(1), 6, (1957).

The results of investigations establishing the possibility of erecting economic solar installations in the sunny regions of the USSR to produce 11 – 13 metric tons/hour of steam (pressure = 30 Atm., t = 400 deg C) are presented. The optical system of the installation consists of 1293 mirrors of 3 X 5 m, mounted on carriages which move on rails, positioned around a boiler shield, on which the solar rays are focussed.

0366

The Solar High Temperature Furnace : Baum V A, Annayev A A, Atlyev K and Kurbangeldyev B, "Applied Solar Energy ", 5(2), (1969).

Description of the furnace and the absolute calorimeter for the investigation of power characteristics of solar power furnaces are given, also power characteristics and the region of temperature in which the thermophysical and radiation properties of refractories can be investigated and presented.

0367

Optical Analysis of Point Focus Parabolic Radiation Concentrators : Bendt P and Rabl A," Solar Energy Research Institute, Colorado (USA) Report SERI/TR – 631 – 336 ", (1980)

A simple formalism is developed for analyzing the optical performance of point focus parabolic radiation concentrators. To account for off-axis aberrations of the parabola, an angular acceptance function is defined as that fraction of a beam of parallel radiation incident on the aperture which would reach the receiver if the optics were perfect. The radiation intercepted by the receiver of a real concentrator is obtained as a convolution of angular acceptance function, of optical error distribution , and of angular brightness distribution of the radiation source. Losses resulting from absorption in the reflector or reflection at the receiver are treated by a multiplicative factor $\rho\alpha$ where ρ = reflectance of the reflector and α = absorptance of the receiver. For numerical calculations, this method is more accurate and less time consuming than the ray tracing method. In many cases there are acceptable approximations whereby the results can be obtained by reading a graph or evaluating a simple curve fit.

0368

High Temperature Studies on $Al_2 O_3 - Lu_2 O_3$ Systems with Solar Furnaces ; Berjoan R, Coutures J P & Foex M, " Proceedings ISES Conference", Los Angeles (USA), (1975).

This type of system for $Lu = L_a, O_e$, Dr, Nd is investigated by high temperature devices generally associated with laboratory solar furnaces. In this case we performed thermal analysis investigations on melting under controlled atmosphere, vapour quenching and splat cooling experiences. Some other techniques like X-rays analysis (at room or high temperature), differential thermal analysis, thermogravimetric analysis and low temperature optical absorption are also used.

0369

Notes on Optical Design Principles, Concentration Ratios and Maximum Temperatures of Parabolic Solar Furnaces : Bliss Jr. R W, " Report No MDC - HDW-TM-56-1, Air Force Missile Development Center", Holloman, AFB, New Mexico (USA), (1956).

The memorandum is intended as a preliminary guide to designers and interested possible users of solar furnaces. Optical design principles are summarised briefly and a simple expression is derived for the solar concentration ratio obtainable from a perfectly reflecting paraboloid of large aperture. The transmission losses necessarily occurring in an actual furnace are considered and estimated, and estimates are presented as to the concentration ratios obtainable with actual furnaces of various aper - tures. The maximum attainable temperatures of such furnaces, when operating under favourable sunshine conditions, is similarly estimated. A short bibliography is included.

0370

Summary of Parabolic Solar Furnace Performance Design : Bliss Jr. R W, "Report No MDC - HDW - TM - 56 -2, Air Force Missile Development Center", Hollman, AFB, New Mexico (USA), (1956).

The memorandum summarises the basic knowledge necessary for the design of a solar furnace to specified performance. The intensity of available south western sunshine is estimated from currently available records. Instruments for the measurement of direct sunshine intensities are described   briefly. The fundamental equations yielding the spot diameter and concentration ratio of a paraboloid are given, and the maximum temperatures attainable by furnaces of given concentration ratio are derived.

0371

Designing Solar Furnaces for Specific Purposes : Bliss R, "Solar Energy ", 1 (2 - 3), 55, (1957).

The basic performance equations of a parabolic solar furnace are manipulated into graphical form convenient for purposes of estimation and design analysis. It is found that slight changes in either the specified performance or the overall efficiency of a furnace can necessitate relatively large changes in design size. A preliminary test program of experimental calorimetric measurements is considered a necessity in furnace design. The essentials of such a program are described briefly.

0372

Plastic - Replica Mirror Segments for a Solar Furnace : Bolin J, Tenukest C J and Milner C J, " Solar Energy ", 5, 99, (1961).
A paraboloidal mirror, 12 ft in diameter and of focal length 47 in, is built up from 162 segments, arranged in six concentric zones. Segments for each zone are cast in epoxy resin, using as matrix one of six glass plates optically worked to convex off-axis paraboloidal form. In order to avoid distortion and wrinkling of the final surface, due to shrinkage and other causes, three successive coatings of resin, each using a different filter and curing schedule, are applied to a stout cast-aluminium backing plate. To part the casting from the matrix a silicone release agent is used, which is effective when applied as an extremely thin film. The final casting is front surface aluminized by vacuum evaporation. A parallel beam of light formed by one such segment (approx. 12 in x 10 in) may be refocussed by another segment, suitably placed, to an image less than 1/20 in in diameter. The surfaces   formed by this process are hard and permanent, and numerous optical and mechanical units are forseen.

0373

Modulation of the Light Flux at the Focal Point of Solar Furnace : Borukhov M Yu., Mavashev Yu. S and Bespal'ko V P, " Applied Solar Energy ", 4 (5), 42, (1968).

The calculation of some light flux modulations in the light flux of a solar oven is discussed. Formulae are deduced for cam shape calculation corresponding to sinusoidal changing of the light flux in the focal point of a solar oven.

0374

Method of Calibrating a Solar Installation with a Paraboloidal Mirror : Bystrov V V, Zhuk V I, Mochuev Yu. I, " Applied Solar Energy ", 11 (1), 20, (1975).

A new method of calibrating a solar installation is proposed. The calorimeter is provided by a solid metal rod with a temperature sensor. Noteworthy features of the method are simplicity of calorimeter . design, lack of water system and an automatic solar tracking system, and simplicity and reliability of measurement. An example is also given.

052

0375

<u>Corradiation Using the Reversible Ammonia Reaction</u> : Carden P O, "Proceedings ISES Conference", Los Angeles (USA), (1975).

A system is described for the large scale generation of power from solar energy in which energy is transferred by means of the reversible chemical reaction $2\,NH_3 \rightleftharpoons N_2 + 3\,H_2$. A multiplicity of pressed steel paraboloidal mirrors is employed each having a focal absorber in which the endothermic forward reaction proceeds. The exothermic backward reaction occurs at a common central plant and the heat energy recovered operates a thermodynamic power plant. The reactants are transferred in small diameter steel piping at ambient temperature. Storage of energy may be catered for providing storage for the reactants.

0376

<u>Design of a Small Thermochemical Receiver for Solar Thermal Power</u> : Chubb T A, Nemecek J J and Simmons D E, " Solar Energy ", 23, 217, (1979).

Capture of solar thermal energy by means of chemical conversion for collection of solar energy is investigated. A converter has been designed for operation at the focus of a 7 m dia. paraboloid. The converter is constructed from a multi-passage ceramic extrusion, which is wound into a spiral form prior to firing. The innermost warp is designed to operate with a cavity facing surface heated to 1000 deg C. In the passages adjacent to this surface SO_3 is catalytically disassociated into SO_2 & O_2. The eight outer warps are used for heat exchange between an inflowing SO_3 rich gas stream and an outflowing SO_2 rich gas stream.

0377

<u>Thermomigration of Silicon Wafers in a Solar Furnace</u> : Cline H E and Anthony T R, " Solar Energy ", 19, 715, (1977).

A 5.5 m, 8.7 Kw solar furnace was designed and constructed to process silicon wafers by thermomigration. Under the intense heat, 160 W/sqcm, of the solar furnace, a grid of aluminium wires was migrated through the wafer in 5 minutes. Helium gas was used to cool the wafer and produce a thermal gradient of 400 deg C/cm. The heat transfer and efficiency of the system are discussed.

0378

<u>Theoretical Concentrations for Solar Furnaces</u> : Cobble M H, " Solar Energy ", 5(2), 61, (1961).

The concentrations for a paraboloid of revolution mirror and for a parabolic cylinder are derived as a function of relative aperture for several target geometries. The expression for optimum concentration for each target mirror combination is developed, and the equation for the ideal target shape for a target of a parabolic cylinder mirror having a relative aperture of 4 is obtained.

0379

<u>Image Quality and Use of the United States Army Quartermaster Solar Furnace</u> : Cotton E S, Lynch W P, Zagieboylo W and Davies J M, " Proceedings UN Conference on New Sources of Energy ", Rome (Italy), 6, 336, (1964).

The general features of US Army Quartermaster's solar furnace are described. Its unique optical characteristics, its performance as an imaging device, and the types of research for which it is suited are also discussed.

0380

<u>Flux Distributions Inside and Thermal Efficiencies of Solar Cavities Heated by Parabolic Dishes</u> : Dasgupta S, Mauk C E and Hildebrandt A F, " Proceedings ISES (American Section) Conference ", Denver, Colorado (USA), 2.1, 840, (1978).

Simple methods have been derived to determine the incident flux inside cavities heated by parabolic dishes. The effect of cavity shape on the flux profiles are analysed. These flux distributions are used to determine the energy loss by reflection and re-radiation through cavity apertures. The results are compared to similar losses from idealised cavities. Convective losses are estimated using correlations available from heat transfer theory. Some design considerations for achieving high efficiencies are discussed.

0381

<u>Simplified Methods for Analysing the Performance of Parabolic Dish Systems</u> : Dasgupta S, Mauk C E and Hildebrandt A F, " Proceedings AIChE Symposium ", San Francisco (USA), 76, (198), 1, (1980).

Accurate analysis of flux profiles inside cavity receivers use complex computer numeric techniques. However, for conceptual design purposes simpler preferably analytic, techniques are desirable. These techniques provide a vital link between preliminary evaluation of receiver concepts and their final design.

0382

<u>Flux Distribution Near the Focal Plane</u> : De la Rue Jr. R E, Loh E, Brenner J L and Hiester N K, " Solar Energy ", 1 (2 - 3), 94, (1957).

In the focal plane of a parabolic concentrator, the assumption of a sun's disk of uniform flux intensity leads to a sharply defined image. However, in planes parallel and adjacent to the focal plane, the heat flux distribution is again more uniform, but is reduced in intensity. For these cases, the effect of a non-uniform flux density across the sun's disk, as well as a uniform solar disk is considered. Typical flux distribution curves are presented.

0383

<u>Computation of a Concentrator - Receiver System with an Equal Heat Stress Receiver</u> : Dudko Yu. A, "Applied Solar Energy ", 3 (4), (1967).

Numerical solution for a differential equation describing a special optical concentration   receiver system is given. The result of the solution is the calculation of the surface form of the concentrator-receiver system which provides a uniform distribution of the radiant flux falling on the receiver.

0384

Calculation of Concentrator-Receiver Systems with a Flat Receiver : Dudko Yu. A, "Applied Solar Energy ", 4(1), 19, (1968).

Numerical solutions of differential equations describing concentrator receiver systems with a flat receiver for different laws of concentration are considered. Results for several systems with flat receivers are given.

0385

Calculation of Concentrator – Receiver Systems with a Flat Receiver : Dudko Yu. A, " Applied Solar Energy ", 4(2), 68, (1968).

The calculation of the concentrator receiver system with a flat receiver consists of successive approximation of the receiver surface shape to one that provides a receiver with the desired f l u x density distribution for a chosen distribution law in the elementary flux.

0386

Designing Concentrating Systems : Dudko Yu. A & Dudko O A, "Applied Solar Energy ", 4(6),60, (1968).

A calculation diagram of the reflecting concentrator based on aberration experiments is attempted for further quantitative estimation of the reflective surface quality and for obtaining the value of radiation flux density on the surface of a receiver of any configuration.

0387

Performance Testing of the General Electric Engineering Prototype Collector : Dudley V E and Workhoven R M, " Sandia Laboratories (USA) Report SAND – 79 – 0514 ", (1979).

This report provides results of tests on the General Electric Engineering Prototype Collector, a 5 m parabolic dish solar concentrator at the Mid Temperature Solar Systems Test Facility.

0388

Plastics for Focussing Collectors : Duffy J A, Lappala R P and Lof G O G," Solar Energy ", 1(4), 9, (1957).

The use of plastics in fabrication of reflectors for focussing collectors offer possibilities for reduction of material and fabrication costs. Fibre glass reinforced polyesters, cellulose acetates butyrate and polystyrene have been considered for reflector shell fabrication, and vacuum   metallized films for reflecting surfaces. Polystyrene shells with aluminized Mylar reflecting lining have been the most promising of the materials considered. Flexible laminates of aluminized Mylar to vinyl or vinyl cloth combinations are considered for flexible reflectors. Application of plastic reflectors to solar cookers is noted.

0389

Utilization of Solar Furnaces in High Temperature Research : Duwez P, Transaction ASME ",79(5), 1019, (1957).

The paper presents a theoretical discussion of the performances of a parabolic type solar furnace. Maximum temperature and maximum heat flux attainable at the focus are presented for furnaces of different diameter to focal length ratios. It is shown that the optical quality of the parabolic reflector i s the most important factor for obtaining high heat fluxes. With present day technology, the heat flux is limited to 600 Btu/sft/sec but this could be raised to 2300 Btu/sft/sec by improving the optics.

0390

Concentration of Solar Energy : Duwez P, " Introduction to the Utilization of Solar Energy ", (Ed.A M Zarem and D P Erway), McGraw Hill Book Co. Inc., NY (USA), p.107, (1963).

Theoretical considerations associated with the design and performance of solar furnaces are discussed. The concentration ratio and concentration efficiency of furnaces are briefly described, and the maximum obtainable temperature for a given furnace is analyzed. The importance of the quality of the focussing paraboloid is discussed, as are factors affecting furnace performance, the basic features of a lens type furnace and the operation of a solar furnace under defocussed conditions.

0391

A New Solar Energy Device for Use in Testing Refractory Materials : Dvernyakov V S and Pasichnyy V V, English Translation from "Dopovidi Akad. Nauk ", UKRSR (Kiev), No. 6, 762, (1966)

Results are described from a study to determine the parameters of a special solar installation (SSI) intended for use in mechanical, spectrographic and chemical investigations of materials in vacuo and various unusual environments. Energy distribution was studied in the focal area for several positions of a regulating cylinder, calorimeter and system of diaphragms. It is emphasised that the purity of the experiment makes possible a synthesis of new materials adopted to the concrete conditions of their use.

0392

Industrial Considerations of Solar Furnaces : Edlin F E," Solar Energy ", 1(2 – 3), 52, (1957).

A solar furnace for research is compared to other furnaces above 3000 deg K with oxidizing or reducing environment. Lower quality reflective furnaces and heat traps are discussed for possible power generation. Plastics are considered as a material or construction for reflectors and heat traps.

0393

A Computational Alignment Method for Paraboloidal Collectors : Edwards B, " Proceedings ISES Silver

054

Jubilee International Congress ", Georgia (USA), (1979).

An alignment method for following collectors is described which has the potential for lowering collector installation costs. Alignment may be regarded as the process which determines constants in a function relating the positions of the actuators on the collector. These constants are called calibration
constants. The function is determined by the type of the collector mount chosen, while t h e constants
are fixed by the actual dimensions of each mount, and the orientation of the mount after installation.
These constants include such items as the direction of the axes of rotation and the position of t h e
zero points of the position encoders. The method described here needs no manual intervention, requires
no adjustments and can be used on any type of mount.

0394

Collector Deflections Due to Wind Gusts and Control Scheme Design : Edwards B, " Solar Energy ", 25 ,231,
(1980).

The relationship between the spectrum of collector deflections due to winds & the composition of the
guidance scheme controlling the collector is considered. The interface between those two factors is
the required rigidity of the collector because rigidity is dependent on both factors. As rigidity i s
an expensive commodity, it is important that the collector is not under or over designed in this respect. The average sun following error is presented as   a function of collector rigidity and guidance
and control scheme type is also presented. This shows that guidance and control schemes which have band
widths encompassing the spectrum of wind gusts and include wind effects as inputs in the servo loop
give an advantage in cost due to lower structural rigidity requirements. There is little difficulty
in designing a guidance and control scheme which will meet both these requirements. The contents of
this paper are applicable to both direct tracking collectors and heliostats.

0395

Design Fabrication and Testing of Three Meter Diameter Parabolic Dish Heliostat System : Engira R M &
Mannan K D, " Proceedings ISES Conference ", New Delhi (India), p. 1278, (1978).

This paper describes the design, fabrication and preliminary performance evaluation of   a parabolic
dish collector and heliostat system. The system was designed to provide a medium concentration ratio
of about 800. The parabolic dish was fabricated using a specially designed reflector element fitted
in the form of a mosaic on a supporting structure. The heliostat was fabricated by using plane mirrors, each 9 m x 3 m. The azimuthal as well as altitude tracking is accomplished by separate tracking
motors based on on-off operations. The electric control system acts in conjunction with the photocell
censors to maintain the pointing error to within an angle of 3 minutes. It has been found experimentally that the system concentrates solar energy on 10 cm dia. disk.

0396

Crystals of High Temperature Materials Produced in the Solar Furnace : Farber E A," Solar Energy" , 8,
38, (1964).

It is demonstrated that a solar furnace can be used to grow crystals of useful sizes by the vapour deposition processes and better still, by the puddle melt method. Rather impure material can be   used
as raw material and can be in powder, rock or crystal form. Therefore a solar furnace could contribute
greatly in helping to reach the objectives of a lunar station.

0397

Analysis of Large Aperture Parabolic Mirrors for Solar Furnaces : Farber J and Davis B I, " Journal
Optical Society of America ", 47(3), 216, (1957).

An analysis is made of large aperture parabolic reflectors as solar furnaces to determine the most effective areas and estimate maximum attainable temperatures for two types of radiation targets. The maximum attainable temperature calculated is 5200 deg K.

0398

Temperature Measurements in the Solar Furnace ; Foex M, " Proceedings UN Conference on New sources of
Energy ", Rome (Italy), (1961).

Two methods are suggested to alleviate the error introduced due to the parasitic reflections during
the optical pyrometry of substances treated in a solar furnace. A method using rapid response pyrometers to study various transitory phenomenon, and in particular, to measure the solidification
points of refractory oxides melting between 1800 and 2700 deg C as well as the temperature of certain
transformations is also discussed.

0399

Concerning Several Devices for the Utilization of Imaging Furnaces : Foex M, " Proceedings First Conference on Thermal Imaging Techniques ", Cambridge, Massachusetts (USA), p. 141, (1964).

Comparison of the operating efficiencies of solar furnaces and classical imaging furnaces is presented. The repective means of measurement, control and regulation of energy and temperature are considered . Devices for the combined heating of substances by high energy light radiation and by plasma are
reported. Arc and solar heating are well adapted for operations of fusion of condensed s u b s t a nces
without contamination, but, due to the poor radiation absorption, are not well suited to the heating
of vapours and gases. The incorporation of a plasma tube or torch thus extends the general utility of
an imaging furnace.

0400

Methode de Mesure des Temperatures des Produits Traites au Four Solaire : Foex M and Coutures J P,
" COMPLES Bulletin ", p. 5, (1969).

A method of measuring temperatures of products treated by a solar furnace is proposed. The information
given by the pyrometer is corrected from interfering solar emission by filters. Other sources of errors in the measurement are also discussed.

0401

Some Remarks on the Design of a Solar Furnace and the Calculation of Concentration Using Spherical - Mirror Elements : Foote J R and Adney J E, " Report No. AFMDC-TR-59-15; Solar Furnace Support Studies", U.S. Air Force Missile Development Centre, Holloman, AFB, New Mexico (USA), (1959).

The problem of replacing a single paraboloidal condenser by an approximate mirror system for a large solar furnace is studied. Ray tracing methods are outlined for approximate and preliminary studies of various mirror types. Formulas are found and methods outlined for calculationg the concentration o f light at the focal spot for spherical mirrors having either of the two local principal radii from the paraboloid. This case is calculated and compared to the performance of the ideal paraboloid having selected the qualitatively best array of spherical mirrors.

0402

Exploitation of Solar Energy Via Modular Power Plants and Multiple Utilization of Waste Heat: Feustel J E and Kraft M, " Proceedings ISES Congress ", New Delhi (India), p. 1696, (1978).

This paper describes the development of a solar farm plant designed on the modular system and having ratings of 15 to about 1000 Kw. Such plants lend themselves particularly to decentralized supply of power in countries with high rates of direct insolation. An initial plant of 50 Kw peak rating and 30 Kw nominal rating also utilizing waste heat is presently being planned and built for a site in Almeria, southern Spain. Sections of the plant will be put into operation in 1978 and the plant will be in full operation by 1979. First tests and optimization have been carried out at MAN's solar test centre i n Munich since 1976.

0403

Obtaining Film Materials with Solar Furnaces : Frantesvich I N, Dvernyakov V S, Kasich-Pilipenko I F , Tikush V L,Rusakov G V, Gaevskaya L A, Suaginyan L R and Birynkova R S," Applied Solar Energy", 15(5), 37, (1979).

It is shown to be possible in principle to produce films of various materials by means of solar ener- gy. Films of aluminium, silicon and silicon monoxide have been obtained on single crystal silicon.

0404

Methode Pour Evaluer L'energiesolaire Fournie par un Insolateur Plan/(Method for the Estimation of the Solar Energy Supplied by a Plane Solar Furnace) :Gicquel R," Rev. Gen. Therm ", 14, (164-165), 581,(1975).

Design criterial for solar furnaces are outlined. The calculation method for energy absorbed by solar heat exchangers are demonstrated.

0405

Lightweight Solar Concentrator Development : Gillette R, Snyder H E and Timer T,"Solar Energy ", 5(1), 20, (1961).

The objective of the present research program is to develop a lightweight solar concentrator that will provide sufficient solar energy for space vehicle power conversion equipment. Solutions to processing problems such as tooling and parting of the lightweight replica mirror are discussed. Solar test data are furnished for performance evaluation. It was found that a 36 in diameter mirror weighing 2 . 9 lbs (.410 psf) has the capability of providing sufficient energy for a 15 watt thermionic generator. These developments demonstrate feasibility of supplying highly concentrated solar heat with light weight co- ncentrators.

0406

Fabrication of Lightweight Parabolic Concentrators From a Glass Master - Final Report : Gillette R B & Snyder H E , " NASA (USA), Report No. NASA - CR - 51732 ", (1963).

Significant refinements are reported in the fabrication of a high optical quality, lightweight solar concentrator by using a 5 ft diameter glass convex master as the reference optical surface. A new spray- ing technique was used to form multiple layers of epoxy resin for the optical substrate, together with aluminium honeycomb and fibreglass cloth. Several 8 in diameter specimens and a 1/3rd scale model con- centrator were obtained with intact substrates. The epoxy filled substrate of a 60 in diameter concen- trator cracked while curing due to the developing tensile stress between the filled epoxy and the glass master . It was concluded that a lower curing temperature might eliminate cooling stresses.

0407

The Design of Solar Concentrators Using Toroidal, Spherical or Flat Components : G i u t r o n i c h J E, " Solar Energy ", 9, 162, (1963).

The difficulty and the cost of making large concentrators for solar furnaces can vary widely depending on the performance required of the furnace. These large concentrators are invariably made f r o m many small elementary mirrors that may or may not have the ideal paraboloidal shape. The author has endeav- ored to supply some of the design data associated with the use of toroidal, spherical and flat mirror components. The magnitude and number of components are determined for mirrors of various angular aper- tures and accuracy requirements. The results indicate that most design requirements can be efficiently fulfilled by the use of these approximations.

0408

Engineering Research with a Solar Furnace : Glaser P E, " Solar Energy ", 2(2), 7, (1958).

Research approaches using a solar furnace for the measurement of properties of matter and t e s t i n g of materials at high temperatures under controlled conditions in the laboratory are discussed. Instru- ments to measure temperatures, heat flux and flux density are described and their uses pointed o u t . Methods for measuring thermal conductivity, thermal expansion, heat content, emissivity of materials are explained, and the analytical and experimental procedure outlined.

056

0409

Algorithm for the Statistical Solution of the Problem of Radiant Flux Distribution in Receivers of Solar Devices with Paraboloidal Concentrators : Girilikhes V A, "Applied Solar Energy", 2(4), 22, (1966).

Development of a method for solving problems that arise in the investigation of radiative transport in solar energy converters. It is shown that using this method it is possible to analyse the effect of various factors on the distribution of reflected radiant fluxes over the solar receiver surface.

0410

New Methods of Manufacturing Solar Concentrators : Girilikhes V A, "Applied Solar Energy ", 4(3), 30, (1968).

Modern methods of paraboloid solar concentrator manufacturing are given. Parameters of the produced mirrors are presented. Ways and means of improving their performances are considered.

0411

Experimental Investigation of Distribution of Irradiance in the Near Focal Region of a Paraboloidal Concentrator : Girilikhes V A,"Applied Solar Energy", 4(4), 68, (1968).

The light stimulation method for investigation of performance of the reflected radiation field in the near focal area of paraboloid concentrators is used. Experimental installation and measuring instruments are described. Test procedure and measurement data are presented.

0412

Methods of Quality Control for Solar Energy Concentrators : Girilikhes V A, "Applied Solar Energy " , 8(4), 71, (1972).

The basic tests for solar energy concentrators conducted during manufacture and operation are examined. Methods for determining the optical precision and optical power characteristics of the reflective surfaces are classified. The instruments and devices used in the Soviet Union are described. The advantages of the different methods and recommendations of various applications for studying the quality of reflective surfaces are given.

0413

Analysis of Radiant Transfer Processes in Cylindrical Cavity Type Receivers of Solar Installations : Girilikhes V A and Obtemperanskii F V, "Applied Solar Energy", 5(2), (1969).

Theoretical and experimental study of the influence of radiation and multiple reflection on the distribution of the resulting radiant fluxes in cylindrical receivers of solar power installations is described. It is shown that, at moderate operating temperatures, the flux density distribution over the reflecting surface is almost unaffected either by the self radiation of the receiver walls or by the wall temperature variations. The main radiant losses in such receivers are due to reflection from the cavity, so that low wall reflection coefficient is desirable. Cavity aspect ratios between 3 & 4 are optimum for minimizing radiant losses.

0414

An Appropriate Method for Calculating Solar Sources of Heat Energy with Circulating Contours : Girilikhes V A, Krasavtsev V P and Matveev V M, " Applied Solar Energy", 5 (5), (1969).

A method is proposed for calculating parameters of solar heat source element, such as concentrators, receivers and heat accumulators. With this method the precision of the concentrator orientation to the mirrors can be determined. This makes the alternation of optimal parameters, throughout the whole system possible.

0415

A Formula for Irradiation Distribution in the Focal Surface of Parabolic Solar Concentrators : Girilikhes V A and Zakhidov R A, "Applied Solar Energy", 7 (4), (1971).

R.Aparisi's model is analysed and the physical and mathematical nature of the assumptions are discussed. A simpler solution for determining the final formula of the radiation distribution in the focal surface of the parabolic solar concentrators is presented.

0416

Universal Power Characteristic for High Temperature Solar Heat Source ; Girilikhes V A and Matveev V M, "Applied Solar Energy", 8 (3), (51, (1972).

The ratio which makes it possible to represent the dependence of the maximum efficiency of a concentrator - receiver system on the entire set of indices which characterize the elements of the system is described. The all purpose power characteristics developed on the basis of this ratio makes it possible to establish a relation between any parameters of the solar heating source. Several e x a m p l e s are illustrated.

0417

Calculation and Analysis of Parameters of a High Temperature Solar Heater with Convective Heat Removal : Girilikhes V A , Matveev V M and Rozhkov I A, "Applied Solar Energy", 8(6), 72, (1972).

The authors outline a design calculation for a high temperature solar heater with convective heat removal and recommend methods for its solution. They also examine some of the specific peculiarit i e s of calculations for gas heaters and outline some basic configurations. As an example, calculat i o n s are made for a SVIT installation with a parabolic concentrator and cylindrical corrugated receiver . The dependences obtained have been analysed to substantiate the choice of optimum values for alternative parameters.

0418

Solar Furnace at Observatory in Algeria : Guillemin J, " Revue de l'Aluminium " (France), 34 (240), 171, (1957) .

Thirty nine feet high Heliodyne furnace was built to study chemical processes. Problems which had to be solved in making parabolic aluminium mirror with diameter of 27 ft 7 in and in achieving sufficient rigidity of light metal frame are discussed.

0419

Conceptual Design of a Parabolic Dish Solar Collector Using Simulation Techniques : Gupta B P and Buchholz R L, (Unknown) .

The development of solar concentrators in recent years has produced a wide variety of collectors for the utilization of solar energy. This paper presents the simulation techniques used to predict the optical and thermal performance of a paraboloid of the revolution type solar collector. Conceptual design of a dish concentrator with a fixed receiver size is obtained by parametrically examining the significant variables.

0420

Distribution of Radiation from Elliptical and Parabolic Mirrors : Hart P J, " Journal Optical Society of America ", 49, 637, (1958).

A method is described for determining the distribution of the radiation reflected from a mirror due to a point source. In particular the distribution from an elliptic cylinder which apparently has not been previously analyzed is investigated and examples are given of the irradiance on surfaces of various configurations. In addition the distribution from mirrors of elliptical, paraboloidal and parabolic cylinder form is considered. The method used may be extended to other types of mirrors.

0421

Method for Fabricating Paraboloidal Mirrors : Hess G and Jenness Jr. J R, " Journal Optical Society of America ", 48 (2), (1958).

Paraboloidal mirrors have been made, using a convex plastic paraboloidal formed by centrifugal casting on a mercury surface as the master for fabricating plastic replica mirrors. Mirrors with disk or confusion diameters less than 1 mm have been obtained by the process.

0422

Status Report on the White Sands Solar Facility : Hayes R, Flores L, Higgins R, Houghton A L, Knasel T M and Liner R T, " Proceedings ISES Conference ", Los Angeles (USA), (1975).

The 25 Kw solar furnace at the White Sands Missile Range (WSMR) in southern New Mexico is the largest operational solar collector in the US and the fifth largest furnace in the world. The operating characteristics of the facility including the measured power and tracking capabilities are described. An overview of the experimental program is presented. Special features such as the ability to track the sun, the use of an ideal light collector at the furnace focus, and the use of flux diverters is discussed.

0423

Theoretical Considerations on Performance Characteristics of Solar Furnaces ; Hiester N K, Tietz T F & Loh E, " Jet Propulsion ", 13,546, (1957).

The theoretical factors affecting the performance of the parabolic type solar furnace are discussed . Calculations of the heat flux and the maximum tamperature obtainable at the focus are presented for furnaces of different diameter to focal length ratios. The analysis is extended to the California Institute of Technology lens type furnace. The results indicate that a paraboloid of relatively low quality is capable of achieving temperatures over 2000 deg K. On the other hand, a research furnace capable of attaining temperatures in the range of 3600 to 4200 deg K would have to have a paraboloid of very high quality.

0424

Solar Furnace ; Its Construction and Performance : Hisada T, et al," Memorandum, Government Industrial Research Institute ", Nagoya (Japan), (1956).

The process of construction and the results of the initial inspection of the first Japanese solar furnace are related in this paper. Based on a design by W M Conn without the auxialiary heliostat, the furnace has an aluminium paraboloidal mirror with an aperture of 2 m. Further development work on the utilization of solar furnace is planned.

0425

Effects of Pointing Errors on Receiver performance for Parabolic Dish Solar Concentrators : Hughes R Q " Proceedings 13th IECE Conference ", San Diego, California (USA), (1978).

The effect of dynamic (moving) pointing errors on the performance of solar thermal receivers is investigated. Only point focussing types of solar collectors are considered. The key element in the study is the analytical derivation of the intercept factor that relates pointing errors to captured energy at the receiver. A detailed example using typical parameter values is modelled on the digital computer and demonstrates the theory and the dynamic nature of the problem.

0426

Effects of Tracking Errors on the Performance of Point Focussing Solar Collectors : Hughes R O," Solar Energy ", 24, 83, (1980).

The author stresses that an important parameter in the design of point focussing solar collectors is the intercept factor which is a measure of efficiency and of energy available for use in the receiver. Using statistical methods, an expression for the expected values of the intercept factor is derived for various configurations and control law implementations. The analysis assumes that a radially symmetric flux distribution is generated at the focal plane due to the sun's finite image and various reflector errors. The time varying tracking errors are assumed to be uniformly distributed within the threshold limits and allow the expected value calculation.

0427

A High Temperature Installation for Calibrating Thermal Problems : Idiatulin Z G, Postnov V K and Kirgzbaev D A, " Applied Solar Energy ", 12 (1), 36, (1976).

A 580.10^3 W/m^2 experimental installation for calibrating thermal probes has been investigated. The experimental data analyzed show that optical furnaces can be used for high temperature calibration. The measuring unit of the system was a highly sensitive calorimeter with a constant effective absorption coefficient of 98 % .

0428

Calculation of the Radiant Flux Distribution in the Working Spot with Three - Dimensional Emitter : Ischenko E F and Lopatina G G , " Applied Solar Energy ", 6 (1), 34, (1970).

A grapho - analytical estimation procedure is presented for calculating emission flux distribution in the active zone of an image - furnace spot. As an emitter, a three dimensional radiating body is used and the aberration of the emitter lens system is taken into account. The procedure covers optical arrangements such as monoellipsoidal, biparaboloidal and biellipsoidal arrangements. Comparison of estimated results and experimental data for a built image furnace is given. The estimating accuracy permits analyses to be made of different factors which influence the distribution law of emission f l u x within the active zone.

0429

Advanced Solar Receiver Development : Jarvinen P O , " Proceedings ISES Silver Jubilee International Congress ", Georgia (USA), (1979).

Ceramic receivers for advanced solar Brayton and advanced solar stirling thermal power systems are considered which utilize impingent jet cooled silicon carbide ceramic dome heat exchanger modules in conjunction with a cavity receiver geometry to heat a pressurized gaseous heat transfer fluid to temperatures in the range of 1800 to 2400 deg F. Conceptual designs of single and multiple dome ceramic cavity receivers are presented and analysed, and a combined analytical and experimental program to develop a high temperature seal for the ceramic dome module is reported.

0430

Geometric Properties of a Modified Whirling Membranes Solar Energy Concentrator : Jerke J M & Heath Jr. A R, " NASA - Report No. NASA - TN - D - 5859 ", (USA), (1970).

The geometry of three modified paraboloidal whirling membrane solar concentrator models of 3.05 m diameter was measured by using an optical-ray-trace technique. The membranes were fabricated of 0 . 01 m m thick aluminized plastic, attached to metal hubs, and rotated at 71 rad/s in a vacuum chamber. Geometric properties such as focal length, mean and standard deviation errors, and geometric efficiency for three models with different metal hub diameters are discussed and are compared with results for a similar whirling - membrane model of an earlier investigation.

0431

Construction of Solar Furnace at Technical Center of Aeronautics, Sao Jose Dos Campus, Brazil: Jorro M A A and Mueller A, " Associacao Brasileira de Metais - Boletim ", 20, 495, (1964).

In this paper a solar furnace unit designed and built for research purposes is described. It h a s a theoretical capacity of 1.5 Kw and can achieve temperatures in the order of 3500 deg K.

0432

The Design of the Heliostat Mirror for a Solar Furnace : Jose P D, "Solar Energy", 1 (2-3), 23, (1957).

An analysis is made of the size and shape required for a heliostat mirror as a function of latitude , hours of operation , angular diameter of the sun, the distances between the parabolic condenser and heliostat, and the arrangement of the axes. All these parameters enter into a general equation from which the shape of the heliostat mirror may be exactly determined. Two arrangements of axes are considered for the altazimuth arrangement. The Trombe mount is discussed.

0433

The Flux Through the Focal Spot of a Solar Furnace : Jose P D, "Solar Energy", 1(4), 19, (1957).

An analysis is made of the flux through the focal spot of a solar furnace, taking into account t h e limb darkening of the sun.

0434

The Design of the Condenser of a Solar Furnace Using Non Parabolic Elements : Jose P D, "Report AFMDC-TR-59-15, Solar Furnace Support Studies ", Air Force Missile Development Center, Holloman, AFB, N e w Mexico (USA), 2, 219, (1959).

This paper discusses the possibility of using non parabolic segments in a solar furnace of large dimensions. A condenser composed of spherical mirrors of different radii is found to be unsatisfactory except when the mirrors are quite small. A great number of small mirrors lead to considerable edge loss. Torroidal mirror segments form a satisfactory approach to the paraboloid. All computations are based on front surface mirrors, although in practice back surfaced mirrors are more desirable. Preliminary consideration is given to the limb darkening of the sun and the resulting energy distribution in the focal plane. Taking this effect into account, it is found that the flux density is about 10% greater than that given by assuming a solar disk of uniform density.

0435

Rigidization of a 1.52 meter Diameter Inflatable Solar Concentrator in Vacuum with Associated Material Studies and Fabrication Techniques - Final Report : Jouriles N, Welling C E and Kryah J C, " N A S A Report No. NASA - CR - 66114, (1966).

Paraboloidal solar concentrator development work was carried out. Precoat foam was used as the rigidizing material for inflatable film type mirrors. 1.52 m diameter units were rigidized in a vacuum chamber utilizing radiant heat as the heat source for initiation of the foaming reaction. The foam h a d been previously applied in sheet form to the back of a commercial aluminized polymide film parabolic membrane. A torus type backup structure was subsequently attached for mounting in ground test equi p-ment. Relatively good quality mirrors were the results of this development program.

0436

The Feasibility of Joining Metal Using a Solar Furnace : Kaddon A F K & Abdul Latif, "Solar Energy", 12(3), 377, (1969).

This note discusses the investigation of the possibility of joining metals with solar radiation. It is expected that in case of employing a large solar furnace, all joining methods, including soldering brazing and welding with their variation may be successfully realized. It is believed t h a t all metals may be joined using the solar furnace and employing any desired atmosphere.

0437

Theoretical Consideration on Energy, Concentration Ratio, Concentration Efficiency and Obtainable Temperature of Solar Furnace : Kamada O, " Bulletin of the Research Institute for Scintific Measures of Tohuko University ", Sendai (Japan), 12, 30, (1963).

By taking the brightness distribution of the solar disk into consideration, the characteristics of a parabolic type solar furnace were theoretically treated. The energy concentration ratio, the concentration efficiency and the maximum temperature obtainable for the flat targets were calculated for t h e furnaces with different aperture ratios. These were then compared with the results obtained by assuming that there was a uniform brightness distribution of solar disk.

0438

Method of Measuring Target Temperature in a Solar Furnace : Kamada O, "Applied Optics ", 3,1397, (1964).

A pyrometric method of measuring the surface temperature of a target in a solar furnace by employing 1.38 m wavelength radiation which is absent in the sun's radiation is described. The measured temperature distribution of graphite exposed to concentrated solar radiation is given, and comparison is made with the theoretical distribution. Illustrations are included showing the optical system of the brightness pyrometer and the relations between the accuracy, the reference temperature and the range in which automatic measurement is possible using the optical null method.

0439

Theoretical Concentration and Attainable Temperature in Solar Furnace : Kamada O," Solar Energy ", 9(1), 39, (1965).

By taking the brightness distribution of the solar disk into consideration, the characteristics of so-lar furnaces provided with paraboloidal concentrator were theoretically treated. The energy concent-ration ratio, the concentration efficiency and the attainable temperature for a flat, a cylindri c a l and a spherical target were calculated for the furnaces with different aperture ratio, the results of which were compared with the results obtained by assuming that there was a uniform brightness distri-bution of the solar disk.

0440

On the Focus Shift in Film Concentrators Under Wind Load : Kamildjanov A, "Applied Solar Energy ", 5(2), (1969)

The effect of wind load on the optical characteristics of a solar concentrator is calculated. The re-flector surface shape is determined. The focal point shape is described by the points of intersection of infinite near parallel rays. If it is necessary to limit the focal point, the given formula mak e s it possible to calculate the required characteristics of a film concentrator.

0441

Film Solar Energy Collector with Concentric Circular Joints : Kamildzhanov A, " Applied Solar Energy", 6(1), 15, (1970).

In this paper the concentrators with parallel, radial and circular concentric seams are discussed. The results of experimental investigations of such concentrators are compared with data from investigation of concentrators without seams.

0442

General Method for Predicting Efficiency of Paraboloidal Solar Collector : Kaykaty G N," NASA - Report No. NASA - TM -X - 1323 ",(USA), (1966).

A technique and working curves are presented for the rapid calculation of the maximum collection effi-ciency of a paraboloidal collector that focusses solar energy into a cavity receiver. Effects of col-lector surface errors, orientation error, and and rim angle are included. Any distribution of surface errors may be specified in the method.

0443

Analysis of the Maximum Performance of a Paraboloidal Solar Collection System for Space Power : Kaykaty G N, " NASA - Report No. NASA - TN - D - 4415 ", (USA), (1968).

This paper presents an analytical study performed to investigate the effects and interactions of the concentrator surface errors and rim angle, collection system orientation errors and cavity receiver operating temperature on the maximum thermal efficiency of a paraboloid collection system operati n g in the vicinity of the earth. The ranges investigated were : standard deviation of surface e r r o r (0 to 18 min.) orientation error (0 to 30 min.) receiver temperature (1110 to 2200 deg K) and concen-trator rim angle (45 to 60 degrees).

0444

Construction and Operation of the Arizona State College Furnace : Kavane C J, "Solar Energy ", 1(2 - 3), 99, (1957).

A description is given of the solar furnace installed early in 1956 on the roof of the science building at Arizona State College at Tempe. In common with many contemporary furnaces a converted searchlight reflector is used in a stationary mounting. The yoke which originally held the searchlight has been converted and serves as a mounting for the heliostat. A photoelectric system is used to make the heliostat follow the sun throughout the day. The furnace is now in use in a program of high temperature studies for which accurate temperature control is needed. This has been accomplished by means o f adjustable sleeves, rotating segments and other devices.

0445

Estimated Precision of the Reflecting Surface of a Paraboloidal Concentrator for Different Angular Sizes of the Radiation Source : Kraisna E A, Nevezhin O A and Rubanovich I H, "Applied Solar Energy ", 10(1), 22, (1974).

The effect of angular size of the source of radiation on the precision parameter and the standard parameter is analyzed. The influences of angular size of the source of radiation on the distribution of the heat flux density in the focal spot is also studied. Results of some typical numerical calculations are represented graphically and discussed.

0446

Berechnung Der Sonnendirekteinstrahlung Auf Ein Spannungkraftwerk Mit Parabolspiegeln / (Computation of Solar Radiation Potential for a Power Plant with Paraboloidal Mirrors) : Kuczera M and Günther R, " Brennstoff - Wärme - Kraft ", 27(11), 418, (1975).

Design guidelines for determination of intensity and absorption of solar radiation to energy collecting mirrors are outlined. The effects of seasons, geographical location and atmospheric conditions on performances of solar energy power plants are investigated and assessed. Design criteria for collectors are discussed.

0447

Design and Development of a Parabolic Dish Solar Collector for Intermediate Temperature Service : Kugath D A, Koenig A A and Drenker G, " Proceedings ISES Silver Jubilee International Congress ", Georgia (USA), (1979).

This paper discusses the design and development of a 7 m diameter parabolic dish solar collector,which is intended for first application in the Solar Total Energy Large Scale Experiment at Shenandoah,Georgia. Each of the four main subsystems of the collector i.e. (i) reflector, (ii) mount,(iii) receiver and (iv) the control, is discussed briefly with major emphasis on the receiver design.

0448

Temperature and Flux versus Geometrical Perfection : Laszlo T S, "Solar Energy ", 1(2 - 3), (1957).

This paper describes the problems involved in determining the energy flux and temperature existing at the focus of a solar furnace. Theoretical values are readily calculated from the geometry and materials of the system but actual measurements are much more difficult to obtain. A method is described by which brightness temperature, as measured by an optical pyrometer, and normal incident solar f l u x, measured simultaneously can be converted into black body temperatures.

0449

On Radiant Energy in High - Temperature Research : Laszlo T S, "ASME Paper 60 - WA - 170 ", (1960).

A new shape for artificial black bodies is conceived, consisting of a truncated right cone with t h e end as the orifice. It is stressed that new approaches should be developed for experimentation while using high intensity radiant energy together with instruments of special design.

0450

New Techniques and Possibilities on Solar Furnaces : Laszlo T S, " Proceedings UN Conference on   New Sources of Energy", Rome (Italy), (1961).

The limitations on experimentation in a solar furnace are discussed and also the improved methods and the new instrumentation.

0451

Investigations of Thermal Imaging Techniques : Laszlo T S and Sheehan Jr. P J, " Proceedings 1st Conference on Thermal Imaging Techniques ", Cambridge, Massachusetts (USA), p.33, (1964).

Method for fabricating large, inexpensive, paraboloidal mirrors for image furnaces is presented. The process uses the principle that a liquid in a rotating dish takes the shape of a paraboloid. A clear, bubble free epoxy resin is poured into a rotating pan with a hardening agent and allowed to cure (2 4 hours). The need for a vacuum chamber is obviated by bonding the required highly reflecting m e t a l lining to the resin with an epoxy adhesive. Tests of the resulting paraboloidal mirror, a sample holder for electrical measurements, and an MgO coated radiometer are described. Methods for meas u r i n g the flux through a solar furnace assembled from these components are illustrated.

0452

Contribution to the Study of a Solar Furnace for Efficient Heating of a Bath to a High Temperature : Le Grives E, Trombe F, Lephat Vinh A, Charron F and Delfolie B, " Revue Internationale Hautes Temp. Refract.", 10, 303 , (1973).

The examination of the curves indicating the energy flux density distribution in the focal area of the paraboloidal mirror on the one hand, and the properties of cavities on the other, have led the solar furnace designers to devise a multicavity arrangement which makes it possible to achieve e f f i c i e n t heating of a bath to high temperatures. This arrangement appears to offer a good compromise between conflicting requirements.

0453

Distribution of Solar Energy at the Focus of a Parabolic Mirror Calculations and Experimental Study : Lephat Vinh A, " Journal Rech. Cent. Nat. Rech. Sci. ", No. 57, 265,(France), (1961).

The flux density of radiation reflected from an idealized parabolic surface is computed, and applied to a calculation of the distribution of radiation received on the walls of a cylindrical cavity placed at the focus of a parabolic reflector. Theoretical results are compared with those obtained by observing the actual distribution of solar energy reflected from an approximately parabolized mirror of focal length equal to 18 meters.

0454

Parabolic Collector for Total Energy System Application : Levine A L, Paradis L R and Truesdal K L , " Proceedings ERDA Conference on Concentrating Solar Collectors", Georgia (USA), p. 4 - 33, (1977).

The Raytheon Company is participating in the ERDA/Sandia Laboratories total energy system development program by designing and fabricating a parabolic, point concentrator solar collector. The point concentrator designed and under construction is a toric parabola 6.7m in diameter, with an effective aperture of 35 sqm. Azimuth and evaluation drive systems are computer controlled and provide maximum aperture utilization over the course of the year. Mirrors are curved glass, hard mounted on an aluminium substructure concentrating the solar energy into a cavity absorber located on the collector optical axis.

0455

Performance and Evaluation of Concentrating Solar Concentrators for Power Generation : Liu B Y H and Jordan R C, " Journal of Engineering for Power (Trans. ASME) ", 87, 8, (1965).

The geometrical accuracy of a real solar concentrator is defined quantitatively in terms of two equivalent parameters, the standard target error and the angular error and the relationship between those parameters and the flux distribution on the focal plane are developed. A general method for determining the optimum size and efficiency of an absorber for any given concentration is described. Specific numerical results are obtained , however, only for the case where the function describing the flux distribution on the focal plane is given by Gauss's normal law of error. Criteria for determining the applicability of the results are proposed. Finally, experimental techniques (both optical and thermal) of evaluating concentrators are briefly described.

0456

Heat Flux Measurements at the Sun Image of the California Institute of Technology Lens Type Solar Furnace : Loh E, Hiester N K and Tietz T E, " Solar Energy ", 1 (4), 23, (1957).

Flux profiles of the sun image were measured at the focus of the California Institute of Technology lens type solar furnace using NRDL water cooled radiometers. A maximum flux value of 220 cal/sec/sqcm was recorded when the direct solar radiation received at the furnace site was 1.04 cal/min/sqcm. Under these conditions the furnace operating efficiency was found to be 47.2%. The losses of 52.8% are ascribed to transmission losses in the lenses, reflecting losses at the mirrors, the geometrical imperfection of the individual lenses and mirrors and lack of perfect superposition of the 19 sets of images.

0457

Thermal Analysis in the Focal Spot of a Solar Furnace : Lorenzini E and Spiga M, " Solar Energy ", 22, 51, (1979).

The paper deals with the analytical determination of time - dependent temperature distribution in a sample placed in the focal plane of the paraboloidal mirror of a solar furnace. Several different heat pulses are considered and some graphs are shown.

0458

Solar Parabolic Dish Thermal Power System Technology and Applications : Lucas J W and Marriott A T , " Proceedings 14th IECE Conference ", Boston (USA), p. 166, (1979).

The activities of two projects at JPL in support of DOE's Small Power Systems Program a r e reported. These two projects are the Point Focussing Distributed Receiver (PFDR) Technology Project, and Point Focussing Thermal and Electrical Applications (PFTEA) Project. The PFDR technology project's major activity is developing the technology of solar concentrators, receivers and power conversion subsystems suitable for parabolic dish or point focussing distributed receiver power systems. Other PFDR activities include system integration and cost estimation under mass production, as well as the testing of the hardware. The PFTEA projects' first major activity is seeking ways to introduce PFDR systems into appropriate use sectors. The second activity is systems engineering and development wherein power plant sytems are analyzed for specific applications. The third activity is the installation of a series of engineering experiments in various use environments to obtain actual operating experience.

0459

Parabolic Dish Technology for Industrial Process Heat Applications : Lucas John W, " Jet Propulsion Laboratory Report ", (USA), (1979).

The point focussing distributed receiver concept utilizes a parabolic concentrator which tracks the sun in two axes across the sky. The concentrator collects sunlight from a large area and reflects & focusses it to a very small area. A receiver, which is mounted at the focal point, captures the concentrated radiation and converts the energy to heat in a working fluid such as hot gas or steam. The working fluid transports the energy via flexible lines to a heat transfer network on the ground to provide process heat. The technology status of the basic concentrator and receiver subsystems is described in the paper.

062

0460

Performance and Economic Risk Evaluation of Dispersed Solar Thermal Power Systems by Monte Carlo Simulation : Manvi R and Fujita T, " Proceedings IECE Conference ", San Diego (USA), p. 1535, (1978).

A preliminary comparative evaluation of dispersed solar thermal power plants utilizing advanced technologies available in the 1985 - 2000 AD time frame is under way at the JPL. The solar power plants of 50 KWe to 10 MWe size are equipped with two axis tracking parabolic dish concentrator systems operating at temperatures in excess of 1000 deg F. The energy conversion scheme under consideration includes advanced steam, open and closed cycle gas turbines, stirling and combined cycle. The energy storage systems include advanced batteries, liquid metal and chemical. This paper outlines a simple methodology for a probabilistic assessment of such systems. Sources of uncertainty in the development of advanced systems are identified and a computer Monte Carlo simulation is excercised to permit an analysis of the trade - offs of the risk of failure versus the potential for large gains. Frequency distribution of energy cost several alternatives are presented.

0461

Flow Systems in the Solar Furnace and the Photolysis of Nitrosyl Chloride : Marcus R J and Wohlers H C, " Solar Energy ", 5 (4), 121, (1961).

This paper reports the design, construction, installation, testing and use of various flow systems in a 2 ft diameter solar furnace. The flow system finally used is more complex than usual, since it was designed to handle a corrosive solution of Nitrosyl Chloride in Carbon Tetrachloride. In this flow system the focal spot is contained in the center of a 15 mm diameter quartz tube through which solution is pumped at the rate of 500 ml per minute. Alternatively the focal spot is in a free - falling stream of solution which impinges on a splash plate immediately after illumination. The flow system was used to study the photolysis of nitrosyl chloride in carbon tetrachloride solution to form nitric oxide and chlorine.

0462

Effect of Transverse Receiver Defocussing on the Energy Characteristics of Paraboloidal Solar Devices: Matveev V M and Teplyakov D I, " Applied Solar Energy ", 2(1), 9, (1966).

The magnitude of the radiant flux reflected by the mirror and incident on the receiver for a paraboloidal solar device with a mirror precision characteristic accurately focusses onto the sun is calculated. Formulas are given for calculating the heat loads existing at the characteristic points of the radiation receiving surface. A double integral is derived which expresses the total radiant energy arriving at the radiation receiving part of the receiver. Calculated curves are presented of the energy distribution in the focal image of a mirror with h = 1.5 in the presence of transverse defocussing.

0463

A Device for Producing Sinusoidal Heat Waves in Solar Furnaces: Mavashev Yu. Z," Applied Solar Energy", 1 (3), (1965).

An apparatus was developed utilizing solar energy to form sinusoidal heat waves for studying the thermal properties of materials at high temperatures, i.e. 3000 to 3500 deg C, regardless of their conductivity. The heat source is a solar furnace two meters in diameter with a heliostat. A periodically changing thermal flux was easily obtained in the furnace by modulating a screen in the form of a disk placed between the parabloid and its focus. The axis of the disk coincided with the principal optical axis of the mirror and the plane of the disk was moved back and forth along the axis, the power of the light flux falling on the specimen was varied.

0464

Determination of Focussing Properties of Solar Collectors by an Integral Formula : Mazur P, " Solar Energy ", 6(1), 23, (1962).

An analytical procedure is derived for determining the distribution of energy at the focal plane of a mirror such as is used in a solar collector system. The method is a general one that permits any shape of emitting and reflecting surfaces, and is an analytical equivalent of the conventional ray tracing procedures of geometrical optics. The advantage provided is that the calculation of intensity of collected radiation at a given point can be performed on a digital computer in less than 15 seconds, whereas one man - week is required for graphical ray tracing. The solution takes the form of an exact double integral which gives directly the intensity of reflected radiation at any given point for an arbitrary reflecting surface. The double integral has been programmed on a digital computer to evaluate the performance of solar collectors.

0465

Solar Concentrator Design and Construction : McCusker T J, " Paper at AIAA / ASME 3rd Biennial Aerospace Power Systems Conference ", Philadelphia (USA), (1964).

The development of fabricating techniques for various types of concentrators is prescribed. Vacuum evaporation of a layer of aluminium onto a thin plastic film provides the reflecting surface, and pressurising this film or assembling properly shaped sections of it produces the desired paraboloid. Problems and methods of ground rigidization of the concentrator are discussed. For large diameter concentrators, shaping and rigidization takes place in space. Methods using a foam of the urethane type have been tried and are compared. Thermal coating of film is mentioned. A cone and column type concentrator has been tested and is considered.

0466

Solar Furnaces ; Development at the New South Wales University of Technology : Milner C J, "Paper at the A N Z A A S Congress, Solar Energy Symposium ", Adelaide (Australia), (1958).

Solar power concentrated with a large high - grade paraboloidal mirror is discussed. Temperatures of 3000 - 4000 deg C are studied. The arrangement of a paraboloid inverted on a tower above a heliostat mirror on the ground, is commended. The influence of size on performance is indicated alongwith the problem of temperature measurement.

0467

High Temperature Phase Studies on the System $Al_2O_3 - Lu_2O_3$ with a Solar Furnace : Mizuno M, Berjoan R, Coutures J P and Foex M, " Proceedings ISES Conference ", Los Angeles (USA), (1975).

The liquidus curves in the $Al_2O_5 - La_2O_3$ and $Al_2O_3 - CeO_2$ systems were determined by the use of a helio-stat type solar furnace. The cooling curves of method specimen were obtained in the black body c e n-trifugal cavity with an optical pyrometer. The quenched specimens from the melt were investigated by X - ray diffraction techniques, chemical analysis, and a petrographic microscopy . X - ray powder dif-fraction measurements on specimens quenched from the melt in air, hydrogen, argon, oxygen and nitro - gen as well as on the samples heated at 1700 deg C for 8 hours in an electric resistance furnace were made using a diffraction unit on the Geigerflex or Philips - Norelco diffractometer with nickel filter-ed $Cu - K\alpha$ radiation . High temperature X - ray diffraction data were also examined. From the liquidus curve data and crystallographic analyses on these systems, tentative high temperature phase diagrams are presented.

0468

High - Flux Low Temperature Solar Collector : Nevins R G and McNall Jr. P E, " Heating / Piping / Air Con -ditioning ", 29 (11), 171 , (1957).

A 5 ft diameter surplus searchlight mirror is used to focus solar energy on 6 in diameter   collector mounted near the focus of the mirror. Solar energy is collected at a rate of 3700 to 4200 Btu/hr and collection efficiencies of 65 to 75% are obtained on clear days with such a collector.

0469

High Temperature Solar Furnace Studies : Noguchi T, Mizuno M, Yamada T and Rouanet A J, " Proceedings ISES Conference ", Los Angeles (USA), (1975).

In a series of high temperature solar furnace works at the Solar Research Laboratory, the high temp-erature phase studies on the $La_2O_3 - Y_2O_3$ and $Ga_2O_3 - Al_2O_3$ systems were summarized in this paper. I n the temperature measurement of a heated specimen at the focal plane of a heliostat type solar furnace of which optical axis is horizontal, a digital computer system has been coupled with the brightn e s s pyrometer to give a higher accuracy.

0470

Adjusting of a Concentrator with Parabolic Facets :   Novicov V V and Skripkar L N, " Applied   Solar Energy ", 5 (1), (1969).

An adjusting method for a concentrating mirror made of a facet mirror, made for a cassegrainian con -centrator is discussed. The optical facet zones are determined giving the maximum light flux.

0471

Optical Analysis of Paraboloidal Solar Concentrators : O'Neill M J and Hudson S L, " Proceedings ISES (American Section) Conference ", Denver, Colorado (USA), p.855, (1978).

Paraboloid solar concentrators are currently being developed for use in high temperature solar ther-mal electric power systems. Since such systems require concentration ratios of 2000 and higher f o r efficient energy collection, the accurate prediction of focal plane flux profiles is critical for ef-fective system design. This paper presents a straightforward and accurate solution techniques, based upon the method of cone optics, for defining paraboloidal focal plane flux profile for both perfect & imperfect concentrator surfaces. In addition, the paper also presents parametric results of the ana-lysis for a variety of cases of practical importance as well as comparisons of the current r e s u l t s with the numerical calculations of other investigators.

0472

Thermal Design and Analysis of an Integrated Sodium Boiler / Receiver for Solar Energy Conversion : Osborn D B, " ASME Paper No. 79 - WA / Sol - 10 ", (1979).

This paper presents the result of the thermal design and analysis of integrated sodium boiler recei-ver used for solar energy conversion. The receiver is a major element of a point focus distribu t e d receiver (PFDR) solar thermal electric system employing stirling engines for power conversion. T h e results of the design / analysis study show that a high temperature cavity receiver, employing p o o l boiling sodium, is an excellent choice for use in dish - stirling PFDR systems. The concept is tech-nically feasible at the present time, employing state of the art materials and technology, and will be a cost effective subsystem when put into production.

0473

Parabolic Collector for Total Energy System Application : Paradis L R, Levine A L and Vallee E C, "Proceedings ISES (American Section) Conference ", Orlando, Florida (USA), (1977).

This paper discusses a parabolic, point concentrator solar collector being designed and fabricated by Raytheon. The point concentrator designed and under construction is a toric parabola 6.7 m in diame-ter, with an effective aperture of 35 sqm. Azimuth and elevation drive of systems are computer contr-tolled and provide maximum aperture utilization over the course of the year. Mirrors are curved glass hard mounted on an aluminium substructure, concentrating the solar energy into a cavity absorber lo -cated on the collector optical axis.

0474

Installation for High - Temperature Studies of Heat - Resistant Aviation Materials and Coatings : Pasi-chnyi V V and Dverniakov V S, " Air Fleet Equipment and Aircraft Design "(ed. In.N. Alekseev), Issue 10, p.77, (1967).

A solar furnace using a 1.5 m diameter reflector with a focal length of 640 mm and an open angle o f 60 is described. The reflector is made of 12 mm thick glass, coated with silver on the inside (reflec-tion coefficient of 87%). The solar furnace makes it possible to study the heat resistance of c o a t-

i n g s a n d similar materials at readily controllable temperatures ranging from 20 to 3200 deg C and
with thermal fluxes ranging from 0 to 3200 Kcal/sqm/sec. The methods used to determine the thermal
parameters and the flux distribution over the heated surface are outlined.

0475

A High Temperature Solar Installation for Studying the Mechanical Properties of Heat Resistant Materials : Pasichnyi V V, Dverniyakov V S,Isakhanov G V, Liashch enko B A and Gashchenko A S, " Applied Solar Energy ", 5 (3), (1969).

The design and operation of an assembly using solar radiation in mechanical tests of heat resistant material at high temperature in air is described. A projector with a parabolic mirror solar radiation
condensing lens, 1.5 m in diameter, is focussed at specimens to provide the maximum radiant energy
density of 1250 Watt/sqm. A specimen loading system of special design is used to minimize the s h a-
dowed area of the working surface of the mirror. Tests at temperatures in excess of 3000 degK can be
performed on this assembly.

0476

The Influence of Shadowing Devices on the Heat Efficiency of Solar Furnaces : Pasichnyi V V, Sergeev
V I, Dverniyakov V S and Gavrilenko A P, " Applied Solar Energy", 5 (6), (1969)

The exact and approximate estimate of relative energy loss is elaborated. The results obtained with
a special device, both experimentally and by calculation coincide with the limits of experimental accuracy.

0477

Determining the Characteristics of the Radiation Field in the Focal Plane of Parabolic Solar Concentrators : Poluektov V P and Grilikhes V A, " Applied Solar Energy ", 4 (6), 68, (1968).

A simplified computation method is proposed for radiant flux density distribution in the peri f o c a l
area of parabolic mirrors. The deduction of the basic computation relations for the perifocal area
is given. Calculation results are given. Comparative analyses are performed with results deduced by
other methods.

0478

Analytic Method of Calculating the Radiation Flux Distribution in the Collectors of High Temperature
Solar Installations : Poluektov V P and Grilikhes V A, " Applied Solar Energy ", 6 (4), 60, (1970).

A suitable method of analytical calculation is presented for investigating radiant flux distribution
at a cavity receiver with any concentrator configuration. The angle dimension of the radiant source
is taken into account.

0479

Modular Electrical Generation Using Parabolic Dish Solar Concentrators : Peterson R and Evans R A,
"Proceedings ERDA Conference on Concentrating Solar Collectors ", Georgia (USA), p.4-1, (1977).

The preliminary design of a modular electric power generation system utilizing parabolic dish shaped
solar concentrators is presented in this paper. Solar Thermal - to - electric power conversion is accomplished at the module level by use of a recuperated open Brayton Cycle gas turbine heat engine,which
drives a 16 KWe generator. A two axis tracking circular paraboloid of revolution dish shaped concentrator 48 ft in diameter collects the solar insolation and redirects it towards the focal point.A high
temperature, cavity type thermal receiver / heat exchanger is positioned with its aperture centered at
this focal point. An alternative system is also discussed that will operate in excess of 2900 hours
per year and will generate 70 MWh (net)of electric energy per year per module.

0480

Research and Development of High Efficiency Lightweight Solar Concentrators : Pichel M A, "NASA Report
No. NASA - CR - 57347 ", (USA), (1962).

The backing and support structure for 5 ft diameter concentrators are investigated and the parameters
related to their application to concentrators of large size were evaluated. Special tooling required
for the fabrication of 5 ft diameter concentrators was designed and produced. A number of supporting
and investigatory studies were carried on relative to the parameters affecting the electroforming of
a variety of materials suitable for use in concentrator fabrication with emphasis on electroformed nickel and copper. Various reflective and protective coatings have also been investigated for use with
concentrators produced by the electroforming process, and their durability under various environmental conditions is being evaluated.

0481

Research and Development Techniques for Fabrication of Light Weight Solar Concentrators (Addendum Re-
Port) Pitchel M A ," NASA - Report No. NASA - CR - 57346 ", (USA), (1962).

Fabrication of a 5 ft diameter copper concentrator was completed, and this unit together with a 5 ft
diameter nickel concentrator was coated with chromium, silicon and aluminium layers. Calorimetric efficiency tests showed a higher efficiency for the copper concentrator with an average of 68% at the
half inch orifice, as compared to the all nickel concentrator with an average efficiency of 64.2%.
It was concluded that modified design concepts for control of the edge effect and parting procedures
will improve the concentrator efficiency.

0482

Optimization of a Point Focussing, Distributed Receiver Solar Thermal Electric System : Pons R L ,
" ASME Paper No. 79 - WA / Sol - 1 ", (1979).

This paper presents an approach to optimization of a solar concept which employs solar to electric
power conversion at the focus of parabolic dish concentrators. The optimization procedure is presented through a series of trade-off studies, which include the results of optical thermal analyses and

individual subsystem trades. Alternate closed cycle and open cycle Brayton engines and organic Rankine engines are considered to show the influence of the optimization process, and various storage techniques are evaluated, including batteries, fly wheels and hybrid engine operation.

0483

Conceptual Design and Analysis of a Dish Rankine Solar Thermal Power System : Pons R L, " ASME Paper No. 80 - C2 / Sol - 10 ", (1980).

This paper presents results of a preliminary design, economic study of a Point Focussing Distributing Receiver (PFDR) solar thermal electric system optimized for application to small community power plants at power levels upto 10 MWe . Power conversion is provided by small Organic Rankine Cycle (ORC) engines mounted at the focus of paraboloidal solar concentrators. The output of multiple power modules (concentrator, receiver, engine and alternator) is collected by means of a conventional electrical system and interfaced with a utility grid.

0484

Reflector Characteristics of a Solar Power Installation with Photoelectric Converters : Rodichev B Ia, and Tarnizhevskii B V , " Applied Solar Energy ", 5(2), (1969).

Discussion of the characteristics of a faceted reflector used in solar power installations with photoelectric converters is presented. The analysis is performed for a reflector consisting of 13,50 mm facets arranged along the half width of a parabola. The efficiency and energy balance of the reflector are determined. Optimization considerations indicate that the reflector efficiency can be increased from 0.75 to 0.8 .

0485

Economic Analysis of Two Sizes of a Point Focussing Concentrating Collector System : Rogers W E, Borton D N, Rice M P and Rogers R J , " Proceedings ISES Silver Jubilee International Congress ", Georgia (USA), (1979).

During the past four years a point focussing concentrating collector system and production techniques have been developed and are currently being evaluated. Two sizes of the system are being built: 23 & 70 Kw thermal units. The units use the same basic components and consist of a structure made of thin wall tubing, gauged reflective columns containing flat, second surface, low iron glass mirrors: a cavity receiver and a sophisticated processor based control package. In order to show a specific energy product superior to another, its value relative to alternatives must be quantified. Cost per unit collector area, or familiar term in the developing solar industry, is only one of many factors which influences the cost of energy delivered to a solar product consumer. This paper presents a method of computing the system costs of delivered energy. Various applications such as process steam, use of photovoltaic cells, air conditioning and heating are considered and specific examples presented.

0486

Determination of the Optimal Dimensions of High Temperature Cylindrical Cavity Solar Energy Receivers: Rubanovich I M , " Applied Solar Energy ", 1(4), (1965).

The operation of the concentrator and receiver of a high temperature cylindrical cavity solar energy receiver is examined for determining the optimal dimensions of such a receiver. The concentrator studied is of paraboloidal type. As part of the optimization process, receiver losses associated with the escape of radiant energy from the receiver inlet and the mean coefficient of irradiation of the cavity inlet from its inner surface had to be determined. Several means for increasing the efficiency of receivers are presented.

0487

Theoretical Investigation of the Effect of Accuracy of Concentrator Guidance System on the Operation of Solar Energy Devices : Rubanovich I M , "Applied Solar Energy ", 2 (6), 13, (1966).

Analysis of the effect of concentrator orientation with respect to the sun on the performance of a solar energy converter is presented. The parameters and pointing inaccuracy of the concentrator , and the density field of the energy flux in the focal plane of the converter at various mismatch angles are investigated as factors affecting solar converter performance. The requirements for precise orientation of the concentrator are studied for energy conversion. The analytical results agree with experimental observations.

0488

On the Objective Estimation of Reflecting Surface Accuracy of Parabolic Concentrators : Rubanovich I M " Applied Solar Energy ", 2 (6). (1967).

The relationship beteen commonly used parameters characterizing the accuracy of solar parabolic concentrator surface is given. The values of parameters for perfect accurate mirrors are presented.

0489

Experimental Manufacturing of Paraboloid Glass - Plastic Collectors : Rubinov E B, Umarov G Ya., Pinkhasov E A, Bashnyak A Ya. and Abduazizov A, " Applied Solar energy", 1 (2), (1965).

The feasibility of manufacturing a glass - plastic paraboloid collector with a reflecting surface obtained by direct vacuum condensation of aluminium vapours is shown. These are to be used for collecting solar energy flux. The methods of manufacture are given, subsequent testing is discussed, and problems associated with these processes are evaluated. Tabulation includes relative temperature distribution in the collector focal spot and average coefficients of reflection under aluminium layer vapourizing conditions.

0490

Fabrication of Paraboloidal Mirror Segments for Large Solar Furnaces : Sakurai T and Shishido K," Applied Optics ", 3, 813, (1964).

Special aspherical grinder constructed for grinding paraboloidal segments for laboratory solar furnace

10 m in aperture and aperture ratio of 3.1 is described ; fronts surface of 181 ground and polished segments were aluminized by vacuum evaporation and mounted on steel frame ; accurary of surface obtained was found to be satisfactory.

0491

Determination of the Power of a Light Pulse in a Solar Device : Sergeev V I, Dverniakov V S, Gavrilenko A P, Denisenko E Tand Pasichnyi V V, " Applied Solar Energy ", 5 (5), 11, (1969).

The design and operation of a solar energy converter with a shutter producing light pulses from 0.1 to 10 secs. is described. A procedure is derived for determining the power of such pulses as a function of the solar light intensity, exposure time, and a set of three certain empirical parameters. G o o d agreement is obtained between analytical and graphical results for a representative converter design. A block diagram is given for a converter with a shutter.

0492

Definition of Light Impulse Energy in Solar Devices : Sergeev V I, Dverniakov V S, Gavrilenko A P, Denisenko E T and Pasichnyi V V, " Applied Solar Energy ", 5 (5), (1969).

Results of the definition of light impulse energy obtained from a solar installation are given. T h e light recorder is blocked by a shutter during the time corresponding to the duration of the light impulse. The light impulse energy is defined with the help of a curve integration circumscribing the light impulse.

0493

Influence of External Reflection on Light Flow Formation by Glass Paraboloid Concentrators : Shakhparonyan V V and Shermanazanyan Ya. T, " Applied Solar Energy ", 7 (3), (1971).

The influence of the reflection of a mirror concentrator 1.5 m in diameter on the flux density dist - ribution in immediate proximity to the outside focal zone is considered. In this distribution the effect of disappearing gaps caused by central shadowing is explained.

0494

Effect of Frontal Reflection on Radiant Flux Formation by Back Reflection Paraboloidal Mirror Concentrators : Shakhparonyan V V and Shermanazanyan Ya. T, " Applied Solar Energy ", 7 (3), 16, (1971).

This paper presents the quantitative effect of central shading of the concentrator by the target and defocus of the receiver on the radiant energy distribution in the near post focal (away from the r e f - lector) region of 1.5 m reflectors. The problem is investigated in terms of the axial and radial components of the radiant energy tranport vector.

0495

Irradiation of Volumetric Receivers in the Near Focal Region of High - Temperature 1.5 m Diameter Solar Furnace : Shakhparonyan V V and Shermanazanyan Ya. T, " Applied Solar Energy ", 8 (1), 40, (1972).

Computerised methods estimate the influx of radiant energy to cylinders, disks, and plates placed within a focal region of a solar furnace.

0496

Application of Optical Pyrometers on a Solar Furnace : Shakhparonyan V V and Stamboltsyan S G," Applied Solar Energy ", 8 (3), 54, (1972)

The peculiarities in measurements of the brightness temperature of an object by an optical pyrometer with a disappearing filament mounted on a high temperature solar furnace are described. Results are presented for measurements of the melting temperature for several refractory oxides which have b e e n made through the application of a rotating radiation flux cut - off device with a reflecting elem e n t. The paper gives the sequence for converting brightness temperature data into true temperature. There is also a description of the cut - off device which is attached to the optical pyrometer for measuring temperatures on a solar furnace and data covering the results are given.

0497

Cost of Paraboloidal Collectors for Solar to Thermal Electric Conversion : Shaner W W and Wilson H S, " Solar Energy ", 17, 351, (1975).

Investment and other costs of paraboloidal collectors with various aperture widths, reflectivities , rim angles, and accuracies of contour and tracking are reported. In studying alternative design-cost performance relationship, emphasis is placed on breadth of coverage, rather than on detailed accuracy. Alternative aperture width led to preferences for different production processes that are n o t particularly sensitive to change in rim angle or reflectivity. Significant cost relationships i n manufacture, installation and maintenance of collectors were found ; other relationships less certain.

0498

On the Calorimetric Measurements of Radiant Flux Density Distribution in the Focus of a Solar Concentrator : Sharafi A Sh., " Applied Solar Energy ", 3 (4), (1967).

A zonal method is described of calorimetric determination of the radiant flux density distribut i o n in the focus of a solar concentrator. Calculations are given. Theoretical analysis of errors arising when different ways of calorimetric measurements are used is also presented.

0499

Approximate Solar Energy Concentrator Consisting of Wedge Shaped Facets of Constant Transverse Curvature : Sharafi A Sh., Umarov G Ya., Abduazizov A, " Applied Solar Energy ", 11 (3-4), 28, (1975).

A high power solar energy concentrator is described and analyzed. It consists of identical n a r r o w wedge shaped reflecting strips with constant transverse curvature. It produces the same average solar energy concentration as the paraboloidal concentrator.

energy concentration as the paraboloidal concentrator.

0500

Large High Temperature Solar Units ; Their Features and Prospects : Shermazanian Ia. F, Umarov G Ya.,
Aparisi R R and Teplyakov D I , " Applied Solar Energy ", 5 (6), (1969).

The necessity of starting home manufacture of large, high temperature solar unit and the difficulties
countered are discussed.

0501

Calculation of the Concentration of Energy at Points Outside the Focal Spot of a Parabolic Condenser:
Simon Alfred W , " Solar Energy ", 2 (2), 22, (1958).

This paper is an analysis of the distribution of intensity and of energy falling outside of the circular projection of the cone of reflected radiation coming from all infinitesimal areas of the surface
of a paraboloidal mirror. The method integrates the energy and intensity falling in the area swept by
the major semi - axis of the ellipses subtended by the focal plane. The results are tabulated & graphed.

0502

Calculation of the Concentration of the Solar Radiation Through the Focal Spot of a Paraboloidal Mirror:
Simon A W, " Solar Energy ", 2 (2), 25, (1958).

This paper is an analysis of the power, or the energy per unit area per unit time, falling on the focal
spot from a paraboloidal mirror. It is unique in that it takes account of the distribution of intensity over the solar disk as a function of wavelength. The results are compared with the w o r k of two
other men who used different assumptions.

0503

The Loss of Energy by Absorption and Reflection in the Heliostat and Parabolic Condensers o f a Solar
Furnace : Simon A W, " Solar Energy ", 2(2), 30, (1958).

This paper analyzes the reflection and absorption losses for back silvered plate glass mirrors such as
would be used in a solar furnace. Taken into account are ; the amount of incident light r e f l e c t ed
from the surface of the glass plate, the amount absorbed by the known thickness of the glass plate, the
amount lost on reflection from the silver surface, and the higher order terms showing the losses a s
the light is reflected and transmitted back and forth within the glass.

0504

Calculation of the Concentration of Energy at Points Outside the Focal Spot of a Parabolic Condenser:
Simon A W, " Solar Energy ", 3(4), 67, (1959).

An analysis of the distribution of intensity and of energy falling outside of the circular projection
of the cone of reflected radiation coming from all infinitesimal areas of the surface of a paraboloidal mirror is presented. A uniform sun is assumed.

0505

Treatment of Molybdenite Ore Using a 2 Kw Solar Furnace ; Skaggs S R, Coutures J P and Renard R,"Solar
Energy ", 22, 367, (1979).

Several 10 gram samples of molybdenite ore containing 5 - 6% as MoS_2 were heated in one of the 2 Kw
solar furnaces at the Laborataire des Ultra - Refractaires, Odeillo, France. The end products showed
excellent separation with pure yellow crystals of 99 + per cent MoO_3 and pure white SiO_2 powder being
the two major components. Impurity levels in the MoO did not exceed 6000 ppm while only a small fraction of the molybdenite was entertained along with SiO_2. All treatment was done in a flowing oxy g e n
atmosphere, and other elements condensed out at specific sets in the gas transport system.

0506

A Simple Technique of Fabrication of Paraboloidal Concentrators : Srinivasan M, Kulkarni L V and Pas-
upathy C S, "Solar Energy", 22, 463, (1979).

A practical and elegant technique of fabricating simple and compound paraboloidal concentrators starting from a plane sheet of material is described.

0507

A Collector of an Asbestos - Cement Base : Starodubtsev S V, Arifov U A, Umarov G Ya., Kordub N V a n d
Akhmedov S, " Applied Solar Energy ", 1 (1), (1965).

Results of experiments conducted on the mirror surface of a solar collector with an asbestos - cement
base to determine its integral reflection coefficient are presented in this paper. The mirror surface
was metalized polyethylene terephthalate film. The integral reflection coefficient was calorimetrically
determined and the procedure is described. The results are tabulated and show that the optical characteristics of the asbestos - cement based concentrator are not inferior to those of cast glass concentrators. The mean value of the integral coefficient of reflection of the concentrator is 0.78. The low
reflection coefficient is attributed to dispersion of radiant energy because of film deformation.

0508

A Solar Vacuum Film Collector 2.7 m in Diameter : Starodubtsev S V, Umarov G Ya. and Kordub N V, " Applied Solar Energy ", 1 (1), (1965).

A vacuum film solar collector 2.7 m in diameter consisting of a film mirror and a vacuum chamber in the
form of a frustrum of a cone with ribbed walls and struts for zenith and azimuthal orientation to the
sun is described. The tracking system consists of an optical sensor using silicon photocells a n d a
relay block. The entire unit is set on a reinforced concrete foundation and the vacuum chamber housing construction has proven sufficiently stable to withstand all wind loads. The film is hermetically
sealed on the open side of the conical vacuum chamber and under even presure assumes a mirror surface

of good reflecting power which concentrates the radiant energy flux. By altering the vacuum in the chamber, mirrors can be obtained which concentrate radiant energy at a given focal length.

0509

Brayton Cycle Solar Collector Design Study : Stewart D E," NASA Report No. NASA - CR - 195 ", (USA), (1965).

This study has developed parametric analysis of 20 and 30 ft one piece electroformed mirrors, conceptual designs and manufacturing methods for these concentrators, predicted concentrator performance, an analysis of structural integrity, and experimental studies of master fabrication techniques and reflectance samples subjected to simulated micrometeorite damage. Experimental studies have demonstrated that either blade grinding or plastic overlay master fabrication techniques are amenable to the fabrication of high accuracy highly specular, collector masters.

0510

Solar Concentration Power and Optimization of the Cavity Type Heater for a Solar Source : Tahakoff W , " Raumfahrtforschung ", 13, 107, (1969).

Review of solar power technology and study of an idealized concentrator-absorber system is presented. The flux distribution in the focal plane is derived, and the opening of the cavity is determined for the maximization of the absorbed energy. The maximum achievable temperature of the cavity is found . A numerical example is presented.

0511

Distribution of Radiant Flux Along the Walls of a Receiver Located Near the Focus of a Solar Radiation Concentrator : Tarnizhevskii B V, Baranov V I and Muchnik G F, " High Temperature", (USA), 3 (5), 666, (1965).

Considers a grapho—analytic method of determining the radiant flux density along the walls of a cavity receiver of solar radiation. The results of a calculation are compared with the experimental measurements of radiant fluxes.

0512

A Study of Optical and Energy Characteristics of Solar Radiation Collectors Formed by Inflation : Tarnizhevskiy B V, Razgovorov A D and Savchenko I G, " Solar Energy ", 1 (5), (1965).

The optical and energy characteristics of solar collectors formed by the inflation of sheet metal were investigated. It was experimentally established that the maximum energy density of a paraboloidal collector 540 mm in diameter fluctuates from 4×10^5 Kcal/sqm/hr. The focal images of the collectors were photographed by the light of a full moon, and it was determined that in addition to the irregular and asymmetric shape of the focal point, the collectors displayed an aureole surrounding the focal point and consisting of radially arranged bands of different illumination. It was also found for a parabolocylindrical reflector 2200 mm x 550 mm, that the focal strip varied arbitrarily in its width along the whole lengths.

0513

Effect of Central Shading on the Energy Characteristics of Paraboloidal Mirrors : Teplyakov D I," Applied Solar Energy ", 2 (1), 15, (1966).

The effect of central shading on the thermal regimes of radiation receivers is examined. In case of paraboloidal mirrors, central shading affects the energy distribution parameters in the concentration area, in addition to limiting the radiation flux power. It is shown that a convenient way of solving the problem may be based on the additivity principle of radiant fluxes which is used for solving ordinary concentrating capacity problems in mirrors. Formulas are given for calculating the energy distribution in the focal plane of a paraboloidal mirror with circular central shading.

0514

Effect of Longitudinal Target Defocussing on the Power Characteristics of a Solar Reflector : Teplyakov D I, " Applied Solar Energy ", 3 (1), (1967).

A general frame of reference for the coordinates, radiant flows, and the intensity of these flows is derived. Application of this frame of reference to the analysis of the effect of longitudinal target defocussing on the power parameters of a solar reflector shows that the reference frame is sufficiently general to characterise by a single family of curves the relatively complex phenomenon of the redistribution of energy in the longitudinal target defocussing of solar reflectors of any accuracy.

0515

Features of Energy Transfers and Distribution in High Temperature Solar Installations : Teplyakov D I, " Applied Solar Energy ", 3 (3), (1967) .

The transport and distribution of radiant energy in the reflection field of parabolical solar reflection is described. This scheme acquires a general character when using a generalized system of measures. Particular features of the thermic regime of cavity targets are discussed.

0516

Transport and Distribution of Radiation in Solar Energy Units with Mirror Concentrators : Teplyakov D I, " Semi - conductor Solar Energy Converters, (Ed. V A Baum) ", p. 143, (1969).

The review and analysis are given of several theoretical solutions for the concentration of direct solar radiation by paraboloid mirrors, which are classified, from the point of view of the results obtained, into ideal and real. Analytic and graphical methods are utilized. Complete allowance for mirror imperfections can be made using equivalent but different methods of solution based on the assumption that the imperfections of the reflecting surfaces of solar energy concentrators are distributed statistically (at random). Energy transport and distribution functions in the concentration field of radiant energy are introduced and commented on : a generalized system for measuring coordinates, radiant fluxes, and radiation density is discussed. The limits of validity of the ideal(simplified) methods are established.

0517

Generation of Initial Power Characteristics of Paraboloid Solar Devices under the Conditions of Angular Defocussing : Teplyakov D I, " Applied Solar Energy ", 5 (1), (1969).

The results are given of employing criterion system of reference, radiant fluxes and their densities, as well as inaccurate following, for generalising the detailed power characteristics of paraboloid solar devices, under statistical conditions of inaccurate sun following. Proposed are the simplest approximation of the specific relationship under these conditions. It is also proposed to base more detailed study on mean numerical values of the distributions used in the theory of probability.

0518

Optical Characteristics of Parabolic Solar Concentrators Based on Analytic Calculation : Teplyakov D I, " Applied Solar Energy ", 7 (5), (1971).

A method for calculating optical characteristics for parabolic solar concentrators is shown. A number of optical characteristics were derived and are commented on.

0519

Optical Characteristics of Parabolic Solar Concentrators Derived from Aberration Theory : Teplyakov D I, " Applied Solar Energy ", 7 (5), (1971).

Applying aberration theory (of the third degree) to the calculation of optical characteristics for parabolic solar concentrators is considered. It is shown that with this theory only approximate results can be obtained. Summary aberrations of high degrees (fifth, seventh etc.) are estimated.

0520

Concentrating Power of Solar Energy Paraboloids in Various Spectral Ranges : Teplyakov D I, " Applied Solar Energy ", 3 (1), 15, (1973).

The paper points out that the spectral contents of radiation in various parts of the focal spot differ basically from the spectral characteristics of the initial solar radiation. The spectral reflectivity of the paraboloid or the nature of the mirror have a decisive influence. The phenomenon is less affected by the distribution of spectral radiation parameters along the solar disk radius.

0521

Similar Generalization in Studies of the Angular Defocussing Regions of Paraboloidal Solar Devices : Teplyakov D I and Poluektov V P, " Applied Solar Energy ", 4 (2), 71, (1968).

Generalisation of theoretical and experimental results is made that are characteristic for non-exact sun tracking of parabolic solar devices. Regime criteria are recommended for angular defocussing that determine their similarity. The modelling of these regimes is based for concentrator - receiver systems. Theoretical and experimental results for different solar devices, are compared.

0522

Defocussing Characteristics of a Solar Device with a Crucial Cavity Receiver : Teplyakov D I, Aparisi R R, Kolos Ya. G, Zakhidov R A and Poluektov V P ," Applied Solar Energy ", 4 (5), 16, (1968).

Found experimentally were the regularities of energy input to the walls of a cubical cavity receiver, depending on the concentrator. For some walls that were not in the aiming position, these relationships were established to have a maximum. The experimental data obtained was processed in a criterial system.

0523

Determination of Effect of Inaccurate Angular Tracking in a Solar Energy Device ; Teplyakov D I and Zakhidov R A, " Applied Solar energy ", 4 (5), 20, (1968).

The method is considered of an experimental investigation of the influence of non exact orientation of solar installation on the receiving characteristics and on energy performance in it. Calculation relations were received that are characteristic for the considered method and corelations are given which one has to consider when treating the experimental data. General practical recommendations are given.

0524

Calorimetric Investigation of a Solar Radiation Concentrator for a Thermoelectric Water Pumping Unit : Teplyakov D I, Aparisi R R, Kolos Ya. G, Zakhidov R A and Annayev A, "Semi - conductor Solar Energy Converters (ed. V A Baum)", p.118, (1969).

The purpose and results are given of a comprehensive investigation of the concentration field of a paraboloid mirror of 5 m diameter, assembled from shaped reflecting aluminium elements. The investigation was carried out using calorimetric methods and special calorimeters. It was possible to determine and find optimum values of the parameters of energy absorption by a semi-conducting thermo - electric generator. The distribution of the radiant flux along the walls of cavity absorbers of various types were determined and an axial asymmetry in the flux distribution was found.

0525

Effect of Central Shading and Longitudinal Defocussing of the Receiver on the Power Characteristics of a Paraboloid Solar Energy Device : Teplyakov D I, Aparisi R R and Kolos Ya. G, "Applied Solar Energy", 6 (5), 4, (1970).

The influence of shading the central part of a solar irradiation concentrator on the distribution of heat load over a receiver's light absorbing surfaces is considered.

0526

Power Parameters of the Solar Concentrator of the S V - 1 Water Lift Unit : Teplyakov D I, Aparisi R R, Kolos Ya. G, Zakhidov R A and Annayev A, " Applied Solar Energy ", 6 (6), (1970).

Calorimeter tests of the reflector for a solar water lifting installation constructed in the Krzhizhanoskiy Power Institute are discussed. The results of these tests serve as a basis for the design of semiconductor thermogenerators.

0527

Combined Solar Arc Optical Furnace for High Temperature Research : Teplyakov D I, Aparisi R R and Kolos Ya. G, " Applied Solar Energy ", 8 (3), 43, (1972).

The design of a unique high temperature furnace which produces high thermal fluxes by utilizing either solar energy, the radiation energy of an electric arc, or a pwerful electric lamp, is described. The efficiency of direct energy use can be increased with this furnace.

0528

60 Inch Stretch Formed Aluminium Solar Concentrator ; Final Report : Thompson R, "NASA Report No. NASA CR - 56469 ", (USA), (1963).

All of the tasks in the development of a 60 in solar concentrator were completed. Of particular importance in the improvement of the concentrator was stretch forming.This improvement is basic to the number of available power conversion systems with which this concentrator concept may be used. The stretch forming, spray coating and optical inspection equipment modifications are discussed in detail. The stretch formed concentrator was improved sufficiently to be used with thermionic conversion system.

0529

Solar Energy Research in France; A Gigantic Solar Furnace is Used for Material Testing : Tornvall T, " Teknisk Tidskrift ", 105, 26, (1975).

The article describes the design, operation and performance of the World's largest solar furnace in the Pyrenees which is also in use for testing thermal properties of various materials,such as aluminium and zirconium dioxide.

0530

Composite Mirrors of Large Area, in Particular, for Concentrating Solar Energy : Trombe F, " U S Patent 2, 707, 903/May 10 ", (1955).

The construction of a large area, composite curved mirror, of the type used in large solar furnaces is described. Each of the elementary mirrors consists of an elastically deformable plate which is given, by mechanical deforming means, the shape of the corresponding portion of the theoretical surface of the large mirror. The theoretical surface is a paraboloid of revolution. The elementary mirrors are trapezoidal in shape and mounted on a frame work, to form substantially the theoretical surface of the large mirror.

0531

Solar Furnaces and Their Applications : Trombe F, " Solar Energy ", 1 (2), 9, (1957).

Problems encountered in focussing the solar energy in the construction of large solar furnaces at Mt. Louis are discussed, and the 1000 KW unit now being built is described. A short review of solar furnace applications and of the economics of their operation indicates increased importance and use of this type of furnace.

0532

Solar Furnaces of 1000 KW in Odeillo, Font - Romeo, France : Trombe F, " Revue des Hautes Temperatures et des Refractaires ", 1 (1), 5, (1964).

This paper presents the principles of construction of big solar furnaces. The orientating systems of solar radiation and focussing systems are described.

0533

Limitations of Solar Collectors for Converters : Trombe F and Le Grives E, " Paper at 6th AGARD Colloquim on Combustion and Propulsion ", Cannes (France), (1967).

Research on solar energy collectors and receivers for space applications is reviewed. Review of various concentrator concepts and definition of receiver configurations leading to most promising performances is presented. Analysis of main factors affecting efficiency (geometrical perfection, optical qualities, structural rigidity, response to meteorite impacts etc.) and receiver efficiency (absorption characteristics, surface radiation), is also given.

0534

Characteristics and Energy Performance of the 1000 KW Solar Furnace at the Centre Nationale de la Recherche Scientifique : Trombe F and Lephat Vinh A, " C R Hebd. Sean Acad. Sci. B. ", 272, (19), 1104, (France), (1971).

The solar furnace consists essentially of an array of automatic deflectors directing a parallel light beam on a fixed 200 sqm reflecting paraboloid. The energy flux at the focus of the system corresponds to the radiation of a 400 deg K black body.

0535

First Results Obtained with the 1000 KW Solar Furnace : Trombe F, Gion L, Royere C and Robert J F, " Solar Energy ", 15, 63, (1973).

The first tests carried out on the 1000 KW solar furnace were conducted with an oxidizing atmosphere at the focus. Two conditions mainly the direct treatment of the material supported by the material itself or treatment on cooled support and treatment in a cavity; either fixed or revolving, were made for the treatment. The conditions were applied to melt various refractory materials, oxides of thorium, zirconium, calcium aluminium, chromium and various compounds formed with these oxides. The efficiencies of treatment were determined. Heat treatment on cooled supports was studied extensively,primarily to obtain refractory materials, solidified in the form of spheres of very small diameters.

0536

The Parabolic Concentrating Collector ;A Tutorial : Truscello V C, " Jet Propulsion Laboratory (USA) Report DOE/JPL - 1060 - 79/1 ", (1979).

This report presents a tutorial overview of point focussing parabolic collectors, optical and thermal characteristics of such collectors are discussed. Data representing typical achievable collector efficiencies are presented and the importance of balancing collector cost with concentrator quality is argued through the development of a figure of merit for the collector. The impact of receiver temperature on performance is assessed and the general observation made that temperatures in excess of 1500 - 2000 deg F can actually result in decreased performance. Various types of two - axis tracking collectors are described, including the standard parabolic deep dish, Cassegrainian and Fresnel, as well as two forms of fixed mirrors with articulating receivers. The present DOE program to develop these devices is briefly discussed as are presented and projected costs for these collectors. Pricing informa t i on is presented for the only known commercial design available on the open market.

0537

Heat and Electricity from the Sun Using Parabolic Dish Collector Systems : Truscello V C and Nash W A, " A I A A Monograph ", 25, 137, (1979).

This study addresses point focus distributed receiver (PPDR), solar thermal technology for the production of electric power and of industrial process heat, and describes the thermal power systems project conducted by JPL under DOE sponsorship. Project emphasis is on the development of cost effective systems which will accelerate the commercialization and industrialization of the plants upto 10 MWe, using parabolic dish collectors. The characteristics of PDFR systems and the cost targets for m a j o r subsystem hardware are identified. Markets for this technology and their size are identified, and expected levelized bus bar energy costs as a function of yearly production level are presented.The present status of the technology development effort is discussed.

0538

The JPL Parabolic Dish Project : Truscello V C and Williams A N," Jet Propulsion Laboratory Report ", (USA), (1980).

This paper discusses a project being managed by the Jet Propulsion Laboratory for the U S Department of Energy for the development of parabolic dish collector systems. The paper summarises the status of the project including various contracts with industry for developing the dish subsystems which include concentrator, receiver, and heat engine. An early market of opportunity for dishes is the dispersed small community market. This market depends heavily on oil to operate diesel or steam turbine plates in order to generate electricity. The present contracts with industry for conducting a series of engineering experiments using the developed dish hardware to demonstrate the technology i n these early opportunity markets is also discussed.

0539

The Small Community Solar Power Program :Truscello V C, Williams A N, Marriott A N and Weisiger J, " Jet Propulsion Laboratory Report ", (USA), (1980).

The parabolic dish system being developed for the DOE small community solar power program is expected to be the most suitable solar thermal electric technology for near term small community use. Its high thermal efficiency and modularity make it most competitive for isolated communities and other communities forced to use high cost conventional energy supplies. Characteristics of this small community energy market through a series of parabolic dish engineering experiments are discussed.

0540

Tests on Paraboloidal Concentrators Using the Leonov Aberrograph : Tveryanovich E V and Madaev V V, "Applied Solar energy ", 10 (3), 16, (1974).

A method is described for testing paraboloidal solar energy concentrators using the Leonov aberrograph. A similar technique devloped for projector type glass reflectors is inadequate for solar concentrators. The proposed method can be used to determine more accurately than the projector method the true focal length,the minimum caustic heck, the root mean square angular error, and the energy distribution in the focal spot.

0541

Experimental Determination of the Shape of an Inflatable Film Collector Reflecting Surface : Umarov G Ya., Kordub M V, Bespalko V P and Gafurov A, " Applied Solar Energy ", 1 (1), (1965).

This paper describes the experiments conducted to determine the shapes of the reflecting surface of an inflatable film collector in its relation to the differential between pressures on both sides of the film. The formulation used,the tabulated data derived, and the discussion of the results are included.

0542

Preparing Aluminium Mirrors on Asbestos Cement by the Transfer Method : Umarov G Ya., Vilkova S N, Ayzenshtal Ya. L, Novikova I A and Sutyagine V M, " Applied Solar Energy ", 1 (3), (1965).

Methods and materials are described for constructing vapour deposited or bonded aluminium thin film mirrors for use in solar power plants. The supporting body of the collector was asbestos cement covered with a layer of epoxy resin which was mixed with a filler to decrease the absorption coefficient.

0543

Geometry and Optical Characteristics of a Hexagonal Reflecting Film Membrane : Umarov G Ya, Solodynikov Yu. A, Kordub N V, Alimov A K, Zarubin V S and Alavutdinov D N," Applied Solar Energy ",1(3), (1965).

Concentrating collector using as a reflecting film, materials made of plastic, e.g. a film of polyethylene terephthalate of the order of $10 - 50\mu$ thick, coated on one side with a thin layer of aluminium is described. To give the film a concave shape it is stretched over a frame with an airtight cavity in which a vacuum is created. If the frame is round the concave surface of the film is close to spherical and in the case of small deflections has the optical properties of a paraboloid.

0544

<u>A Solar Furnace of Polyethylene Terephthalate Film</u> : Umarov Ya. G and Kordub N V, " Applied Solar Energy ", 1 (4), (1965).

A high temperature solar furnace made by stretching a roll of metalized polyethylene terephthalate film back and forth over a dome shaped aluminium frame is described. The film was hermetically sealed to the open face of the aluminium basis. Different pressure on the film gave mirrors of various focal lengths for concentrating the sun's rays. A 92 cm mirror surface had a focal length of 120 cm & reached temperatures of 950 deg to 1000 deg C.

0545

<u>Manufacturing Feasibility of Long Focus Vacuum Film Collector</u> ; Umarov Ya G, Alavutdinov D N and Alimov A K," Applied Solar Energy ", 1 (4), (1965).

The feasibility of manufacturing a large reflecting film surface vacuum collector was evaluated b y considering the collector surface as spherical and computing the collector sagitta by geometric method. Film tension was determined for measurements of focal spot diameter and absolute black body temp e r ature in the center of the focal spot. It was developed that manufacturing processes for large collectors require the tightening of the metalized film against an optically accurate frame. Long fo c u s collectors with adjustable focal length from 1.5 to 10 m regulate black body temperature in the focal spot from 1200 to 300 deg C.At a focal length of more than 10 m, wind stresses have a negative effect on the optical characteristics of the collector.

0546

<u>Rotation Method of Manufacturing Paraboloidal Solar Collectors</u> : Umarov Ya. G, Alimov A K,Suleimanova F N and Alavutdinov D N, " Applied Solar Energy ", 2(6), 41, (1966).

Technology of copies manufactured from 1.2 mm diameter matrices is considered. Epoxy resins are used for preparation of matrices. It is shown that manufacturing of large solar collectors from matrices by the rotation method is feasible.

0547

<u>Contactless Method of Concentrator Shape Measuring</u> : Umarov Ya.G, Alavutdinov D N and Muzaforov H , " Applied Solar Energy ",3 (2), (1967).

A method of measuring the shape of concentrators is given by use of distance gauge. The description of the device is given.

0548

<u>Investigation of Film Concentrators Deformation Field under the Influence of Air Pressure</u> : Umarov Ya. G, Jadraev U J and Djalalov Z, " Applied Solar Energy ", 3 (4), (1967).

Formulae are deduced with the theory of absolute flexible membranes, for the dependence of radial and peripheral deformation on the relative radius air pressure and mechanical features of the film. Results are given on experimental investigations of radial and peripheral deformation determination.

0549

<u>The Relation Between the Air Pressure of the Toroid and the Pressure of Film Concentrator Chambers</u> : Umarov Ya. G, Jadraev U J and Djalalov Z, " Applied Solar Energy ", 3(6), (1967).

Formulae of dependence between the air pressure of the toroid & film concentrator chambers for t h e case when the change of presuure in the chamber does not influence the pressure rate in the t o r o i d are given. Theoretical results are compared with experimental data.

0550

<u>Computation of Facet Solar Concentrators on the Basis of a Rotating Paraboloid</u> : Umarov Ya. G, & Sharafi A Sh., " Applied Solar Energy ", 3 (6), (1967).

Presented are calculation of facet coordinates, rates of concentration yield and efficiency of concentrator area. The concentrator consisted of equal hexahydral and round facets, which are located o n the surface of the rotation paraboloid.

0551

<u>Asbestos Cement Film Concentrator Fabrication Technology</u> : Umarov Ya. G and Abduazizov A," Applied Solar Energy ", 4 (6), 105, (1968).

In this paper the manufacturing technology of parabolic and parabolocylindrical concentrators and the energetical and optical characteristics are discussed.

0552

<u>Investigation Results of Film Pump Out Concentrators</u> : Umarov Ya. G and Gafurov A M, "Applied Solar Energy ", 5 (2), (1969).

Experimental investigation of film pump out concentrator of 2.7 and 5 m diameter is discussed. T h e mean value of integral coefficient of reflection is determined. As the diameter of the film pump out concentrator increases, the stability against wind load decreases. This fact makes it possible to manufacture film concentrators of great diameter.

0553

On the Investigation of Wedge Plates Deformation : Umarov G Ya., Suyushkaliev N H and Abduazizov A, " Applied Solar Energy ", 5 (3), (1969).

Formulae for the determination of glass wedge plate shape difference from parabolic is discussed i n the case when they are fastened at several points on the meridian on a parabolic rigid substrate. Calculation is given based on the formulae. The results are used for the manufacturing of a 5 m diameter concentrator.

0554

Concentrators with the Focal Image in the Form of a Ring : Umarov G Ya. and Sharafi A Sh., "Applied Solar Energy ", 5 (4), (1969).

The article considers solar radiation concentration by mirror reflectors. The radiation is received by the rotation of a parabolic generatrix around an arbitrary axis.

0555

The Technology of Manufacturing Cheap Round Facet Concentrators of Window Panes : Umarov G Ya., Alimov A K, Alavutdinov D N and Ovechkin N F, " Applied Solar Energy ", 5 (6), (1969).

This article describes the optical and power characteristics of paraboloid concentrators assem b l e d from round window pane facets with a double curvature. Manufacturing technology is also discussed.

0556

Light Film Solar Concentrators: Umarov G Ya., Alavutdinov D N, Alimov A K, Muzafarov H and Kuznetsov E P, "Applied Solar Energy ", 5 (6), (1969).

This article presents the technology of manufacturing light weight concentrators of solar energy from metalized film by inflation of paraboloid shells after thermal treatment.

0557

A Highly Productive Thermoevaporator of Aluminium for Producing a Mirror Coating : Umarov G Ya, Gaziev U Kh., Trukhov V S, Klucheverky Iu. E and Orda E P, "Applied Solar Energy", 6(3), (1970).

This article describes a thermoevaporator of aluminium which allows the deposition of a mirror coating on the substrate in a continuous flow. The evaporator considered has a large front evaporat i o n unit, supplies uniform wear on the pivot-evaporator ; and does not need water cooling. This d e v i c e can operate for 8 - 10 hours without vacuum breaking in the evaporation box.

0558

A Concentrator made from Parabolic Glass Facets : Umarov G Ya., Alimov A K, Alavutdinov D N and Abdu - rakhamanov A, " Applied Solar Energy ", 7 (4), (1971).

Results of an investigation into the optical power characteristics for parabolic glass facets with aluminium covering are described.

0559

Long Focal Length Concentrator Made from Wedge Shaped Window Glass Elements : Umarov G Ya., Alimov A K, Alavutdinov D N and Abduazizov A, " Applied Solar Energy ', 10 (4), 20, (1974).

A single solar energy unit designed for high power levels in which to reduce the cost of the concentrator, the reflecting area was made from window - type glass and the body from ferroconcrete is considered.

0560

Fabricating Paraboloidal High Temperature Solar Concentrators from Mollified Sectors : Umarov G Ya., Alimov A K, Abduazizov A and Usmanov M, " Applied Solar Energy ", 10(6), 73, (1974).

Preliminary calculations of the cost of the reflecting element and the supporting structure for a sectoral concentrator are made. The solar energy concentrator fabrication cost depends on the sector dimensions ; the smaller the sector size, the more expensive the concentrator. Experience in fabricating paraboloidal sectoral concentrators shows that high temperature concentrators for power generation and processing purposes can be fabricated by mollifying individual paraboloid of revolution segments.

0561

Integrated Precision Parameter of the Heliostat Paraboloid System and its Connection with Mirror Errors : Umarov G Ya., Zakhidov R A and Vainer A A, " Applied Solar Energy ", 11 (3 -4), 24, (1975).

A statistical model is put forward and is used to investigate the effect of mirror errors on the energy and temperature precision parameters of the heliostat paraboloid system. The errors of the two mirrors must satisfy the same conditions (exactly from original deviations and approximately from sagit - tal deviations).

0562

A Mosaic Solar Energy Concentrator : Umarov G Ya., Alimov A K, Alavutdinov D N and Abduazizov A, " Applied Solar Energy ", 12, (1), 52, (1976).

A concentrator 2 m in diameter and using a 50 X 50 mm^2 aluminium mosaic has been made. It has the following parameters. Aperture 52.5°, focal length 1.05 m, focal spot diameter 7 cm, power in focal spot 1300 W (for incident solar radiation of 784 W/m^2).

0563

Concentrating Power of Paraboloidal Facets : Umarov G Ya., Zakhidov R A and Khodzaev A Sh., " Applied Solar Energy ", 12 (4), 11, (1976).

The formation of the image figure of an off - axis infinitely remote point source by a paraboloidal

074

mirror is considered. Formulas are obtained for calculating the coordinates of the points at which the reflected rays intersect a plane perpendicular to the rays incident on the mirror.   Intersection point coordinates calculated by means of a M 222 computer are used to construct the image figures and aberration diagrams on various planes for mirrors differing in flare angle at different ray angles of incidence.

0564

Concentrating System of a Self-Contained Solar Power Installation : Umarov G Ya.,Alimov A K, Khodzhaev A S, Abduazizov A and Khasanov R B, " Applied Solar Energy ", 15 (2), 90, (1979).

This article reports the results of tests on concentrators composed of square paraboloidal glass facets. At a radiation input of 760 W/m^2 it was found that a large 2.2 m focal length installation composed of 24 paraboloidal glass facets, each 60 X 60 cm in size generated a power of 4.5 - 5 KW with the mean temperature of an absolute black body at the focal point reaching 1170 deg K.

0565

Experimental Determination of Optimal Size of a Heat Exchanger for Solar Furnaces with Large Aberrations : Vannucci S N, " Solar Energy ", 15, 51, (1971).

A method is presented for experimentally determining the optimal area of a heat exchanger to be u s e d in solar furnaces with large aberrations and relatively low coefficients of reflection. This process is intended to be used in domestic solar furnaces with low power requirements. The method first determines the intensity distribution of radiation in the focal plane by measuring it with a photovoltaic cell which has a linearized output and which has been calibrated with a standard radiometer. I t is then possible to determine the area which will collect the maximum energy. An example is given where 100 Watt of net power are obtained by means of a small solar furnace.

0566

Geometrical Interpretation of Mathematical Model of Energy Distribution in the Focal Plane of a Paraboloid : Vartanyan A V, " Applied Solar Energy ", 2 (5), 6, (1966).

A geometrical method of computing energy in the focal plane of a paraboloid in a collector-receiver system for the case of uniform energy distribution in the elementary reflected beam is examined. The b e a m energy distribution is divided into elements, solved element by element by integration, and the whole pattern of energy distribution in the focal plane at a given regulator position is derived by summing the curves. The method is also useful in solving problems associated with designing high temperature receivers in that part of the calculation involving irradiance of opaque cavity walls with regard t o the condition of visibility, and for determining the concentrating capacity of non circular partial paraboloids.

0567

Automatic Regulator of Radiant Flux for High Temperature Solar Installations : Vartanyan A V, Shermasanyn Ya. T and Agasyan Ya. M, " Applied Solar Energy ", 5 (1), (1969).

A system of automatic regulation of power parameters of solar installations ray flow is proposed.Precision of regulation of 3 - 5% is provided with the automatic power regulator of paraboloid installa - tion with a diameter of 1.5 m accomplished according to given scheme.

0568

Measuring the Distribution of Light in the Focal Spot of Solar Collectors : Veinberg V B, Konayeva G Ya. and Sattarov D K, " Applied Solar Energy ", 1(2), (1965).

A fibre optical micrometer for direct measurement of irradiance in the focal spot of solar r a y collectors is described.

0569

Reduced Drag, Paraboloid Type, Solar Energy Collectors : Vermeulen P J, Badari A and Elfner P, " Pro - ceedings ISES Conference ", Winnipeg (Canada), 2, 264, (1976).

The design and development of two novel paraboloid type solar energy collectors of reduced wing drag is presented. The designs are aimed at reducing drag focus from winds nominally from (a) the lateral direction and (b) the frontal direction. The design consists essentially of conical frusta mounted in (a) to produce a front focus mirror and in (b) a neat focus mirror. The frusta are mounted so that annular gaps for air flow are presented partially in (a) and frontally in (b). In design, (a), interference by the focal plane energy receiver devices produces the usual collector area loss, but i n design (b), there is no such interference loss. Wind tunnel models of both designs are tested and the resulting drag force data are presented and discussed. The data show that significant drag reductio n is possible. A 0.914 m diameter prototype of the front focus design is constructed and the results of calorimetric measurements are presented.

0570

Solar Furnace of 1000 KW : Vicariot H, Courtot P and Fondeville F, "Institut Technique du Batiment et des Travaux Publics Annales ", 19 (228), 1387, (1966).

The plant and equipment including directing mirrors, large parabolic mirror and furnace building are described. The principle and method of construction of heating surfaces, problems of rigidity a n d stability ; anchorages of building, balancing of upper parts and new criteria for form of layouts o f cables and problems related to geographical parameters of site are also discussed.

0571

The Protection of Aluminium Mirrors by Hard Polymer Substrate : Vilkova S N, Gunner E A and Sutyagina V M, " Applied Solar Energy ", 5 (6), (1969).

Results of tests on mirrors with aluminium coatings protected from external influence by varnish with induced photostabilisers are reported.

0572

Opto - Mechanical Section of a Thermoelectric Solar Water - Pumping Unit : Vladimirova L N and Garf B A,
" Semiconductor Solar Energy Converters (Ed. V A Baum) ", p.111, (1969).

A description is given of a reflector and a mechanism for 24 hr. rotation of a solar unit. An original
feature is the system for controlling the rate of rotataion of the reflector by a drive, based on the
uniform flow of sand through a small aperture. The special features of the reflector, the control
mechanism and of the mounting are described. A method for designing the opto - mechanical section of a
SV - 1 solar water pumping unit is discussed in detail from the point of view of strength and rigidity.
Design formulas are given, as well as the results of wind tests on an aerodynamic model of the reflector.

0573

Thermal Performance Trade - offs for Point Focussing Solar Collectors : Wen L, " Proceedings 13th IECE
Conference ", San Diego (USA), 2, 1548, (1978).

Solar thermal conversion performance is assessed in this paper for representative point focussing distributed systems. Trade - off comparisons are made in terms of concentrator quality, solar receiver
operating temperature and power conversion efficiency. Normalised system performance is presented on
a unit concentrator area basis for integrated annual electric energy production.

0574

Effect of Optical Surface Properties on High Temperature Solar Thermal Energy Conversion : W e n   L ,
" Journal of Energy ", 3 (2), 82, (1979).

The effect of thermal surface properties on the performance of representative point focussing s o l a r
power plants are assessed in this paper. The trade - off relationships are presented in terms of normalized system performance as a function of thermal optical design parameters. Crucial surface pro -
perties include solar reflectance, specular spreading due to microscopic roughness, surface error due
to manufacturing slope tolerance or waviness, and concentrator pointing accuracy. Two representative
power conversion systems, a Rankine steam cycle and an open - air Brayton cycle are considered.

0575

Comparative Study of Solar Optics for Paraboloidal Concentrators : Wen L and Huang L, " ASME P a p e r
No. 79 - WA/Sol - 8 ", (1979).

Different analytical methods for computing the flux distribution on the focal plane of a paraboloidal
solar concentrator are reviewed. An analytical solution in algebraic form is also derived for an idealized model. The effects resulting from using different assumptions in the definition of o p t i c a l
parameters used in these methodologies are compared and discussed in detail. These parameters include
solar irradiance distribution (limb darkening and circumsolar), reflector surface specular spreading,
surface slope error, and concentrator pointing inaccuracy. The type of computational method selected
for use depends on the maturity of the design and data available at the time the analysis is made.

0576

Geometric Efficiency of an Electroformed Nickel Solar Concentrator : Willis C M and Houck O K, " N A S A
Report No. NASA - TN - D - 4072 ", (USA), (1967).

Geometric efficiency of a 152 cm diameter paraboloidal solar energy concentrator is calculated f r o m
optical ray trace data. The data is obtained by reflecting a beam of light incident parallel to the
concentrator axis, from the concentrator to the focal plane image plate and measuring image displacement from the concentrator axis. Two of the three methods used to calculate efficiency provide results
that are within 0.025 of the measured calorimetric efficiency for energy - absorber apertures l a r g e r
than 14 solar image diameters. The data is also used to calculate the slope errors of the concentra -
tor reflective surface. The values of one standard deviation of slope error are 1.45 milliradians for
the radial component and 2.32 milliradians for the circumferential component.

0577

A High Temperature Solar Furnace 2 Meter in Diameter : Yagudayev M D, Bashnyak A Ya., Nechayev Yu. Ye.,
Mavashev Yu. Z and Rudshteyn R L, " Applied Solar Energy ", 1 (1), (1965).

A high temperature solar furnace consisting of a parabolic reflector with a mirror two meters in diameter and an orientation mechanism, each with its own system of azimuth - zenith axes is descr i b e d.
The device is designed for direct aiming of the reflector at the sun and working the heliostat with
the optical axis of the paraboloid horizontal. The reflector design, tracking system, method for determining heat flux at focus, and characteristics of high temperature solar furnaces are discussed.

0578

Calculation of the Energy Distribution in the Radiation Field of Reflector - Type Solar Energy Devices:
Zakhidov R A, " Applied Solar Energy ", 1 (2), (1965).

An analytical method is presented which permits the solution of important practical problems by determining the characteristics of the radiation field at each point on the walls of the cavity receiver.

0579

Experimental Study of Effect of Inaccurate Orientation of Solar Mirror Device on its Power Characteristics : Zakhidov R A and Teplyakov D I, " Applied Solar Energy ", 1 (2), (1965).

This article presents the results of an experimental study of the effect of inaccurate solar orientation of the paraboloid - mirror solar devices on effective concentrating capacity under static conditions. The effect is experimentally evaluated from the apparent displacement of the sun in the sky
when the device is stationary. Radiation flux densities near the geometric focus of the mirror are
determined by calorimetric methods.

0580

Computation and Optimization of the Thermal Regime in Concentrator Type Solar Devices : Zakhidov R A and Teplyakov D I, " Applied Solar Energy ", 2 (2), (1966).

The energy distribution over randomly oriented elements of the radiation receiving surface of the hollow collector of a concentrator type solar device is computed. The energy distribution characteristics in the specular reflection field of a real paraboloidal concentrator are fully established by a structural solution of the energy transfer and distribution problem. Expressions for the radial and axial components of the reflected radiation vector are given, together with expressions which enable the more general laws of radiation field specularly reflected from the paraboloidal mirror of a solar device. Radiant flux measurements necessary to the investigation of the structural characteristics of the concentration field are made, using the calorimetric method. The level and distribution of energy over the elements of the collector wall constitute the prime factor in thermal regime optimization.

0581

The Analysis of Heat Regimes of Solar Furnace Cavity Receivers :Zakhidov R A and Teplyakov D I, " Applied Solar Energy ", 3 (4), (1967).

An analysis of heat regimes of cylindrical cavity radiation receivers for solar devices with parabolic reflectors is discussed. The geometry of the receiver is optimized to obtain equal average heat distribution loads. A curve showing the distribution of local heat loads over the walls of the receiver is given.

0582

A Method for Determining Local Radiant Fluxes in Cavity Type Radiation Receivers : Zakhidov R A and Teplyakov D I , " Applied Solar Energy ", 4 (6), 64, (1968).

The relationships which enable the proposal of new means for determining local radiation fluxes were obtained, which departed from the physical concept of a radiation field in a diathermal environment. The principal distinctive feature of the proposed method is the possibility to establish local densities of the incident radiant fluxes in cavity receivers, by measuring the total radiant flows through normal cross section of the cavity. The method may be applicable to cavities of various shapes (cylinder, cone, sphere etc.).

0583

Thermal Conditions in Cylindrical Cavity Absorbers of High Temperature Solar Energy Convertors : Zakhidov R A and Teplyakov D I , " Semiconductor Solar Energy Converters (Ed. V A Baum) ", p.134, (1969).

The problem of the optimization of the geometry of cavity absorbers from the absorption parameters of the radiation (density level and nature of the distribution along wall surface) reflected by a concentrator is studied. The problem is related to the thermal conditions in the absorber i.e.to the possibility of heat transfer by conduction along the walls. An analysis, based on certain assumptions about the structure of the radiation field and on the results of specially conducted experiments, show that, to obtain optimum average intensities of the radiation incident along the bottom and side walls of a cylindrical absorber, the optimum ratio of depth and the diameter of the absorber (l/d) should be equal to unity. The distribution of the local radiant flux, incident on various parts of an absorber,is investigated.

0584

Distribution of Radiation Produced by a Paraboloidal Concentrator : Zakhidov R A and Vainer A A, " Applied Solar Energy ", 10 (3), (1974).

A calculation is given of the distribution of the radiant vector in the field of a paraboloidal concentrator using a previously determined brightness distribution in the reflected beam. The dependance of the parameter 'h' on the meridional and sagittal macroimperfections is investigated.

0585

Optimum Distribution of Facets on a Large Concentrator : Zakhidov R A and Dudko Yu. A, " Applied Solar Energy ", 11(5), 66, (1975).

A paraboloidal faceted concentrator is considered. The facets are flat polygons, whose corners lie on the paraboloid. The shape of the facets is determined under the condition of dense packing along horizontal belts on the paraboloid.

0586

Standard Size Facets for the Reflecting Surface of a Solar Concentrator : Zakhidov R A and Dudko Yu.A, " Applied Solar Energy ", 12 (2), 13, (1976).

This paper reviews the use of square facets of a single standard size on horizontal belts on a paraboloidal solar concentrator.

0587

Energy Computation of Concentrating Capability of Paraboloidal Facets : Zakhidov R A & Khodzhaev A Sh., " Applied Solar Energy ", 12 (5), 20, (1976).

The distribution of the radiation vector in the concentration field of long focus paraboloidal mirrors (facets) is investigated for various ray angles of incidence. Formulas are obtained for calculating the distribution of radiant flux density on planes perpendicular to the incident rays. The calculated results are given as graphs.

0588

Analysing Multifacet Concentrating Systems : Zakhidov R A and Khodzhaev A Sh., " Applied Solar Energy", 13 (2), 26, (1977).

Various methods of designing multifaceted concentrating systems are analysed. The distribution   of irradiance in the focal volume of long focus mirrors (facets) is investigated. It is shown that for the general case, the plane in which maximum irradiance is reached does not coincide with the plane of the mirror. The position of the maximum irradiance plane depends on the relationship between the angular dimensions of the mirror and the radiation source (the sun). A method is proposed for calculating the focal parameters of facets as a function of the position of the supporting base and the function of the multifacet concentrating system.

0589

Sodium Heat Pipe Use in Solar Stirling Power Conversion Systems : Zimmerman W F, Divakaruni S M a n d Won Y S, " ASME Paper No. 80 - C2/Sol - 13 ", (1980).

Sodium heat pipes were selected for use as a thermal transport method in a focus mounted, distributed concentrator solar stirling power conversion system. Heat pipes were used to receive thermal power in the solar receiver and to transmit it to a secondary heat pipe containing both latent heat salt (for upto 1.25 hours of thermal storage) and the heat exchanger of the stirling engine. A brief description is given of the system, and of the performance testing of the primary heat pipes experiment a p - plicable to this unique solar power conversion system.

0590

9.50 Foot Diameter Master and Mirror : " NASA Report No. NASA - CR - 64204 ", (USA), (1964).

The design fabrication and optical inspection of a plastic spincast female master mold for producing a 9.5 ft diameter male paraboloidal master are described. The male master is used to produce l i g h t weight high quality paraboloidal solar concentrators with collectors' efficiencies of 75% when operating a 2000 deg K black body cavity.

0591

Stretch Formed Aluminium Solar Concentrator : TRW Equipment Laboratories, Cleveland, " NASA Report No. NASA - CR - 66069 ", Ohio (USA), (1966).

Results of the investigation of the solar concentrator fabrication technique and mounting ring design are presented. Fabrication of a 60 in diameter concentrator is described which resulted in an optical accuracy of 1.59 mins and 2.08 mins radial and circumferential standard deviation of surface e r r o r s respectively. Measured specular solar reflectivity is 88% to 90% & the concentrator weight is 10.75 pounds.

Central Tower Receiver Systems

0592

Production D' Electricite Par Conversion Thermodynamique De L' Energie Solaire Project 10 MWe / (Production of Electricity Through Thermodynamic Conversion of Solar Energy - 10 MWe Project): Abatut J L, " Revue Energie ", 28 (293), 230, (1977).

A combined research program for the development of solar energy (PIRDES) organised in France by t h e National Scientific Research Center (CNRS) in association with the French Electricity Board (EBF) and a group of industrialists is described. The concept of tower type electro solar power plants, their optimum design, and the design of heliostats is explained in some detail.

0593

A New Method for Collector Field Optimization ; Abdel Monem M S, Hildebrandt A F, Lipps F W and Vant- Hull L L, " Proceedings COMPLES International Conference ", Dhahram (Saudi Arabia), p.372, (1975).

A comprehensive computer simulation program has been used to optimize heliostat locations in a collector field and to compute the fraction of the solar energy which actually strikes the receiver. T h e new optimization method makes a 10% improvement over previous approximations.

0594

Develpment of the Advanced Components Test Facility : Altman R F and Brown C T, " Proceedings Seminar on Testing of Solar Energy Materials and Systems ", Gaithersburg (USA), p.50, (1978).

The Francia solar powered steam generator and its solar tracking and receiver systems are described. The DOE facility incorporating the Francia Design has been installed at Georgia Tech. for the purpose of testing innovative solar receivers and systems. A brief description of the facility and its auxiliary equipment is given.

0595

Update on the Advanced Components Test Facility : Altman R F, Brown C T and Teague H L, " Proceedings ISES (American Section) Conference ", Denver, Colorado (USA), 21, 835, (1978).

Installation of the U S DOE's Advanced Components Test Facility has been completed. The facility is operated by Georgia Tech.'s Engineering Experiment Station and is located at Georgia Tech. Campus in Atlanta, Georgia. The principal feature of the facility is a hexagonal array of 550 mirrors or heliostats that are mechanically driven so that they focus sunlight at a point 21.3 m above the center o f the field. The facility is designed to serve as a test bed for solar components that require concentrated solar energy for their operation.

078

0596

Reflector Designs for Tower - Top Solar Collector : Ansari J S and Krishnaprasad K, "Proceedings ISES Congress and Exposition ", New Delhi (India) p.1296, (1978).

In earlier studies on tower top focus solar plant, the receiver has a bowl like shape. For such a receiver, the flux density of incident reflected solar radiation is more or less uniform over its entire surface. It is pointed out in this study that if the receiver contains an evaporator and a super heater, the optimum flux densities for the two are different ; the evaporator should receive higher power per unit area than the super heater. Hence, a modified geometry is suggested. Furthermore a bowl shape does not conform with the sun's image formed at the receiver if rectangular mirrors are used.

0597

Analysis of the Terminal Concentrator Concept for Solar Central Receiver Systems : Athavaley K and Vant - Hull L L, " Solar Energy ", 22, 493, (1979).

A terminal concentrator design is considered which produces 90% of the maximum concentration and reduces the size of the conical reflector by 5 - 6 times. The effectiveness of this compromise design permits one to conclude that a practical terminal concentrator of the conical variety can almost double the concentration without any appreciable loss of total power. There will be losses due to reflectivity but not due to beam spillage because of the reduced aperture.

0598

Influence of the Working Fluid on Heat Transfer and Layout of Solar Tower Receivers : Bammert K, " Studies in Heat Transfer : A Festschrift for E R G Eckert"(Hemisphere Publications), Washington, (USA), p.383, (1979).

Extensive experience has been obtained from the design and operation of fossil fired heaters for closed cycle air turbine plants. Based on this experience, the layout of solar heated receivers for gas turbines is described. To make optimum use of the calculation methods developed for fossil-fired radiation parts, the receiver is of cylindrical or polygonal design. The essential equations for calculating the heat transfer and the heat flux distribution on the heating surface are given and discussed. The receiver for a solar tower plant with a helium turbine is designed. The layout data of this plant were chosen to make possible a comparison of the helium receiver with the receiver of an air turbine plant of the same output. Favourable values of pressure and geometric dimensions are given for both working fluids. The influence of the working agent on these data is discussed. Corelations between the receiver design and the layout of the other components of the plant are shown.

0599

Design Method for Optimizing Colector Systems for Small Solar Central Receivers : Bannerot R B and Laurence C L, " ASME Paper No. 79 - WA/Sol - 14, (1979).

The design methodology for the determinaion of the optimal heliostat field designs is presented in detail for a small solar central receiver. The optimization process is reviewed. Cost and performance models are discussed. To illustrate the design processes, a representative small solar central receiver system is optimized. Cost factors were developed from current prices. The individual heliostat design and cost data were taken from the design of the 10 megawatt electric Barstow Pilot Plant design. A north field configuration, steel guyed lower and a tilted circular aperture cavity receiver were utilized. It is demonstrated that solar central receiver systems are more cost effective at higher power level, above those considered here. But this fact has nothing to do with relative cost effectiveness of small, stand alone, power systems.

0600

Structure of Beam Reflected by Heliostat : Baranov V K, " Applied Solar Energy ", 13(6), 1, (1977).

The structure of the beam reflected by a heliostat is examined. It is shown that at a distance of about 107 times the heliostat light aperture diameter there is a central zone in the form of a converging cone, in the limits of which the flux density is uniform and differs very little from the flux incident on the mirror . Beyond this is a segment, the central zone of which has the form of a diverging cone. With increasing distance from the mirror the flux density in this zone decreases as the square of the distance.

0601

Optimum Concentration Ratio for a Solar Central - Receiver Electric Power Plant : Baughm J W and Bergquam J B, " Journal of Engineering Power (Trans. ASME) ", 99, 490, (1977).

In this note a simple equation for the optimum design concentration ratio is derived. This optimum concentration ratio is found to be a function of a non dimensional receiver temperature and a non dimensional receiver number (Nr) which is introduced in this note.

0602

Feasible Thermophysical Conditions for Gas Receiver Tubes in Solar Power Stations : Becker M, " ASME Paper No. 79 - WA/HT - 37 ", (1979).

Heat transfer to air in turbulent pipe flow is analysed and numerically computed. This is applied to estimate the energy transfer processes in a solar thermal reciver of a central tower facility. The general condition of uniformly heated pipe flow will be discussed with dependence on e.g. heat flux (10 to 1000 KW/m^2), inlet pressure (4 to 86 bar) and tube radius (10 to 40 mm). For technical and economical optimization, the combination of high heat flux and high pressure is recommended. Thus for a solar power station with gas as heat transfer medium, the closed thermal cycle is favoured.

0603

The Helios Model for the Optical Behaviour of Reflecting Solar Reflectors : Biggs F and Vittitos C N, " Sandia Laboratories Report No. SAND 76 - 0347 ", (SA), (1976).

The helios model simulates the optical behaviour of reflecting concentrators. The model follows the
incident solar radiation through the system (including the atmosphere) and includes all t h e factors
that influence the optical performance of a collector. An important output is the flux density pattern (W / sqcm) at a grid of points on a surface such as the absorbing surface of a receiver and its
integral (power in watts) over the surface. The angular distribution of sun rays for the radia t i o n
incident on a concentrator is modified by convolution using the fast Fourier transform, to incorpor -
ate the effects of other non deterministic factors such as sun tracking errors, surface slope errors
and reflectance properties. The analytic   methods used for statistics, the off-axis reflecting optics, the atmospherics effects, and the various coordinate systems are described and illustrated.This
model forms a basis for the simulation code HELIOS as well as for other codes under development. Some
of the HELIOS routines are described, a few of its capabilities are discussed and illustrated, a n d
comparisons of data with calculations are presented. These capabilities have been used for perform -
ance predictions, safety studies, design trade - offs, data analysis problems and specification a n d
analysis of concentrator qualities, and for the general understanding of solar concentrator technology.

0604

Mathematical Modelling of Solar Concentrators : Biggs F and Vittitoe C N, "Proceedings I S E S Conference - Sharing the Sun ", Winnipeg (Canada), (1976).

A computational capability that models the operation of any solar energy collector that uses flux concentrators is a valuable aid in the planning, design, construction , calibration, safety analysis a n d
operation of the system. In addition to the usual optical consideration the model should treat such
imperfections as reflecting - surface slope errors, sun tracking and alignment errors and mirror focussing errors. It should properly account for the angular distribution of incoming sun rays and the
effects of atmospheric transmission on this distribution. A model with these capabilities is described and two computer programs for implementing it are illustrated.

0605

Helios : A Computational Model for Solar Concentrators : Biggs F and Vittitoe C N, " Sandia Laboratories Report SAND - 77 - 0642 C ", (USA), (1977).

Helios is a computer code for mathematically simulating the behaviour of the flux pattern from   t h e
concentrator field for a solar central receiver power station. Statistical methods are used to incorporate non-deterministic factors. This code is described & some examples of its output are given.

0606

A Parameter Study for a Critical - Receiver Power Station : Biggs F and Vittitoe C N, "Sandia Laboratories Report SAND - 77 - 0667 C ", (USA), (1977).

The interaction between alignment and focussing strategies and heliostat errors are described and illustrated. Some descriptions of astigmatic aberrations are developed and are used to suggest an evaluation criterion for concentrators. Finally, an analysis of measurements for evaluating helio s t a t
reflectors is given.

0607

Unfolding Concentrator Slope Errors from Reflected - Beam Measurements : Biggs F, Vittitoe C N & King
D L, " Proceedings ISES Silver Jubilee International Congress ", Georgia (USA), (1979).

A measured flux density pattern in the reflected beam of a heliostat is analyzed to unfold the distribution of surface slope errors for a heliostat reflecting surface. It is found that two dimensional
effects in this error distribution are within the resolution of the measurements.

0608

Design Fabrication and Test of a Heliostat for a Central Receiver Solar Thermal Plant; Project Technical Report May 1974 - September 1975. : Blackmon J B, " Rep. (PB - 252667), (MDC - G - 6043), (1975).

A full scale central pedestal mounted elevation / azimuth heliostat was designed, fabricated and tested. The heliostat consisted of a flat reflective surface composed of two coplaner lights of black silvered (second surface) glass, each 3.7 m (12 ft) X 1.85 m (6 ft) X 6 mm (1/4 in) bounded to a steel frame
structural support. The structural support was mounted on a drive housing equipped with a linear act-
uator for control of the elevation (tilt) axis, and a harmonic drive for azimuth control.D.C. gear motors (1/30 hp) were used in both drive axes. A quadrant photosensor mounted on a separate pedestal &
aligned along the target mirror of sight provided properly apportioned steering command to the D. C.
motors on the elevation azimuth movement tracking. The beam dispersion characteristics were investigated at 6 typical positions in the field ranging from 300 to 3000 ft horizontally from a screen target suspended between two 360 ft high towers.

0609

Desert Tests of a 3.7 m Square Heliostat : Blackmon J B, Vant - Hull L L and Hildebrandt A F, "Proceedings ISES Conference ", Los Angeles (USA), (1975).

In support of the objectives of demonstrating technical and economical feasibility of the heliostats,
an approximately full scale pedestal mounted heliostat supporting two coplaner sheets of backsilver-
ed float glass, each 3.7 m X 1.85 m X 6 mm has been built and field tested in a desert environment.

0610

Baseline Design of Commercial Central Receiver Solar Power Plants : Blake F A, " Proceedings I S E S
(American Section) Conference ", Orlando (USA), (1977).

The commercial plant with a rated output of 150 MWe during sunlight operation and output of 105 Mwe
from storage, consists of fifteen solar collector modules, a thermal storage field and an electrical
power generation unit. Each of the collector modules features 1718 focussing heliostats which focus
sunlight into tower mounted cavity receiver steam generators. Basic technical characterists of t h e
solar plant designs and current subsystem research experiments being performed to substantiate t h e
design approach are discussed in the paper.

080

0611

Update on the Solar Power System and Component Research Program : Blake F A & Walton J D, "Solar Energy", 17, 213, (1975).

Preliminary design configuration features of the point design 100 MWe solar energy conversion power system and user applications for the direct solar / hydroelectric installations are presented. The 405 acre (1.64 Km^2)plant consists of a central power station surrounded by eight solar energy collection/ concentration / conversion fields, which supply the required 1250 psi, 950 deg F (510 deg C) superheated steam for turbine generator operations. Boiler and superheater assemblies in each field are mounted on a 400 ft (121.9 m) high tower at the center of the southern border of the field. In each field, 1840 mirrors perform the collection / concentration function with a single reflection of the solar energy. The user application analysis examines the potential contribution of the 100 - MWe plant to a south - western utility based on the local demand and solar weather patterns of 1973. An overview of the component testing for a bench model boiler / superheater and associated turbine generator being planned in collaboration with CNRS France is presented.

0612

One MW(th)Bench Model Cavity Receiver Steam Generator : Blake F A, Tracey T R, Walton J D and Borner S, " Proceedings ISES Conference ", Los Angeles (USA), (1975).

Design of a bench model steam generator having geometric and thermal characteristics of the full scale steam generator at the 100 MWe solar energy conversion power plant and the necessary scaling and adaptation features required for operation in the CNRS laboratory furnace at Odeillo, France, was a major element of the Solar Power System and Component Research Program. The planned follow-on program to fabricate the bench model steam generator, to provide required instrumentation - control- adaption equipment, and to perform check out testing, is being performed during 1975. Solar operation of the steam generator is expected to take place during 1976. Thermal losses and resultant cavity efficiency values established by the thermal analyser program for the finalized full scale design are calculated.

0613

Utility Problems of Central Tower Solar Plants in Italy : Borgese D, Cefaratti G and Floris R, "Proceedings ISES Silver Jubilee International Congress ", Georgia (USA), (1979).

A review of both technical and economical considerations,in order to single out the best choice among different components of the 1 MW solar power plant in course of construction in Sicily, Italy, is presented. Some consideration has been made about store system sizing which allows covering load peaks (e.g. evenings peak) in case of correct operation. It will result in an implementation of the energy use produced from solar plants. Finally, as far as operating is concerned, a dynamic mathematical model of the cycle receiver system has been worked out, by means of which it is possible to define both regulation system characteristics and running procedures.

0614

A 400 KW High Temperature Solar Test Facility : Brown C T, "Proceedings ERDA Conference on Concentrating Solar Collectors ", Georgia (USA), p.4 - 13, (1977).

A 400 KW Solar Thermal Test Facility owned by ERDA and operated by Georgia Tech., has been completed on the Georgia Tech. Campus. Its features include a hexagonal array of 550 mirrors driven mechanically to track the sun, a central tower for test apparatus and a computerised data acquisition system . Maximum radiation flux density is expected to be about 300 W / sqcm. By March 1978, a second tower, capable of supporting heavier loads will be built south of the mirror field. The first test, that of a 350 KW Francia type receiver is presently under way and will be completed by January 1978. Other users are being encouraged to schedule tests at the facility.

0615

U S DOE Advanced Components Test Facility : Brown C T, " Proceedings 4th National Conference (Technology Energy Conservation) ", Albuquerque, New Mexico (USA), p.79, (1979).

The US DOE Advanced Components Test Facility (ACTF) is a 400 KW(th)solar thermal test facility operated by Georgia Tech's Engineering Experiment Station for the US Department of Energy. The A C T F provides a place for private industry, universities and government laboratories and qualified individuals to evaluate experimental solar thermal conversion components such as steam generators, chemical reactors, or other heat exchanger devices.

0616

Results of a Tilt - Low Profile Heliostat Test Program : Brown G L, " Proceedings International Conference on Alternative Energy Sources ", Miami (USA), (1979).

Honeywell designed and built a unique four facet, low profile tilt - tilt gimballed heliostat as a candidate for the solar central electrical power pilot plant. It has an entirely open loop control scheme for which one central processor could control 25000 heliostats. This paper addresses the tracking accuracy and optical performance achieved related to an error budget along with the environmental effects of wind, atmosphere alternation etc. Test instrumentation developed for a program is described, with emphasis placed on the flexibility of the calibration array and the wind tunnel test conducted. In Summary, the design demonstrated that it could meet the end to end 2 mr. redirected image tracking accuracy requirements even under operational wind loads using an open loop control mechanism.

0617

High Temperature Solar Power Tower Plants Concept Considerations and Operational Criteria: Cafaratti G and Gretz J, (Unknown).

In the present work some concept considerations and operational criteria are described for solar power plants, central receiver type. These considerations are mainly based on the experiences obtained with Eurelios, the EEC 1 MW plant, now under construction in Italy. A description of the plant is made in the paper. Some economic and material requirement considerations are made, indicating the fact, however, that published figures still vary considerably.

0618

<u>Unique High Concentration Solar Test Facility</u> : Cape J A and Deyhmy I, " Solar Cells ", 1, 29, (1979 / 80).

A high concentration solar test facility is described in which a 60.96 cm X 66.04 cm plane mirror heliostat made from a commercial (Celestron - 14) telescope mount reflects the insolation to a fixed f/4 focussing mirror 53.34 cm in diameter in an enclosed laboratory. Tracking is provided by a feedback system in which the unconcentrated sunlight is pinhole imaged onto a 1 cm quadrant detector. Photocurrent differences from opposite segments of the quadrant drive the right ascension and declination controls of the Celestron - 14 mount. The tracking accuracy is better than $0.01°$ and the total s y s t e m cost is less than $11000.

0619

<u>Centrales Solaires Realisations et Projects Francais / (Solar Power Plants, France ; Realizations a n d Projects)</u> : Clemot M, Dersus B, Etievant C and Pharabod F, " Proceedings 10th World Energy Conference", Istanbul (Turkey), (1977).

The paper discusses solar thermal power plants and project applications in France. Among the projects discussed is the THEM project, which is scheduled to produce 3500 KWe by 1980. This plant will operate with a thermal system at 350 deg C. The following step, two years later, will use a high efficiency thermal system now under development. A prototype solar tower plant (62 KWe) has been constructed in Odeillo in 1976 for the purpose of demonstration and component development. Several projects are now under study for power production in the 100 - 1000 KWe range.

0620

<u>Solar Powered Steam Generator Heliostat Final Report</u> : Cottingham J G, " Brookhaven National Labora - tory Report BNL - 50974 ", (USA), (1978).

A small size central receiver type solar energy collecting system delivering commercial grade s t e a m is analysed and a wind avoidance type heliostat designed, built and successfully tested. The h e l i o- stat design effort is described, including reflecting surface materials and measurements, optic considerations and mirror field arrangements, mechanical analysis and fabrication techniques, and economics and cost effectiveness. Measurements of normal incident solar energy at Upton, New York, are reported and a method is proposed for estimating this input parameter for other locations proposed.

0621

<u>Distributed Computer Control and Data System for the Solar CRTF</u> : Darsey D M, " Proceedings National Electronics Conference ", Chicago (USA), 33, 246, (1979).

The Central Receiver Test Facility (CRTF) is the World's largest solar experimental facility, capable of collecting 5 million watts of thermal power for testing purposes. As an experiment facility, t h e CRTF requires more versatility and precision in controls and a large amount of data for experiment evaluation which are not required in an operating plant. The key to these special requirements is the Master Control System (MCS) consisting of a modular network of mini computers and associated equipment. Real time data acquisition, analysis and presentation and human engineered software, allow operator visibility and control of experiment in progress. Programmed test sequences assure repeatability ; online modifications with transaction logging permit real time changes as required. Control s y s t e m design philosophy, components and operation are discussed.

0622

<u>Colour Graphic Controls for the Solar Central Receiver Test Facility</u> : Darsey D M, " ISA Transactions", 19(2), 65, (1980).

The World's largest solar thermal facility uses a single operator to control and supervise experiments. The critical man - machine interface with the computerized control and data system is an intelligent refreshing colour graphic system with seperate consoles for the operator and the experimenter. The graphic hardware, integration software and application program are discussed as they apply to t h e real time control and data presentation of the testing facility.

0623

<u>Master Control System for the Central Receiver Solar Power Plant</u> : Darsey D M, Routree R C, Sheahan R R, Soderstrand M A and Winarki C P, " Proceedings Joint Automation and Controls Conference ", Philadelphia (USA), p.287, (1978).

A Master Control System (MCS) is described in terms of its operation and evaluation role for the nation's first major solar thermal electric power venture. The specific application referred to t h e 10 MWe solar thermal central receiver pilot plant being built at Barstow, California. This p i l o t plant represents the initial step (involving complete power generation subsystems) being performed to demonstrate feasibility of eventual commercial central receiver solar applications. Accordingly, an MCS has been conceptually designed to direct and coordinate the subsystems and to provide a s o u n d capability for evaluating feasibility. This paper describes the MCS concept and the basis for its adoption and is the first publication regarding the Pilot Plant MCS made available to the technical community at large.

0624

<u>Developing Master Controls for a Central Receiver Solar Power Plant</u> : Darsey D M, Routree R C, Sheahan R R, Winarski C P and Soderstrand M A, " In. Tech. ", 26 (3), 41, (1979).

A distributed control system, comprising conventional and computerized modules, is being developed for use in a pilot central receiver solar power plant. The configuration has capability to gather data for evaluation as well as to provide central supervision and coordination.

0625

<u>A User's Manual for DELSOL : A Computer Code for Calculating the Optical Performance, Field Layout and Optimal System Design for Solar Central Receiver Plants</u>: Dellin T A and Fish H K, " Sandia Laboratories Report SAND - 79 - 8215 ", (USA), (1979).

082

DELSOL is a quick and accurate computer program for calculating the collector field performance, field
layout and optimal system design for solar central receiver plants. The code consists of a detailed
model of the optical performance, a simpler model of non optical performance, an algorithm for field
layout and a searching algorithm to find the best system design. The latter two features are coupled
to a cost model of central receiver components and an economic model for calculating energy costs. The
code can handle flat, focussed and / or canted heliostats, external cylindrical, multi - aperture cavities and flat plate receivers. The program optimizes the tower height, receiver size, field layout &
power level.

0626

DELSOL ; A Code for Central Receiver Performance and Optimization Calculations : Dellin T A and Miriam
J F, " Sandia Laboratories Publication ", (USA), (Unknown).

DELSOL is quick and accurate computer program for calculating the field performance, field layout and
optimal system design for solar central receiver plants. The code consists of a detailled model o f
the optical performance, a simpler model of the non optical performance, and algorithms for field layout and finding the best system design. The latter two features are coupled to a cost model of central receiver components and an economic model for calculating energy costs. The code can handle flat,
focussed and/or canted heliostat and external cylindrical, multi - aperture cavity, and flat plate r e -
ceivers. The program optimizes the tower, height, receiver size and field layout as a function o f
power level.

0627

Cost Model and Optimization of the Combined Structural Components of Large Power Solar Central Receiver Systems : Dundev V, " Proceedings ISES Silver Jubilee International Congress ", Georgia (USA)(1979).

This paper focusses on the cost optimization of the structural system for the central solar receiver
concept. The components of the structure system are the solar receiver (including absorber) t h e
tower supporting it, and the foundation on which it stands. The design of each of the three compon -
ents is determined by the environmental loadings that it is subjected to as well as the forces on it
due to the other two components. Because of the latter, the optimum design of the three   components
might not yield the optimum for the entire structure. The only environmental loadings considered are
wind and earthquake forces since it has already been determined that conditions such as rain, ice and
snow are not important, and tornadoes, floods and other natural disasters are generally not designed
for. It is necessary to determine the effect of the main design parameters in relation to the relative cost of each component on the cost and behaviour of that component as well as the cost consequences of interactive effects on the other two components, for a variety of loading conditions.

0628

French CNRS 1 MW Solar Power Plant : D'Utruy B, Blay D and Coeytaux M, " Proceedings ISES Conference",
New Delhi (India), p. 170, (1978).

This report details the principle of the electrical solar power plant tested in the French CNRS 1 MW
facilities of Odeillo as well as the main results obtained. The plant was realized in 1976 and operates on a Rankine cycle with steam production at 270 deg C under 27° bar. A 30 m^3 storage tank c o n -
taining a thermal fluid at 335 deg C permits delivery of thermal energy of 2 MWh.

0629

Evaluation of Central Solar Power Plants : Easton C R, Hallet Jr. R W, Gronich S and Gervais R L," Proceedings International Conference on Syst. Man and Cybern. ", Dallas,Texas (USA), p. 77, (1974).

A baseline design for heliostat, receiver, tower and energy transport subsystems for a solar t o w e r
power plant is defined down to the major component level. Preliminary manufacturing plans are given
for component fabrication assembly, and installation capital costs are estimated for the baseline design. The major cost element, other than the power plant, is the heliostat and its supportive equipment. The complete collector system is estimated to cost between $3 and $4 / sft ($30 to $40 / sqm)ins-
talled. This corresponds to a levelized, fixed charge cost for generating energy of 15 to 20 mills /
Kwh_e exclusive of operation maintenance and plant parasitic load costs, which is competitive with current oil prices and may become competitive with coal at prices projected to the midpoint of operation
of the solar power plant.

0630

Heliostat Designs for a Central Receiver Solar Power Plant : Easton C R, Applebaugh C R and Nourse J H,
" Proceedings ISES Conference ", Los Angeles (USA), (1975).

As with other concepts , the cost of collecting solar energy is the deciding factor in the economic viability of a system using solar energy. The concentrator (heliostats and related equipment) has been
estimated to comprise approximately 80% of the cost of collecting the solar energy. The helio s t a t s
therefore, become the economic driver of the central receiver solar power plant. In this paper t h e
dependence of the heliostat cost per unit surface area on physical size has been studied. It has been
found that heliostats normally optimize at sizes which are too large for convenient transportation .
The primary interface for the heliostats is with the receiver. The increase of cost with tracking error and the increase in receiver size has also been studied.

0631

Collector Field Design for a Central Receiver Solar Thermal Power Plant : Easton C R, Raetz J E a n d
Vant Hull L L , " Proceedings ISES Conference - Sharing the Sun ", Winnipeg (Canada), 5, 374, (1976).

A description of the pilot plant collector subsystem is presented. The design embodies (i) first surface mirrors (ii) elevation / azimuth gimbal axes, (iii) a central pedestral support, (iv) a digital
field controller servicing 24 heliostats on a time - sharing basis and (v) a beam sensor directly measuring beam errors and providing error signals to the field controller. The designs of the heliostats
and the controls system are presented together with cost estimates for the commercial collectors.

0632

Collector Field Optimization for a Solar Thermal Electric Power Plant : Easton C R, Raetz J E & Vant -

Hull L L, " Proceedings 11th IECE Conference ", Nevada (USA), (1976).

A procedure for optimizing a central receiver solar thermal electric power plant is presented . Key parameters of the optimization are identified and examined to show the degree of interdependence and to derive a secondary set of variables more amenable to optimization. The procedure is detailed with numerical examples and the results for a 100 MWe commercial plant are shown.

0633

Economic Implications of a Low Cost Heliostat Design : Easton C R and Blackmon J B, " Proceedings ISES Silver Jubilee International Congress", Georgia (USA), (1979).

Design and costs of a prototype heliostat for the central receiver solar thermal power system are presented and comparisons made with commonly manufactured items. Central receiver thermal energy costs are shown to be competitive with fossil fuels.

0634

Subsystem Research Experiments on a Central Receiver Collector : Easton C R, Blackmon J B and McCornick R E, (Unknown).

McDonnell Douglas Astronautics Company has just completed a series of tests on representative hardware for the collector subsystem of a central receiver solar thermal power unit. These tests comprised a portion of the preliminary design of a central receiver system preliminary design under contract to the Energy Research and Development Administration (ERDA). Summary descriptions of these tests and key results are presented in this paper.

0635

Distributed Micro Computer Based Control System for a Large Scale Solar Total Energy System : Farrell J O and Reska R S, "Proceedings IECE Conference", Boston (USA), p.77 – 81, (1979).

A unique computer based digital control system is being designed by General Electric to control, monitor and acquire data from the World's largest industrial solar thermal energy system, the US DOE's solar total energy large scale experiment at Shenandoah, Georgia. The control system employs distributed microcomputer architecture, a human factors engineering control console and large scale integrated (LSI) electronics hybrid signal conditioning to control solar energy collection, storage, conversion and utilisation. The total energy system can thus operate under either automatic or manual control to produce electricity, process steam and provide space heating and cooling for a large knitwear factory.

0636

Dispersed Solar Electric Power ; A Small Power System Program : Ferber R R, Marriott A T and Truscello V C, " Proceedings ISES (American Section) Conference ", Denver, Colorado (USA), 2.1, 822, (1978).

The Small Power Systems Applications (SPSA) Project has been established to develop and commercialise small solar thermal power plants. The technologies of interest to this project include all distributed and central receiver technologies which are potentially viable in power plant sizes of 1 – 10 MWe . This paper presents an overview of the JPL managed SPSA project and the technologies which are being considered for application.

0637

The Cost / Performance of Large Central Power Systems as a Function of Heliostat Design Parameters : Fish M J and Dellin T A, "Proceedings ISES Silver Jubilee International Congress", Georgia (USA),(1979).

This paper addresses the topic of system cost / performance as influenced by heliostat design variables. The study quantifies overall performances and energy cost for systems that are individually optimized for each heliostat design perturbation considered . The cases tested derive from the current "best" DOE sponsored heliostat designs. The results are then analysed to determine high versus low leverage design parameters (e.g. what is the change in system cost / performance for a change in heliostat size and how does this compare with comparable changes in other design parameters). Break even costs for design trade – off's are also given (e.g. how much cheaper must a poorer performing tracking motor be in order to keep the cost of energy constant).

0638

High Concentration Solar Receiver : Floris R and Francia G, " Proceedings Solar Thermal Power Generation Conference, ISES (UK Section)(C – 16) ", London (U.K.), p.101, (1978).

The authors present some considerations about the design of solar receivers for high concentration solar plants based on their own experiences and especially on the results of the theoretical and experimental activities of G Francia in the St. Ilario Facility. This facility was built in 1963 and four different receivers were installed and tested since then. The last one, presently tested, is a scaled down model of a receiver which will be built by Ansaldo for the 1 MW (el) solar power plant .

0639

Heliostat Field Design for the ERDA 5 MW Solar Thermal Test Facility : Fourakis E M, Curtner K L, and Mitchell P D," Proceedings ISES Conference – Sharing the Sun", Winnipeg (Canada), 5, 385, (1976).

The working heliostat field for the ERDA 5 MW solar thermal test facility is designed to meet the solar thermal power requirements for testing a different receiver subsystem research experiments . Requirements for the field are specified. In terms of early solar thermal energy available from the heliostats and instantaneous power level delivered by the heliostat field to the experiment. A ray trace simulation model of the subsystem research experiment receivers and heliostat field is the principal tool used in this conceptual design. Field deployment, focussing strategy, number of heliostats and total mirror area required to accomplish the test objectives of the facility are established.

0640

Low Profile Heliostat Design for Solar Central Receiver Systems : Fourkis E and Severson A M, " Solar Energy ", 19, 349, (1977).

Heliostat designs intended to reduce costs and the effect of adverse wind loads on the device   were developed. Included was the low profile heliostat consisting of a stiff frame with sectional focussing reflectors coupled together to turn as a unit. The entire frame is arranged to turn angularly about a center point. The ability of the heliostat to rotate about both the vertical and horizontal axes permits a central computer control system to continuously aim the sun's reflection onto a selected target. An engineering model of the basic device was built and is being tested. Control and mirror parameters, such as roughness and need for fine aiming, are being studied. The results of economic studies on this heliostat are also presented.

0641

Power Plants for Solar Steam Generating Stations : Francia G, " Solar Energy ", 12, 51, (1968).

We describe here a way to use honeycomb panels (anti radiating cellular structure). In this research, investigators have used the plant for thermodynamic conversion of solar energy into electric energy at high temperatures. We refer to the boilers at Monseilles and Genoa, and we describe briefly the plant being built at St. Ilario or Nervi, Genoa. This plant has a maximum production of 140 Kg / Hr of steam at 150 atm., in the temperature range 500 - 650 deg C.

0642

Quality Assurance Program for Solar Thermal Test Facility (STTF) Heliostat Production : Freeman F P and Hillman J T, "Sandia Laboratories Report SAND - 78 - 1303 ", (USA), (1978).

The Quality Assurance (QA) Program followed during heliostat production for the Solar Thermal Test Facility is described. Problems encountered as well as the corrective action taken are discussed. Brief descriptions of the validation of processes, the Martin Marietta Quality Control Inspections, and the data package and computer storage of record of assembly data are included.

0643

Central Receiver Solar System Applicable to Central Power Stations : Gervais R L, Friefeld J M and Mc Kenzie A W, " Proceedings ISES Conference - Sharing the Sun ", Winnipeg (Canada), 5, 325, (1976).

The design is distinct in that it utilizes (i) First surface mirrored Heliostat, (ii) an external single pass to super heat receiver, and (iii) a sensible heat thermocline type thermal storage. A 100 MWe commercial power plant, designed for integration into a utility network, is described. A 10 MWe pilot plant project whose purpose is to establish technical feasibility and provide an indication of system economics for the commercial system is then given. The problems in thermal storage are also addressed.

0644

Closed Cycle High Temperature Central Receiver Concept for Solar Electric Power : Gintz J R," Electric Power Research Institute Report EPRI - ER - 629 ", (USA), (1978).

The technical feasibility of a high temperature central receiver in a solar plant employing closed cycle helium as a heat transport fluid was examined in terms of system life, efficiency, cost and technology requirements. These considerations have been implemented in the conceptual design of a receiver and its components, for utilization in a solar plant of 100 megawatts of electrical power output. The rationale is provided that supports the configuration, equipment arrangement and material choices. Thermal cycling tests simulating the 30 year lifetime of a receiver's heat exchangers at temperatures to 816° C and at 3.45 MN/sqm (500 psi) helium pressure confirmed material choices. Preliminary design considerations are presented for a 1 MW thermal test receiver and for a 10 MW electrical pilot plant. This report also contains system / supporting subsystem definition for employing the central receiver design in a solar plant. This includes conceptual design of several thermal energy storage devices and their integration into plant operation.

0645

Testing Procedures and Test Results for the EPRI / Boeing Gas Cooled Receiver Tested in the Central Receiver Solar Thermal Test Facility : Gintz J, Zentener R, Marshall B and Matthews L, " Proceedings ISES Silver Jubilee International Congress ", Georgia (USA), (1979).

An initial series of central receiver solar thermal tests was completed at the US DOE Central Receiver Test Facility in February, 1979. These first production tests in the new facility consisted of experiments with a 1 MW Gas Cooled Receiver for Brayton Power cycle applications. The receiver was designed and fabricated by Boeing Engineering and Construction under contract to EPRI. The testing was jointly sponsored by EPRI & DOE. This report describes the Gas Cooled High Temperature Receiver, the planning required for accomplishment of a major solar / thermal test, and preliminary test results including solar power measurements and experimental results.

0646

Infra - red Television Measurement of Heliostat Images : Gray D," Proceedings SPIE Conference ", 85, 95, (1977).

The current development of large focussing heliostats to concentrate solar energy on a boiler or central receiver requires a method for measuring the overall efficiency / performance of the heliostat. A flat plate calorimeter, using the temperature rise in water at a measured flow rate, has been used to test a 22.3 sqm heliostat focussed at a distance of 31 m. These measurements represent the first test of a large area concentrating heliostat in this country. This type of testing also established the requirement for varying the accuracy of the collected calorimeter data. One of the methods used to accomplish this verification is described.

0647

Solar Thermal Conversion to Electricity Utilizing a Central Receiver, Open Cycle Gas Turbine Design : Grosskreutz J C, " Proceedings ISES Conference ", Los Angeles (USA), (1975).

Combining an open cycle gas turbine with the central receiver concept for concentrating solar radiation offers an attractive design concept for electric power generation. The high concentration ratios available at the central receiver make possible the generation of temperatures in the range necessary for efficient gas turbine operation i.e. 1500 - 2000 deg F or even greater. The principal design problems associated with the high temperature receiver have been discussed, and also the cost estimates have been made to compare it with conventional designs.

0648

A Central Receiver Solar Thermal Power System : Hallet Jr. R W and Gervais R L , (Unknown).

This paper presents an overview of a central receiver solar thermal electric system presently being developed for the US DOE. The design is distinct in that it uses (i) centrally supported low cost heliostats, (ii) an external single pass to superheat the receiver and (iii) a sensible heat, thermocline type thermal storage. The system is capable of producing 100 MWe with water-steam as its working medium. The system is designed for a 30 year operational life with a 90% plant availability goal. The definition of a 10 MWe Pilot Plant is also presented. The purpose of the Pilot Plant is to operate a system which demonstrates technical feasibility and provides an indication of system economics for the commercial size system.

0649

The Solar Ten Megawatt Pilot Plant : Hallet Jr. R W and Gervais R L, " Proceedings ISES Silver Jubilee International Congress ", Georgia (USA), (1979).

The requirements and characteristics of the Solar Central Receiver Ten Megawatt Pilot Plant presently under construction by the US Department of Energy in Barstow, California,is described. The system requirements, characteristics, operating modes and site features are presented. The system can be characterised as a surround field configuration with radial stagger heliostat layout, capable of generating, 10 MWe net directly from insolation and 7 MWe net from thermal storage. Water - steam is the primary working fluid where a single pass - to - superheat external receiver (steam generator), located atop a 91 meter tower,is used to convert solar radiation to thermal energy. Thermal storage is provided by a dual media, rock and oil, thermocline type, sensible heat storage system. Control of the plant is primarily automatic with the capability of operator override. The plant has capabilities for both sun and load following operation, possessing eight steady state operating modes.

0650

Optimal Module Sizing for Solar Central Receiver Thermal Electric Power Plants : Hankins J D," Proceedings ISES (American Section) Conference ", Denver, Colorado (USA), 21, 815, (1978).

The turbine generator set in a solar central receiver thermal electric power plant can be fed by one or by several heliostat - receiver groups or modules. The cost of the electricity produced by solar plants will depend on plant and the module size. A number of factors must be balanced to produce minimum cost electricity. Among them are the cost of piping between modules and the turbine generator, the cost of towers, the amount of redirected power which misses the receiver. The paper is a report on work in progress on plant and module size optimization for several central receiver technologies.

0651

White Sands Solar Test Facility : Hays R A, " Proceedings Seminar on Testing of Solar Energy Materials and Systems ", Gaithersburg (USA), p. 54, (1978).

The facility is operated by the Army's Nuclear Weapon Effects Laboratory primarily for nuclear weapon thermal effects testing but is also used for solar energy research by numerous government agencies and universities. A description is given of the four main parts of the facility; the heliostat, attenu - ator, concentrator and test and control chamber.

0652

Solar Tower Concentrator : Hildebrandt A F ,   Vant - Hull L L and Easton C R, " Proceedings ISES Conference ", Los Angeles (USA), (1975).

Optical concentration and thermal conversion of solar energy using the solar tower concept has been evaluated as a function of heliostat size and dispersion, heliostat spacing, tower height and boiler configuration. The progress to date will be summarised. Studies show no sharp design maxima exist but excellent performance is predicted for heliostats of 4 - 6 m flat silvered mirrors that provide cover - age over 40% of the area of a 2 Km X 2.4 Km rectangular area with a 300 m tower near the centre. This baseline configuration with 3 - 6 hr thermal storage could provide 100 MW electric power for intermediate and peaking loads. The most attractive storage fluid candidates identified are a low vapour pressure high temperature hydrocarbon and a liquid metal.

0653

Power with Heliostats : Hildebrandt A F & Vant Hull L L, " Science ", 197, 1139, (1977).

The possibility of producing large amount of power using solar power tower system is explored. The paper deals with the following aspects (i) plan of solar thermal power system based on optical transmission (ii) the history of the solar tower (iii) the receiver subsystem (iv) the design of the heliostats and their placement in a field (v) thermal storage (vi) environmental concerns and (vii) economics.

0654

Survey of Power Tower Technology : Hildebrandt A F and Das Gupta S, " Journal of Solar Energy Engineering (Trans. ASME) ", 102(2), 91, (1980).

Recent technology has shown advances in receiver design, in heliostat fields surrounding the power tower, heliostat cost and storage. Calculations indicate an energy amplification factor (EAF) of 20 for solar energy plants i.e. the ratio of electric energy produced over the lifetime of a power plant to the thermal energy required to produce the plant.

0655

Definition of Two Small Central Receiver Systems : Holl R J, " Sandia Laboratories Report SAND - 78 -7001", (USA), (1978).

The performance and costs of two small central receiver systems are defined and hardware development required to deploy such systems in the solar total energy program is recommended. The two solar energy systems are described starting with the receivers, then covering the heliostat field and tower.

086

0656

Central Receiver Test Facility (CRTF) Experiment Manual : Holmes J T, Matthews L K, Seamons L O, Davis D B and King D L, " Sandia Laboratories Report SAND - 77 - 1173 (Revised) ", (USA), (1979).

The central receiver test facility is operated by Sandia Laboratories for the US Department of Energy. The CRTF is being used for component and subsystem evaluation within the solar thermal large power systems program. This experiment manual provides users of the CRTF detailed information about (i) implementation of testing at the CRTF, (ii) details of the CRTF capabilities and interfaces, and (iii) requirements of experiments.

0657

Circumsolar Radiation Data for Central Receiver Simulation : Hunt A, Grether D and Wahlig M," Proceedings ERDA Workshop on Methods for Optical Analysis of Central Receiver Systems ", Houston (USA)(1978).

The circumsolar measurement project is being carried out to improve data to assess the effects of circumsolar radiation on the operation of solar thermal conversion system using solar concentrators, especially central receiver system. Four circumsolar telescopes have been constructed and are providing detailed intensity vs angle profiles of the solar and circumsolar region, as well as other solar and climatological data. These measurements have been underway for more than one year at several locations. The current program emphasis is on reducing the data and making it available to groups analysing the performance of central receiver systems. The results of this project provide the detailed type of input data for central receiver simulation codes that are necessary for determining these losses , optimizing the receiver or field size, and determining the distribution of stray flux due to circumsolar radiation.

0658

Optical Analysis of Solar Facility Heliostat : Igel E A and Hughes RL, " Solar Energy ", 22, 283, (1979).

A method for estimating the approximate size of the solar image cast by individual heliostats is developed. As a consequence from the Coddington equations, a simple analysis of astigmatization, the major observation which describes optical off axis image behaviour of a focussing collector system has been made. These predictive equations agree well with experiments performed with spherical mirror over a range of angles of incidence exceeding 60°. Other than flat mirrors, several heliostat configurations were proposed a n d explored and were found amendable to the same analysis. All designs which attempt to superpose t h e reflected energy from several mirrors mounted and tracked together on the same frame are subject to the same sample rules. Although this analysis is not rigorous, it does give useful insight into the cause of the large image spread found under some operating conditions and it does indicate methods for optimizing the system. Conclusions reached are explicitly related to the off axis performance of the mirror corelates of a central receiver solar station.

0659

Model Analysis of the First Production Design Heliostat Used at the Solar Thermal Test Facility(STTF) : Janczy J R, "Sandia Laboratories Report SAND - 77 - 1393 ", (USA), (1977).

The experimental study sought to characterize the dynamic model characteristics of a heliostat to be used at the Sandia operated Solar Thermal Test Facility. Three model studies were p e r f o r m e d. Two studies were conducted of the yoke and one on the facet assembly. During the course of the test, Power Spectral Density studies were performed using wind loading and transport environmental data. The modal (frequency, damping and stiffness) data is presented. The data obtained through the various studies is compared favourably. The data indicates a possible structural modification and this modification i s discussed.

0660

Designs of a Sodium Cooled, Central Receiver Solar Power Plant : Johnson J L and Thomson W B, "Proceedings 12th IECE Conference ", Washington (USA), 2, 1203, (1977).

The design of a 100 MWe sodium cooled, central receiver solar power plant is presented. A sodium cooled system has advantages of good heat transfer, stable coolant flow and high efficiency. Sodium components either are available now or will soon be available as a result of the soldium cooled reactor programs. A description of three liquid sodium heat transport systems are presented. Several alternative storage options are examined. The sodium cooled plant's initial cost is slightly less than an equivalent water cooled plant. However, levelized energy cost is 18% lower for the mid 1980's.

0661

Environmental Reflectance•Degradation of CRTF Heliostats : King D L and Myers J E, " Proceedings S P I E Symposium ", Los Angeles (USA), (1980).

A laboratory bi - directional reflectometer and a field portable reflectometer have been used to determine the effects of environmental exposure and consequent dust buildup on the reflectance of mirrors located in the heliostat field of the central receiver test facility (CRTF). Surprisingly l o w rates of reflectance degradation have been observed in 64 weeks of sample mirror exposure and in two and one half years of CRTF heliostat exposure.

0662

Concepts for Central Solar Electric Power Generation : Kintigh J K, " Proceedings Annual Front of Power Technologists Conference ", Stillwater (USA), (1974).

The authors concluded that the best conceptual design concept for central solar electric generation is the central receiver system in which solar radiation is redirected by an array of heliostats and collected at one central receiver which is supported by a tower. The solar energy is absorbed as heat by this central receiver and is transferred into the working fluid. The uncertainties associated with the production of central solar electric power are not those of technical feasibility but involve e c o n o mics, and the compatibility of solar power plants with existing electrical systems. It is consequently difficult to predict the impact of central solar electric generation on the future of the electric utility industry. The major economic issue related to solar electric generating plants is their high costs.

0663

Ingenieuraufgaben bei der Errichtung von Solarthermischen Kraftwerken (Engineering in the Construction of Solar Thermal Power Plants) : Kleinrath H, " Elektrotech Maschinenbau", 96, (9), 406, (1979).

Two systems show prospects of an early large scale technical application for the production of electricity from solar energy by way of heat. For the output under 1 MW it is the farm concept with distributed collectors, for the output upto 100 MW per unit it is the tower concept with central steam generation. Using the examples of a 10 KW experimental power plant, of an already designed 500 KW power plant, and a 3200 MW large scale unit, the development tasks for the engineers are presented in detail.

0664

The Tower Reflector as an Alternate to the Tower Boiler Concept for a Central Receiver Solar Thermal Electric Conversion Plant : Knasel T M, Liner R T, Simmons J and Higgins R, " Proceedings SPIE Conference ", 68, 85, (1975).

A second generation central receiver solar thermal power conversion system that may offer more cost effective plant operation than plants presently under study for early DOE development is described. It uses formed focussing heliostats more extensively, secondary reflection to a boiler at ground level, a flux concentrator at the boiler entrance, transpiration cooling of the boiler heat exchanger, and high energy density storage devices to smooth plant output. Tests of some of the critical subcomponents in an actual intense solar environment are reported. Finally, the potential system advantages for each element of the second generation plant are summarised.

0665

Central Receiver Solar Thermal Power System : Lang W R and Riedesel R G, " Proceedings American Power Conference ", Chicago Illinois (USA), 40, 448, (1978).

In 1975 a team was formed by McDonnell Douglas Aeronautics Co. Hutington Beach, California to respond to a request for proposal (RFP) issued by Energy Research and Development Administration (ERDA). The RFP Phase 1 effort required Subsystem Pilot Plant for a central Receiver Solar Thermal Power System. In August 1977, ERDA, now the Department of Energy (DOE), announced that the 10 MWe pilot plant to be constructed in the desert at Dagart / Barstow would be essentially patterned after the concept submitted by the team. This is a summary of the concept.

0666

Concentration du Rayonnement Solaire Par des Heliostats Convergents / (Concentration of Solar Radiation by Convergent Heliostat Mirrors) : Le Phat Vinh A and Le Phat Vinh M, " Revue Générale Thermologie ", 16, 192, 883, (1977).

The possibilities of concentrating sun's rays using several heliostat mirrors are discussed. A description is given of an arrangement with both flat and convergent mirrors controlled by a heliostat during the daylight hours.

0667

Performance Calculations for a High Temperature Solar Energy Collection System : Leonard Jr. C M, "Sandia Laboratories Report SAND - 74 - 8031 ", (USA), (1975).

A high temperature, central receiver solar energy collection system is modelled with the DAZZLE computer program. Results for a large hypothetical system located near Albuquerque, New Mexico, are given for clear weather on spring equinox, and summer and winter solstice days and for a year of tropical Albuquerque weather as described in the US Climate Atlas.

0668

Analytical Integration of the Solar Flux Density Due to Rectangular Mirrors : Lipps F W, " ASME Paper No. 74 - WA / Sol - 10 ", (1974).

The solar power tower concentrator requires roughly 5×10^4 flat mirrors arranged in a horizontal plane array and oriented to reflect sulight to an elevated central receiver, where a concentration of roughly 1000 X can be utilised. The flux density on the receiver is developed analytically. The most important input parameters concern the diurnal motion of the sun, the size of the solar disk, the solar constant, and the limb darkening phenomenon. The analytic approach given here has been implemented on a UNIVAC 1108 and some results shown.

0669

A Numerical Approach to the Flux Density Integral for Reflected Sunlight : Lipps F W, " Proceedings ISES Silver Jubilee International Congress ", Georgia (USA), (1979).

The new approach assumes isotropic Gaussian guidance errors. Hence the flux density integral reduces to several iterated single integrals which can be precalculated and stored in a table for interpolation as needed. The LBL solar telescope data is fed into a convolution integral which represents the guidance errors. Aureole effects can be switched on or off at this point. A vector of convoluted solar data is input to another integration which gives the table of normalized flux contributions. The tabular values depend upon the position of the flux point with respect to an edge of the helio - stat as seen in the image plane. Effects due to canting and dishing are included by a ray trace of the vertices of the heliostat. A larger number of ray traces is not required because the image map of the heliostat is linear unless ripples or irregularities occur.

0670

Shading and Blocking Geometry for a Solar Tower Concentrator with Rectangular Mirrors : Lipps F W & and Vant Hull L L, " ASME Paper No 74 - WA / Sol - 11 ", (1974).

This paper provides an analytical formulation of the shading and blocking phenomenon for rectangular mirrors. Shading is the loss of illumination on a given mirror due to the interception of the incident sunlight by a neighbouring mirror. Blocking is the loss of illumination on the central receiver due to the interception of reflected sunlight by another neighbouring mirror. These phenomena play a major role in limiting the effectiveness of a tower top solar concentrator.

0671

A Statistical Approach to the Solar Flux Density Calculation for a Tower Top Solar Concentrator : Lipps F W, Vant Hull L L and Walzel M D, " Proceedings ISES Conference ", Los Angeles (USA), (1975).

A simulation procedure is developed which enables the determination of flux density distribution on the receiver of a tower type solar concentrator. The effect of the guidance errors and flatness of the heliostats as well as the heliostat geometry and the solar brightness distribution may also be s t u d i e d using this procedure.

0672

Four Different Views of the Heliostat Flux Density Integral : Lipps F W, " Solar Energy ", 18, 555, (1976).

The image due to a single heliostat is represented by its flux density, which can be formulated as an integral over the solid angle of the incoming rays. The initial formulation is transformed into three alternative representations, each having some particular utility. The incoming ray formulation leads to analytic results for flat heliostats with polygonal boundaries. The mirror plane formulation leads to a numerical integration over the mirror plane which can be used to study effects due to distributions of the mirror. The pin hole view leads to an approximate expression for the flux density integral as a convolution of the image due to a point sun with respect to the brightness distribution of the real sun . This formulation allows us to treat the sun size as though it were a source of guidance errors or alternatively, we can introduce a degraded sun which includes the guidance errors.

0673

Parametric Study of Optimized Central Receiver Systems : Lipps F W and Vant Hull L L , " Proceedings ISES (American Section) Conference ", Denver, Colorado (USA), 2.1, 793, (1978).

The RCELL computer program provides a " Cellwise Method for the optimization of Large Central Receiver Systems ". This program contains an adequate model of the solar central receiver system and can output nearly complete performance data for an economically optimized collector field. The collector field utilizes radial stagger neighbourhoods (i.e. the peak through feature) and adjusts the radial and azimuthal spacing coordinates in each cell to balance the losses in an optimum way.

0674

An Analytic Evaluation of the Flux Density due to Sunlight Reflected from a Flat Mirror Having a Polygonal Boundary : Lipps F W and Walzel M D, " Solar Energy ", 21, 113, (1978).

Computer algorithms for the flux density of reflected sunlight from a heliostat become an e s s e n t i a l part of the optical simulation problem for the central receiver system. An exact analytic r e s u l t i s available for heliostats having polygon boundaries. An analytical method for round heliostats is given in appendix A, which is extremely complex and requires quartic roots. A useful numerical method is given in appendix B for heliostats of arbitrary shape. A comparison is made between the analytic method and the Hermite function method which is much faster but less accurate. The analytic method provides a b a s i s for evaluating all other flux density calculations.

0675

A Cellwise Method for the Optimization of Large Central Receiver Systems : Lipps F W and Vant Hull L L, " Solar Energy ", 20, 505, (1978).

The total number of heliostats in the collector field determines the approach to the optical simulation problem. For large central receiver systems, it is desirable to introduce a cell method w h i c h establishes an array of representative heliostats. An arsenal of computer programs has been developed which allows to optimize the arrangement of heliostats in the collector field subject to the approximations of the cell model. Each cell contains an arbitrary regular two dimensional a r r a y of heliostats. Four categories of heliostat arrangement are considered (i) radial cornfield, (ii) radial staggers, (iii) N – S cornfields and (iv) N – S staggers. Some of the important results from the 100 MWe commercial model optimization study are described.

0676

Status Report on the Solar Thermal Test Facility : Marshall B W, " Proceedings IEEE Annual Conference ", Oklahoma (USA), (1978).

A description is given of the solar thermal test facility (STTF) which is made up of a field of mirror assemblies (heliostats), which concentrate sunlight onto experiments located on a central tower. The facility is designed to deliver upto 5 MW of thermal power to experimental equipment a n d successful operation at partial power of approximately 1.8 MW(t).

0677

Les Centrales Electrosolaires a' tour, Optimization due Champe de Reflecteurs et Application a' l'effect de taille du Champ : Mersier Cl., " Revue De Physique Applique ", 14(1), 1, (1979).

Author describes a method to determine heliostat fields allowing optimal energetic efficiency taking into account the receivers. The method is used to compute the local density and the shape of the fields. It is shown that when the receivers have negligible thermal losses, the two main characteristics of the fields are shadowing results and the cosine effect. In the opposite case i.e. important thermal losses from the receivers, the resulting field is very dependent on these losses. In the two cases, the technology of the heliostat gives some limitation in the theoretical optical configuration.

0678

DOE 5 MW Solar thermal Test Facility : Matthews L, " Proceedings Seminar on Testing of Solar Energy Materials and Systems ", Gaithersburg (USA), p. 45, (1978).

The solar thermal test facility in Albuquerque, New Mexico, is operated for the Department of Energy Sandia Laboratories. Initial Testing at the Solar Thermal Test Facility will involve components scheduled to be used for solar central receiver power plants. The heat rejection system, heliostat array subsystem, and data acquisition system are described.

0679

Optical Systems for Large Scale Solar Power Plants : McFee R H, " Paper at Electro Optics International Laser Conference ", California (USA), (1975).

This paper presents an approach to the solution of some of the optical problems associated with a central tower receiver system. The optical problems are related to the behaviour of the array of mirrors as a complex optical system. The total radiant power incident on the receiver as well as the power density distribution on the absorbing surface of the receiver is determined. The effect of mirror surface irregularity, shading and blocking by adjacent mirrors, sun position and tracking error on the power collection efficiency is also studied.

0680

Power Collection Reduction by Mirror Surface Non - flatness and Tracking Error for a Central Receiver Solar Power System : McFee R H, " Applied Optics ", 14, 1493, (1975).

The effect of random waviness, curvature and tracking error of plane mirror heliostats in a rectangular array around a central receiver solar power system are determined by subdividing each mirror into 484 elements, assuming the slope of each element to be representative of the surface slope average at its location, and summing the contributions of all elements and then of all mirrors in array. Total received power and flux density distribution are computed for a given sun location and a set of array parameter values. Effects of shading and blocking by adjacent mirrors are included in the calculation. Altazimuth mounting of the heliostats is assumed. Representative curves for two receiver diameters and two sun locations indicate a power loss of 20% for random waviness, curvature and tracking error of $0.1°$ rms, $0.002\ m^{-1}$ and $0.5°$, 3σ, respectively, for an 18.2 m diameter receiver and 0.3 rms, $0.005\ m^{-1}$ and greater than $1°$, respectively for a 30.4 m diameter receiver.

0681

Solar Tower - Thermal Collection Energy Component - 10 MWe Pilot Plant : Meyers A C and Hildebrandt A F, " Proceedings ISES (American Section) Conference ", Orlando (USA), (1977).

Net energy analysis concepts are developed and applied to the thermal collection segment of the 10 MWe solar tower central receiver pilot plant to determine the amount of energy this sub-system component represents with respect to construction and operation of the facility.

0682

Central Receiver Test Facility, Albuquerque, New Mexico (Review Paper) : Moeller C E, Brumleve T D, Grosskrentz C and Seamons L O, " Solar Energy ", 25, 291, (1980).

The basic concept philosophy is outlined for the Central Receiver Test Facility (CRTF) and defines the capabilities of CRTF. A general description of the facility is given with details of all support systems, the tower, the heliostat array, the control building with its computer control and data acquisition systems, and the metereological station and tower. Maximum energy delivery is 65 MW(th) so that 5 MW can be provided under a reasonable range of conditions. Operating tests with a working receiver absorbing over 1 MW of thermal energy are summarised. Present and future tests are outlined with a listing of proposed high temperature experiments by university and experimental investigators. Tests completed include evaluation of 1 MW air receiver and a 3 MW steam generating receiver.

0683

Development of the Solar Power Central Receiver Concept : Murphy L M and Skinrood A C, " Sandia Laboratories Report SAND - 76 - 8677 ", (USA), (1976).

The main incentive for the development of the solar central receiver concept is to examine the collection and efficient use of solar energy to produce electricity : the particular concept being developed to accomplish this goal by various contractors are described and Sandia Livermore's role as technical manager discussed. In addition, the ERDA plan for eventual commercialization of this concept as well as the role and use of the Sandia solar thermal test facility is described. Furthermore details of particular subsystem components as well as potentially fruitful research area for Sandia and other development laboratories to pursue, which can lead to performance improvement, are discussed.

0684

Structural Design of a Superheater for a Central Solar Receiver : Narayanan T V, Rao M S M & Gupta G D, " Journal of Pressure Vessel Technology ", 101, (1979).

One of the critical components of a receiver in a central solar thermal power system is the superheater. A solar superheater is subjected to non - axisymmetric radiant heating and is usually required to operate at elevated temperatures. Unlike those in conventional and nuclear power systems, solar super heaters are subjected to thermal and stress cycles associated with start - ups and shut - downs due to diurnal solar variations and cloud covers. Because of these features solar superheaters warrant special treatment with regard to structural analysis and design. The paper describes the structural analysis and design considerations and proposes a set of evaluation criteria for a superheater in a central solar receiver.

0685

5 MW Solar Thermal Test Facility Heliostat Focus and Alignment System : Oldham L P, " Proceedings SPIE Conference ", 114, (1977).

A unique system has been developed to focus and align the nearly 8000 mirrors for the nation's first Central Receiver Solar Thermal Test Facility, located at Albuquerque, New Mexico. The computer controlled system utilises a laser light source which is scanned over a 61 in diameter collimating mirror to simulate a collimated solar beam. The collimator and laser source are mounted on a microprocessor controlled precision pointing mount with 0.1 milliradian pointing accuracy. Control of the system is achieved through a vehicle mounted terminal plugged directly into the control electronics of the heliostat undergoing alignment. A simulated solar beam is reflected from the heliostat mirror and imaged on to retro - reflective target screen. This work gives an overview of the solar thermal test facility and a detailed description of the focus and alignment system.

0686

Preliminary Design of the Solar Total Energy Large Scale Experiment at Shenandoah, Georgia : Poche A J, Hass S A and Hunke R W, " Proceedings 13th IECE Conference ", San Diego (USA), 2, 1508, (1978).

The paper discusses a preliminary design study for the Solar Total Energy Systems (STES) experiment. The STES will supply electric power, process steam, and hot water for heating and (absorption) a i r conditioning demands of a 42,000 sft building operated by a knitwear manufacturer. In addition to the preliminary design of the STES, the study included experiment site layout, definition of s i t e specific system requirements and development of preliminary plans for the operating phase.

0687

Comparison of Heat Exchanger Designs for Sodium Cooled Solar Control Receivers : Pomeroy B D & B o n d J A, " ASME Paper No 80 –C2 / Sol – 12 ", (1980).

The major constraints on the design of a sodium cooled central receiver are not associated with t h e sodium side heat transfer, but rather with ; thermal losses to the ambient ; sodium flow distribution and control ; mechanical design and manufacturability and thermal stresses and creep / fatigue failure. In this paper two receiver heat exchangers are analysed and evaluated to select the more appropriate concept for receivers in the 400 MW(thermal) size range.

0688

Testing of Central Receiver for Central Thermal Power Systems : Puckert R P, Tobin R D & Friefeld J M, " Proceedings of Seminar on Testing Solar Energy Materials and Systems ", Gaithersburg(USA),p.39,(1978).

Central thermal power systems which utilize a large heliostat field focussed on a tower mounted solar receiver steam generator, have been extensively investigated. These investigations have included sub- scale testing of key components such as the central receiver. The purpose of this paper is to summa- rize the test approach and results for one such central receiver concept which was subsequently selec- ted by DOE as the baseline for the 10 MW pilot plant near Barstow, California.

0689

Flexed Beam Foundations for Central Receiver Heliostats : Raser W H, " Proceedings ISES (American Sec- tion) Conference ", Denver, Colorado (USA), 2.1, 1017, (1978).

Horizontally flexed beams in the form of large leaf springs are considered as a means for both support- ing and positioning reflectors in central receiver systems. Their use promises to reduce significant- ly both, the quantity of structural material required and the number of azimuth tracking drive motors needed to control a given number of heliostats. Aiming accuracy is discussed in connection with wind velocity.

0690

Theory of Concentrators of Solar Energy on a Central Receiver for Electric Power Generation : Riaz M, " ASME Paper No. 75 – WA / Sol – 1 ", (1975).

The modelling of the performance of large area solar concentrators for central receiver power plants is formulated using a continuum field representation of ideal heliostat arrays that accounts f o r two governing factors ; the law of reflection of light rays, imposes steering constraints on mirror orien- tation ; the proximity of mirrors creates shadow effects by blocking the incident and/or r e f l e c t e d solar radiation. The results of a steering analysis which develops the space time characteristics of heliostats and of a shadow analysis which determines the local effectiveness of mirrors in reflecting solar energy to a central point are combined to obtain in closed analytical form, the global charac - teristics of circular concentrators.

0691

Solar Flux Density Distributions on Central Tower Receivers : Riaz M & Gurr T, "Solar Energy", 19, 185, (1977). (1977).
An analytical formulation of the solar flux density distributions produced on the surface of a central tower receiver by large mirror fields is developed which accounts for dispersion, shading and screen- ing effects of mirrors and for degradation of insolation levels. In the case of symmetrical geomet- ries involving circular mirror fields and vertical cylindrical receivers, a general method of calcul- tion yields closed form solutions for the concentration ratios in terms of normalized parameters des- cribing the mirror field configuration, the receiver dimensions, the insolation level, the m i r r o r chracteristics and the time of the day. Aiming strategies of mirror focussing are devised to reshape the solar flux in accordance with desired distributions. Mirror field asymmetries created b y t h e configuration itself or by operational conditions blanking a portion of the field due, for instance , to maintenance or cloud cover are shown to set up flux gradients around the receiver which c a n be computed using a flux superposition technique. The methodology elaborated in the case of the verti- cal cylindrical receiver for simplicity and insight of treatment is applicable to many other geomet - ries presently envisioned in receiver studies for solar power tower systems.

0692

Heliostat Array Computer Simulation Model : Rubeck M, " Proceedings ISES Conference ", Los Angeles (USA), (1975).

A Heliostat Array Computer Simulation model has been developed in order to evaluate the performance of a heliostat field used in a central tower receiver system. This model will take the input description of t h e field , including the heliostats and the collector and calculate several performance parameters. These re- sults can be combined with cost data to determine solar flux patterns at the cavity aperture, which are needed in the thermal analysis of the collector interior. This paper describes the mathematical analyses, as well as the computer algorithms, which form the model.

0693

Theoretical Considerations on Solar Concentrations by Central Receiver Systems : Sakurai T, " Proceed- ings ISES Silver Jubilee Conference & Congress ", Georgia (USA), (1979).

This paper presents the theoretical concentration of solar radiation by a number of heliostats distributed in a circular field in radius onto a focal point at height 'f' from the center of the field. In the calculation it is assumed that plane heliostat mirrors have suffciently small dimensions given by dividing the field into differential rings and subdividing them radially into segments to cover the whole field without clearance.

0694

10 MW Solar Central Receiver Pilot Plant : Schweinberg R N and Rasband J L, "Proceedings IEEE Power Engineering Society Summer Meeting Prepr.", Los Angeles (USA), (1978).

Through a cooperative effort of government and private industry, a 10 MW solar thermal pilot plant is to be constructed in the 1980 - 81 time frame. The pilot plant will utilize the central receiver concept of generating electricity from incident solar energy. Sunlight is reflected from tracking mirrors to concentrate on a receiver boiler that produces steam. The plant design for the collector field calls for approximately 2000 mirror modules called heliostats. The collector field will surround the receiver tower with the lower set off center in the southern position of the field. The field is an elliptical arrangement about 2600 ft at the major axis and 2300 ft at the minor axis. The total land area for the pilot plant has been set at 130 acres. The receiver boiler will be a once - through design with rated steam conditions of 960 deg F and 1515 psia. A thermal storage system will be provided to extend the plant's usefulness at night.

0695

Solar One Project - A 10 MW Solar Thermal Central Receiver Pilot Plant : Schweinberg R N & Reeves J N, " Proceedings 14th IECE Conference ", Boston (USA), p.181, (1979).

Through a cooperative effort of government and the private industry, a 10 MW solar thermal pilot plant is to be constructed in the 1980 - 81 time frame using central receiver technology. The pilot plant will be tested for five years and information derived will be used for future commercial solar generating stations.

0696

Solar Thermal Central Receiver Systems : Sibers D L, Abrams M and Gallagher R J, " ASME Paper No 79-WA/ HT - 38 ", (1979).

Research needs in the thermal sciences relevant to the development of the solar central receiver concept are identified. The primary need is the creation of theoretical and empirical tools to predict energy loss from cavity and external receivers due to combined natural and forced convection heat transfer. The possibility of using boundary layer suction to reduce the convective energy loss is explored. The technique is found to be advantageous if the energy in the withdrawn air is recovered. In an unrelated problem area, a test to quantify boiling induced thermal fatigue in a solar receiver is described.

0697

Issues and Methodology for the Selection of a Conceptual Design for a Solar Central Receiver Pilot Plant : Skinrood A C, " Sandia Laboratories Publication ", (USA).

This paper describes the selection process used by Sandia Laboratories to recommend a conceptual design for the solar central receiver pilot plant to be built at Barstow, California by the US Department of Energy. Included are the key selection issues and their significance.

0698

Central Collector Solar Energy Receivers : Sobin A, Wagner W and Easton C R," Solar Energy",18, 21, (1976).

This paper discusses the application of the compact steam generator technology to the design and fabrication of central receivers for solar energy powered electric power plants. Receiver designs are discussed for tower mounted applications where size and weight are important. The heat flux rates necessary for central solar receivers are nearly identical to the design heat fluxes for the compact steam generator. Fabrication of the central receiver is discussed as well as design details and applicable materials.

0699

European Community 1 MW(el) Solar Power Station : Stephens W H, " Solar Thermal Power Generator Conference (ISES, UK Section), (C 16) ", London (UK), p.45, (1978).

A 1 MW capacity central receiver power plant is to be built in Sicily by an industrial consortium. This paper summarises the main technical parameters of the plant, outlines the research program and presents a preliminary conceptual design with specifications.

0700

The European 1 MWe Solar Tower Plant : Straub A and Toth G, " Proceedings ISES Silver Jubilee International Congress ", Georgia (USA), (1979).

The European community decided two years ago to launch an ambitious program : designing and building a solar tower plant by a multinational industrial consortium. The contract was signed on Nov.15, 1977. It foresaw two phases : detailed studies to end by Nov. 15, 1978 and erection of the plant to end by Nov. 15, 1980.

0701

Solar Central Receiver Program : Tallerico L N and Skinrood A C, " Proceedings 14th IECE Conference ", Boston (USA), p.176, (1979).

A number of solar central receiver programs are being conducted with the goal of developing a system capable of producing low cost electricity. The programs are concentrating on the heliostat, receiver and storage subsystems. The emphasis on the heliostat programs is to develop designs which are amenable to mass production and minimum maintenance. Both glass and plastic designs are being evaluated. The receiver studies have addressed the use of water (steam), salt (sodium), air and helium as the

092

heat transport fluid in the receiver. Air, rock, salt and sodium storage subsystems are being investigated. The water / steam, sodium and salt receivers are coupled to a Rankine cycle turbine and the
air and helium receivers to a Brayton cycle turbine. Extensive parametric studies were conducted to
obtain the information required to evaluate the systems. An overview of these systems, the scope of
parametric studies, the preliminary conclusions and the future direction of the program are presented.

0702

Identification of Alignment and Tracking Errors in the Open Loop, Time Base Heliostat System of the
400 KW Advanced Components Test Facility : Teague H L, Swafford C J, Brown C T and Lewis G C, "Proceedings ISES Silver Jubilee International Congress ", Georgia (USA), (1979).

This paper describes the original tracking system hardware, installation procedures & alignment methods for the heliostats of the Advanced Components Test Facility (ACTF) operated by the Georgia Institute of Technology. The sources of tracking errors are discussed and modifications necessary to minimise their effects are described.

0703

Tower Type Solar Power Plant : Configuration and Thermal Regime Stability of Receivers & Steam Generators : Teplyakov D I and Aparisi R R, " Applied Solar Energy ", 13 (2), 31, (1979).

A graph analytic method is presented together with results obtained in calculating the irradiances at
the generatics of cylindrical receiver steam generators formed by flat fields of reflectors in tower
type solar power plants of various configurations (axisymmetric and eccentric). The ' thermal bias'
appearing at the ray collecting surfaces of the receivers in the course of the working day is demonstrated; it results from changes in geometry and in the amount of direct solar radiation.

0704

Real – Time Computer Control of 5 Megawatts of Solar Thermal Energy : Thalhammer E D, " ISA (Trans.)",
18 (4), 3, (1979).

The Central Receiver Test Facility (CRTF), operates under the control of a 9 – machine distributed mi-
nicomputer network. The prime functions of this network are heliostat controls, heat rejection system
controls, and data acquisition. This paper describes the control computer. This computer's main tasks
are (i) the sun position calculation, (ii) automatic heliostat command execution, (iii) graphic display
of heliostat status and selected items of tower and boiler control data, (iv) operator control of the
heliostat command cycle, (v) heliostat alarm handling and (vi) the permanent recording of all test parameters that fully describe the experiment performed. The control system is capable of directing 5 MW
of thermal energy at any point within 327 m of the solar receiver tower. Also the procedures & check-
ing performed by the computers to ensure personnel and facility safety are described here.

0705

Measurement of Heliostat Performance Characteristics : Thornton J P, " Journal of Solar Energy Engineering (Trans. ASME) ", 102, 22, (1980).

A key in the development of any solar collector is described as the comparison of actual performance
data with predicted data. During a series of tests at the Martin Marietta Solar Test Facility during
1976 and 1977, the performance of four full size, prototype heliostats with different types and sizes
of glass mirrors were measured and subsequently shown to compare closely with the predicted performances.

0706

Experiments with Heliostats for Central Receiver Power Plants : Thornton J P and Waddington D," Journal
of Energy ", 3(2), 69, (1979).

Two new applications of standard techniques are providing more accurate measurements of heliostat
performance over longer ranges. Four experimental heliostats have been fabricated and installed in the
Solar Test Facility at Martin Marietta, Denver. Performance testing of these heliostats has been completed, and the resulting data are being used to support the design of commercial central receiver solar power plants. Both the design of the experimental heliostats and the results of the performance
testing are described.

0707

Low Cost Central Receiver Solar Power Plant Using Molten Salt as a Heat Transfer and Storage Medium :
Tracey J R, " Proceedings 6th Energy Technology Conference ", Washington (USA), (1979).

The primary objective of the proposed alternative approach to solar thermal power plants is to signif -
icantly reduce the cost of electricity relating to the first generation water steam systems. By using
a molten salt which is a eutectic mixture of 60% sodium nitrate and 40% potassium nitrate as the heat
transfer and storage media. It is believed that substantial cost reductions can be achieved. The single phase molten salt temperature is increased typically from 550 deg F to 1050 deg F in the receiver.
A series of system level trade studies were conducted including plant size, amount of storage, collector field configuration, and receiver type in order to define the most cost effective system. The performance improvement of the alternative system over the first generation system operating from the
tower is 25% . The performance improvement while operating from storage is about 50%. The cost of
storage in the alternate system is significantly reduced from the first generation system. Thermal
storage costs below $ 5.00/KW appear achievable. Studies show that electricity cost of less than 40
mills/Khe is a reasonable goal.

0708

1 MW Solar Cavity Steam Generator Solar Test Program : Tracey J R, Blake F A, Royere C and Brown C T,
" Proceedings 12th IECE Conference ", Washington (USA), 2, 1224, (1977).

The test article is a cavity type receiver with an aperture of 1.03 m (40.3 in) the inner walls consist
of blackened steel tubes designed to deliver steam at 8619 KPa (1250 psig) and 510 deg C (950 deg F). The
solar testing consisted of a total of 161 hours of which 71.5 hours were at rated operating conditions
8619 KPa (1250 psig) 510 deg C (950 deg F) 105 hours were at full boiler pressure. The maximum power in-

put was 738 KW. The measured thermal efficiency of the cavity was 92%. During the program the receiver was cycled to rated conditions 26 times and the boiler was cycled to rated conditions 30 times. In addition, many other transients were encountered including cloud interruptions.

0709

Solar Thermal Power Systems Based on Optical Transmission (A Feasibility Study) : Vant Hull L L and Easton C R, " McDonnell Douglas Astronautics Company Report (NP - 20493) ", California (USA), (1974).

The objective of this project is to perform a preliminary technical and economic feasibility study of a solar thermal power system based upon optical transmission of collected solar energy to a central absorber and boiler unit. The study includes system definition, modelling and primary component design and testing. Tasks to be completed include studies of (i) the solar flux concentration system including system geometric analysis of the optical characteristics, guidance and control system studies and reflector designs,(ii) the receiver and energy transfer system including conceptual designs of the tower, (iii) system definition and evaluation with particular attention to economic analysis of the selection, energy transfer, energy conversion system and (iv) a scale model heliostat & collector with hydraulic steering.

0710

Educated Ray Trace Approach to Solar Tower Optics : Vant Hull L L , " Proceedings SPIE Conference ", 85, 111, (1976).

The author describes an approach to the analysis of the optical system of the solar tower concept, which is designed to provide maximum design information. By inputting all quantities in terms of angles or dimensionless ratio, the results remain relatively scale independent. First the power redirected from the heliostat field to the central receiver is computed. Then solar images are projected from each heliostat to the central receiver and interception factors are obtained. This information permits sizing the central receiver, to enhance performance by redistributing heliostats in the field and finally to define the rim angle of the field.

0711

Development of Solar Tower Program in the United States : Vant Hull L L," Optical Engineering ", 16 , 575, (1977).

The author has traced the history and development of the solar tower from an idea in 1969 through first Federal funding in 1973, to a program to initiate pilot plant construction in 1976. Outlines of some of the future programs are also given.

0712

Liquid Sodium Cooled Solar Tower System : Vant Hull L L, " Proceedings ISES (American Section) Conference ", Orlando (USA), (1977).

A low pressure liquid sodium cooled central receiver system is described. The high thermal conductivity of liquid metals leads to a more compact and efficient receiver while the generation of turbine steam from hot stored sodium decouples the turbine from the receiver and leads a very efficient trouble free operation. Safety and economics of the sodium system are comparable to a water steam system with hot oil and rock thermal storage.

0713

Central Receiver System and Heliostat Field Optimization : Vant Hull L L , " Proceedings DOE Workshop on Systems Studies for Central Solar Thermal Electricity ", Houston (USA), (1978).

Extensive codes have been developed for the optimization of heliostat fields for solar central receiver power systems. Before they can be implemented, a conceptual design of the entire system must be generated so that the component costs and losses can be developed for use in the economic optimization of the heliostat field. In addition, turbine efficiency (and hence the receiver temperature) plays an important role in establishing the thermal power required. After a discussion of these considerations the operation of optimization codes is summarised.

0714

Liquid Metal Cooled Solar Central Receiver Feasibility Study and Heliostat Field Analysis : Vant Hull L L, Coleman G, Springer T and Friefeld J, " University of Houston Report ORO 5178 - 78 - 1, Final Report Part - 1 ", (USA), (1977).

A conceptual design for a 100 MWe solar tower system employing liquid sodium as a heat transfer fluid and as a storage medium is generated. This design intentionally parallels the current commercial base line design for a water / steam transfer fluid in order to enhance comparisons between the two approaches and to maximise the the application of the limited funds in this study to the unique features of the liquid sodium system. Heat transfer at the receiver and the steam generator were considered in detail. Components for the liquid sodium loop currently under test on development were identified for a 10 MWe pilot plant and a 10 MWe demonstration plant. Economic comparisons with the A D A C water / steam receiver are favourable.

0715

Analysis of the Kinematics of Heliostats : Vartanyan A V, Shermazanyan Ya. T and Arutyunyan V V, " Applied Solar Energy ", 10(3), 32, (1974).

Kinematic formulas are investigated for different mountings of the heliostats with two mutually perpendicular axes and arbitrary orientation of the principal axes in space. The resulting analytic expressions for the angles of rotations of the heliostat about the axes, and the corresponding angular velocities, are applicable to vertical, horizontal and inclined optic axes.

0716

HELIOS ; A Computer Program for Modelling the Solar Thermal Test Facility : Vittitoe C N, Biggs F and Lighthill R E, " Sandia Laboratories Report SAND - 76 - 0346 ", (USA), (1977).

Helios is a flexible computer code for evaluations of proposed designs for central tower solar energy collector systems, for safety calculations on the threat to personnel and to the facility itself for determination of how various input parameters alter the power collected and for design trade-offs. Input variables include atmospheric transmission effects, reflector shape and surface errors, sun tracking errors, focussing and alignment strategies, receiver designs, placement positions of the tower and mirrors, time of the day and the day of the year for calculation. Plotting and editing computer codes are available. Complete input instructions, code structure details and output explanations are given. The code is in use on CDC 6600 and CDC 7600 computers.

0717

Comparison Between Results of the HELIOS and MIRVAL Computer Codes Applied to Central Receiver Solar Energy Collection : Vittitoe C N, Biggs F and Leary P L, " Sandia Laboratories Report SAND - 79 - 8266 ", (USA), (1979).

The Sandia Computer Codes HELIOS & MIRVAL were developed to predict the optical performance of reflecting solar concentrators and to model power collection by central - receiver solar energy power plants. HELIOS is an analytic code, whereas MIRVAL uses Monte Carlo ray tracing techniques. The objective of this study was to verify that HELIOS & MIRVAL give the same performance predictions. The sample problem for comparison consists of a rectangular target and altazimuth heliostats deployed in a north field. The results indicate that HELIOS & MIRVAL closely agree on predictions of field performance and of power density on the target plane.

0718

HELIOS and Reconcentrators : Vittitoe C N, Biggs F, Matthews L K and Seamons L O, "Proceedings ISES International Silver Jubilee Congress ", Georgia (USA), (1979).

The HELIOS computer code for modelling solar energy collection has been evolving over the last 2 years. The code was originally developed for solar power tower applications. It has been extended for large parabolic dish systems having a variety of receivers. Further extensions are now being developed to include reconcentrators that may be used to boost collected energy for certain applications. One type of reconcentrator is mounted close to the receiver to redirect spillage back to the receiver. Such measures are necessary for Central Receiver Test Facility (CRTF) tests of receivers with small transverse dimensions that require enhanced power densities. Another extension is being developed to allow evaluation of heliostats of CRTF. Data are also becoming available for more detailed comparisons with HELIOS predictions. Such predictions are compared with measurements taken by the Real Time Aperture Flux System (RTAF) and by a cross configuration of gauges deployed for a set of 23 heliostats at the CRTF. Comparisons are also made with data from a single facet experiment. The single facet was chosen partly because of distortion that was apparent visually after the facet was adjusted to be approximately flat. Calculations have been performed for two aperture and three aperture reconcentrators attached to the CRTF at the edges of a vertical 0.89 X 12.5 m flat rectangular receiver facing north & mounted on top of the cover. The method used to model secondary reconcentrators are described along with several sample calculations.

0719

Georgia Tech's 400 KW(th) Solar Thermal Test Facility : Walton Jr. J D, Bomar Jr. S H and Poulos N E, "Proceedings ISES Conference - Sharing the Sun", Winnipeg (Canada), 10, 245, (1976).

Under ERDA sponsorship, Georgia Tech is constructing a 400 KW(th) solar test facility for high temperature solar energy research and development. Designed by Prof. G Francia, it is an enlarged version of a central receiver facility which he developed near Genoa, Italy. The Georgia test facility utilizes 550 round mirrors 111 cm in diameter which may be operated flat or focussed to provide radiant heat fluxes from 25 to 200 W /sqcm to a test area centrally located above the mirror field. It is used first to heat a boiler - superheater to deliver 365 Kg/hr of steam at 150 atm and 600 deg C. Other uses include evaluation of experimental receivers utilizing such heat transfer fluids as steam, air, helium, oil, molten salts and liquid metals and basic research in the area of metals, ceramics and coatings.

0720

Preliminary Results from the Georgia Tech. 400 KW(th) Solar Thermal Test Facility : Walton Jr. J D, Brown C T, Bomar Jr. S H and Poulos N E,"Proceedings ISES Conference", New Delhi (India), p.1706, (1978).

This paper describes the initial operation and evaluation of Georgia Tech's Solar Thermal Test Facility. Operating characteristics of the facility are discussed with respect to (i) performance of the heliostat system and (ii) performance of the receiver, boiler, super heater including steam temperatures & pressures provided by the thermal cycle set.

0721

A Solar Flux Density Calculation for a Solar Tower Concentrator Using a Two Dimensional Hermite Function Expansion : Walzel M D, Lipps F W and Vant Hull L L, " Solar Energy ", 19, 239, (1977).

The calculation of flux density on the central receiver due to a large number of flat polygonal reflectors having various orientations is a basic part of the system simulation program for the tower concept for solar energy collection. A two dimensional hermite function expansion is adapted to the simulation problem and numerical results are contrasted with an analytic integration of the solar flux density at specific nodes on an image plane. Various measures of error in the flux density calculation are monitored versus distance to the image plane and orientation of the reflector. The flux densities predicted by the statistical method compare favourably with those of the analytic model and required approximately one tenth the computer time.

0722

Comparison of Central Receiver Systems Optimized for Advanced Receivers : Walzel M D and Vant Hull L L, " Proceedings ISES (American Section) Conference ", Denver, Colorado (USA), 2.1, 799, (1978).

A conceptual design analysis has been carried out for an advanced commercial 100 MWe control receiver system using FORTRAN code which optimizes a particular system by seeking the lowest system cost per unit of annual thermal energy produced. Cost models for the heliostats, land, wiring, pipe network, tower and foundation and fixed costs were included. With optical and thermal losses models for the collector field and receiver in the production of the optimized collector field and system figure of merit. The scope of the analysis included external cylindrical reveivers with surrounding collector fields, cavity receivers with north biased fields and single and triple tower configurations for each.

0723

An Investigation of Optimum Heliostat Spacings for the Sub Tower Region of a Solar Power Plant : Walzel M D, " Proceedings ISES Silver Jubilee International Congress ", Georgia (USA), (1979).

The RECELL system of computer programs developed at the University of Houston can give information concerning the optimal placement of heliostats for a solar tower. The information is taken from a cellular structure of the mirror field which has a representative heliostat with neighbours at the center of each cell. The optimization is usually done on an annual basis by a simulation of the optical behaviour of the system cost, and loss models are also employed to determine the optimal deployment of glass of given fixed sizes for the heliostats, receiver and tower constraints such as design point power requirements and peak flux limits can be satisfied by examining the extensive annual performance summaries and determining the changes needed in certain system parameters. The figure of merit (the system's initial capital cost divided by the annual thermal energy produced) can be monitored during a sequence of outputs to determine the cost / benefit difference for variations of system parameters that are within a range capable of satisfying the design constraints.

0724

General Formula for the Incidence Factor of a Solar Heliostat Receiver System : Wei Li Y, " Applied Optics ", 19, 3196, (1980).

A general formula is derived for the effective incidence factor of an array of heliostat mirrors for solar power collection. The formula can be greatly simplified for arrays of high symmetry and offers quick computation of the performance of the array. It shows clearly how the mirror distribution and locations affect the overall performance and thus provide a useful guidance for the design of a solar heliostat receiver system.

0725

Design of a 10 KW Photovoltaic 200/1 Concentrator : Wilkening H A, " Proceedings 13th IEEE Photovoltaic Spec. Conference ", Washington (USA), p. 669, (1978).

An optical solar concentrator coupled with a compatible array of Solarex Corporation 200 X photovoltaic cells uses a fluid loop heat transfer system for cooling the cells. The concentrator consists of a mounting frame, reflectors mounted in a pivoting gimbals and a receiver / photovoltaic cell assembly. The tracking system consists of mechanical drives and linkages, and a control system which provides for tracking the elevation and azimuth motion of the sun. An emergency fail safe secondary reflector system prevents damage to the photovoltaic receiver array in any type of malfunction.

0726

Dynamic Simulation of a Sodium Cooled Advanced Central Receiver Solar Electric Power Plant : Willcox W and Beckman J, " Proceedings 14th IECE Conference ", Boston (USA), p.118, (1979).

A numerical dynamic simulation model of the thermal hydraulic characteristics of a sodium cooled advanced central receiver solar electric power plant has been developed. The model was used to determine the optimum receiver temperature control system configuration. Independent feed back control of each of the 24 receiver panels was found to be an adequate method of receiver outlet coolant temperature control. However, integral and derivative control modes, as well as proportional, are required for satisfactory response speed and acceptable temperature off set. The use of simple independent feed back loops for control does not cause any unusual problems even though the panels are hydraulically coupled by common feed and return lines.

0727

Analysis of Convective Heat Loss from the Receiver of Solar Power Plants : Yao L S & Chen F M, "A S M E Paper No. 79 - WA / HT - 36 ", (1979).

The receiver of solar power plants is modeled as a semi infinite vertical cylinder. The analysis is carried out to investigate the convective heat loss. Constant surface temperature is selected as the idealized condition for the receiver. Heat loss from the receiver is due to the combined mode of the free convection and the forced convection. The problem is studied under the condition that a steady horizontal breeze passes the receiver. The forced convection is treated as a perturbed effect. The solution shows that the free convection dominates the heat loss along the bottom of the cylinder.

0728

Calculation Principles for the Heliostat Field of Solar Power Plants : Zakhidov R A, " Applied Solar Energy ", 15 (3), 20, (1979).

The article offers a brief analysis of the current state of theoretical research into the concentrating systems of tower type plants. A need for developing a general design model is demonstrated. The basic framework of a theoretical model are presented, and formulas are derived for the irradiance on an arbitrary radiant energy receiver from heliostats of arbitrary shape.

0729

Design and Testing of a Cavity Type Steam Generating Central Receiver for a Solar Thermal Power Plant: Zoschak R J, Wu S F and Gorman D N, " Journal of Engineering for Power (Trans. ASME)", 102, 486, (1980).

This paper focusses on the design and operating aspects of a 10 MWe cavity type natural circulation, steam generating receiver for a central receiver thermal power plant. The development of the receiver concept and the basic design features are described. The solar energy input analysis, thermal/hydraulic performance, and structural design of the receiver are discussed alongwith its control concept & transient operation. The design, construction and testing of a 5 MW(th) scaled down version of the 10 MWe receiver are summarised with emphasis on test objectives, scaling criteria and design similarities to the full scale receiver.

0730

Mid Temperature Solar Systems Test Facility Experiment Manual and Test Plan :" Sandia Laboratories Report SAND – 79 – 0379 ", (USA), (1979).

This manual describes the Mid Temperature Solar Systems Test Facility (MSSTF), outlines the procedures for testing items to be evaluated there and defines the prerequisites for installing such items, describes typical tests performed and lists the reports generated and disseminated from the facility.

0731

Summary of Results of Solar Power Arrays for the Concentration of Energy Study : " Sheldahl Inc. Report ", (USA), October 24, (1974).

Topics cover heliostat array design and performance alongwith evaluation of reflective membrane materials having potential application in the fabrication of heliostat mirrors. Design and test requirements for a representative section of a solar thermal energy receiver designed for use in a 10 MWe solar thermal energy, proof of concept, system are discussed.

Two – Stage Solar Concentrators

0732

Two Mirror Multifacet Solar Concentrator for Agricultural Use : Alavutdinov D N, Umarov G Ya, Sadykov M S, Enileev H H and Saifullina Ya, " Applied Solar Energy ", 4(6), 99, (1968).

An original design of a two mirrors facet concentrator is considered. It is supposed that the new concentrator may be used for irradiation of different seeds before sowing, with the help of impulse concentrated solar light. Use of this new concentrator model makes it possible to place the object to be irradiated at any required height over the ground.

0733

Two Mirror Unit with Plane and Hyperboloidal Counter Reflectors : Alimov A K and Alavutdinov D N," Applied Solar Energy ", 15 (3), 42, (1979).

A brief description is given of a direct tracking unit with a principal paraboloidal diameter of 20 m and a power of 150 Kw, employing plane or hyperboloidal counter reflectors mounted in the prefocal region of the concentration zone.

0734

Calculation of the Optical Characteristics of High Power Two Mirror Solar Furnaces : Azimov S A, Mallaeva Kh. M, Pinnatov I I, Riskiev T T and Suleimanov S Kh., " Applied Solar Energy ", 15(2), 23, (1979).

A calculation scheme for large solar furnaces is considered. The results can be used in designing large solar furnaces and solar electric power plants of tower type.

0735

Parabolotoric Focone as Secondary Solar Energy Concentrators : Baranov V K, "Applied Solar Energy", 13 (3), (1977).

The optical system for solar energy concentration is studied. It consists of paraboloidal mirror and a parabolotoric focone (focussing cone). It is shown that introduction of focone into the system makes it possible to increase the irradiance into the focal spot by tens of times in those cases where the mirror included angle is less than 20 to 25°. For included angle of 25 to 65° the illumination in intensity increases to a lesser degree. The basic relations connecting the focone and focline (focussing wedge) parameters with those of the primary concentrators – the mirror and the lens system, are given.

0736

Optical Analysis of Cassegrainian Concentrator Systems : Bars Jr.A H, Schrenk G L, Poon P T Y and Higgins S N, " Proceedings ISES Silver Jubilee International Congress ", Georgia (USA), (1979).

Cone optics procedures have been developed to treat optical systems employing multiple reflectors. The objective of the procedure is to develop efficient techniques to treat multiple reflections of cones when the secondary reflecting surface is curved ; in fact, curved differently in different directions. The particular optical system under consideration is the cassegrainian concentrator system, a system consisting of a paraboloidal primary reflector and a confocal hyperboloidal secondary reflector. Multiple reflection cone optics methodology as well as preliminary results for the flux distribution and intercept factors based on it are presented in this paper. It is shown that neglecting local curvature effects of the secondary reflector i.e. even in the differential solid angle within the sun's cone, may over estimate the peak of the flux distribution at the receiver aperture by a factor of 2 or more.

0737

Cone Concentrator with Point Focussing Secondary Reflector : Buzin E I, " Applied Solar Energy ", 4(2) 63, (1968).

The parallel flux concentration by means of reflection from two surfaces of rotation with a common axis is considered.

0738

Cylindric Concentrators with an Auxiliary Reflector : Buzin E I, " Applied Solar Energy ",4(3), 23,(1968).

The concentration of parallel ray flux by reflection from the surfaces is considered. The first essential reflecting surface is a cylindrical surface. To determine its shape, a nonlinear differential equation of the first order was compared. The shape of the supplementary reflector was determined by a nonlinear differential equation. An example is given in which the main reflector is a parabolic cylindrical surface and the supplementary reflector is an algebraic surface of rotation of the fourth order.

0739

Analysis of a Cassegrainian Solar Furnace : Cobble M H, Hull W C and Hays R A, " Proceedings S P I E Conference – Optics Applied to Solar Energy IV ", p.55, (1978).

The solar furnace consisting of a paraboloid of revolution which tracks the sun and reflects radiation to a hyperboloid of revolution having a common focus with the paraboloid is analysed to determine the concentration available, using various eccentricities. The hyperboloid in turn, reflects relation to a focal plane placed at various distances from the vertex of the paraboloid. The ideal concentration is determined using the largest radius of all the rays from the sun falling on the paraboloid, this radius being the distance from the pierce point of the ray, in the focal plane, to the hyperboloid focal point. The concentrations can be augmented using a compound paraboloidal concentrator, and the ideal augmented concentration is developed for various combinations of eccentricity and vertex distances. The effect of scattering angles for the paraboloid and hyperboloid o n t h e ideal concentration is shown separately and jointly.

0740

Overall Efficiency of a Cassegrainian Solar Collector Using a Thermoelectric Module to Generate Electric power : Cobble M H & Thacher E I," Proceedings 14th IECE Conference ", Boston (USA),p.1831, (1979).

A Cassegrainian solar collector is analysed to determine the concentration and augmented concentra – tion. Experiment results of the concentration are shown graphically and compared to the theoretical concentration for a specific design. The efficiency of a thermoelectric module is developed where there are convection and radiation losses to the surroundings at the hot plate. The concentration is also shown as a function of the hot plate temperature.

0741

Lens Mirror Combinations with Maximum Concentration : Collares Pereira M, Rabl A and Winston R, " Applied Optics ", 16, 2677, (1977).

By the addition of suitable reflectors the concentration of a lens can be increased to the thermodynamic limit, which is equivalent to an f–number of one half. Such lens mirror combinations are useful whenever concentration rather than image formation is important, for example, in radiation detectors and solar energy collectors. The design of lens mirror combinations with maximum concentration is described. To the approximation that the lens has sharp focal points at off axis incidence, t h e solution for the reflector is readily found to be compound hyperbolic. With proper choice of the f–number of the lens, the hyperbolic reflector reduces to a V – trough or cone, an arrangement which offers considerable advantages for fabrications. The 2 – D case (line focus lens) suffers from aberrations due to focal length variation with non planar incidence. The optical performance of a 2–D lens mirror combination at non planar incidence is analysed and evaluated for its suitability in solar energy applications. A prototype Fresnel lens plus V–trough has been built and test data are presented.

0742

High Temperature Solar Collector with Optimal Concentration ; Non Focussing Fresnel Lens with Secondary Concentrator : Collares Pereira M, " Solar Energy ", 23, 409, (1979).

A non evacuated collector consisting of a linear Fresnel lens and a second stage concentrator of the Compound Parabolic Concentrator (CPC) type is described and tested in detail. Use of a Fresnel lens accomplishes two different objectives simultaneously : it allows for the design of a nearly ideal light collector (of the CPC type) of high concentration and height to aperture ratio close to 1, and plays the role of a cover, making the collector less sensitive to the new environment than one with exposed reflector surface. The geometric concentration is 15.56 and the acceptance half angle is 3°.

0743

Non Evacuated High Temperature Linear Fresnel Lens and Second Stage CPC Type Collector w i t h Large Tolerance for Tracking Errors : Collares Pereira M, O'Gallagher J, Rabl A and Winston R, " Proceed – ings ISES Silver Jubilee International Congress ", Georgia (USA), (1979).

A non evacuated collector with a linear nonfocussing Fresnel lens and nonimaging maximally concen – trating second stage reflector designed for tracking about the polar axis has been built and tested. The geometric concentration ratio of this collector obtained is 16. The test results confirm t h e large acceptance angle of the design. By virtue of this large acceptance angle this collector c a n collect all of the circumsolar radiation in addition to the rays from the solar disk itself.

0744

Second Stage Concentration with Tapers for Fluorescent Solar Collectors : Goetzberger A & Schirmer O, " Applied Physics ", 19(1), 53, (1979).

Light concentration of fluorescent sheet collectors can be enhanced by providing the output e d g e s with a taper of a high refractive index with reflecting surfaces. Concentration ratios achievable by these means are computed for two shapes of tapers, one with a plane boundary towards the collector and one with an additional cylindric lens. The general limitation of concentration ratio is given by Louiville's theorem or Abbe's sine condition. Computed second stage concentration ratios are (i) for plane boundary taper and $n_1 = 1.5$, $n_2 = 2$ and three reflections in the tapers : C = 1.49 & (ii) for the lens taper combination under the same conditions : C = 1.76, nearly reaching the Louiville limit 1.79.

0745

Performance of A Two Stage Solar Concentrator : Kandpal T C, Mathur S S & Singh R N, " Applied Energy ", 7, 191, (1980).

The geometric concentration characteristics of a two stage system employing a composite parabolic trough as a primary and a compound parabolic concentrator as a secondary concentrator are discussed. The relative merits of such a two stage system over a single stage CPT are brought out from simple thermal performance considerations.

0746

A Fresnel - Winston Tandem Concentrator System : Kandpal T C, Mathur S S and Singh R N, " International Journal of Energy Research ", 5, 77, (1981).

In this paper, some geometrical concentration characteristics of a solar concentrator system employing a Fresnel reflector concentrator and a CPC in tandem have been discussed.

0747

Analysis of Performance of Double Reflector System for Collecting Solar Energy : Kaykaty G, " N A S A Report No. NASA - TN - D - 3534 ", (1966).

A double reflector system employing a paraboloid for a primary and either an ellipsoid or a hyperboloid for a secondary reflector is analyzed with regard to its performance to provide an appraisal of its applicability to a solar Brayton Cycle power system. The maximum efficiency expected under the most favourable conditions was predicted and compared with that of a collection system consisting of a single paraboloid with the receiver aperture at the focus.

0748

Energy Distribution in the Concentration Field of a Solar Installation with a Hyperboloidal Counter Reflector : Kirgizbaev D A and Zakhidov R A, " Applied Solar Energy ", 10(4), 71, (1974).

A computational method for a coaxial system consisting of a paraboloid of revolution and a hyperboloidal counter reflector is developed. An expression is obtained for the relation between precision measures of the reflecting surfaces and an estimate is made of the influence of the precision of the concentrator and counter reflector on the energy distribution in the system.

0749

Energy Distribution in the Concentration Field of a Two Mirror Device with a Paraboloidal Back Reflector : Kirgizbaev D A and Zakhidov R A, " Applied Solar Energy ", 10 (5), 16, (1974).

The effective disposition of the heating plane in the region of convergence of twice reflected rays of a two mirror device with a paraboloidal back reflector is studied and the energy distribution over the heating spot is determined. A comparison of the system with that consisting of a hyperboloidal back reflector is also made.

0750

Optical and Thermal Analysis of a Cassegrainian Solar Concentrator : Mauk C E, Prengle Jr.H W & Eddy C S, " Solar Energy ", 23, 157, (1979).

For the storage of solar energy in a chemical system, a Cassegrainian type solar collector is investigated. With generalized geometry, a nonskew ray from the edge of the solar disc is followed through the Cassegrainian configuration to the final image to define the image height. Parametric results are presented in a form which can be used to predict performance based on off - axis optics. Equations and graphs allow the incorporation of angular error into calculations. Slope of 5 for the limiting tangent to the hyperbolic secondary appears near optimum. A set of design equations is developed, and their use is demonstrated. Design parameters are calculated for chemical reactor loads of 1.0, 0.5 & 0.1 MW.

0751

Optical Analysis of Paraboloidal Concentrators with Secondary Reflectors : McDanal A J, " Proceedings ISES Silver Jubilee International Congress ", Georgia (USA), (1979).

While the paraboloidal concentrator / secondary reflector configuration has found widespread acceptance within the antenna industry, a generalized optical analysis is currently available for solar applications. The objective of this study is to present the results of a generalized solar concentrator for a family of secondary reflectors. Solar thermal electric power systems are currently under development which require actual flux concentration ratios in the order of 2000 or greater to achieve the high temperature required for efficient heat engine operation. As a point of comparison, optically perfect parabolic dish concentrators produce high concentration ratios (11500) but require the bulky receiver/ heat engine to be placed at the focal point of the parabola. The optical perfect paraboloidal concentrator with an optimized secondary reflector yields a concentration ratio of 9200 and has the distinct advantage of having the receiver / heat engine located at the vertex of the paraboloid.

0752

Secondary Concentration for Linear Focussing Systems ; A Novel Approach : McIntire W R, " Applied Optics ", 19, 3036, (1980).

A novel approach to secondary concentration for linear focussing system is described. It provides a receiver with the appropriately changing aperture size as viewed from different points on the primary concentrator. It also provides a sizeable gap to accomodate a convection suppressing cover tube between a round absorber and the secondary concentrator with no loss of radiation through the gap. The approach will be described as it applies to a 70° rim angle parabolic trough concentrator, but variations of the designs presented here are applicable to primary concentrators with other rim angles.

0753

Preliminary Testing of a Scale Model Secondary Concentrator for the Sandia Solar Thermal Test Facility: Mulholland G P and Matthews L K and " Proceedings Annual Technical Meeting of Institute of Environmental Sciences ", Washington (USA), p.179, (1979).

A qualitative test of a three dimensional Winston type concentrator at the White Sands solar furnace is discussed. The concentrator has an entrance aperture of 15.25 cm, an exit aperture of 7.62 cm, a length of 19.8 cm and a reflective surface of polished aluminium. Nine individually controlled channels have been cut into the wall of the concentrator for cooling water. Wall temperatures at eleven positions and flux densities at the exit aperture have been obtained.

0754

Cassegrainian System Modified for Solar Concentrators : Novikov V V, Barnov V K, Vasilyeva L V and Protasov N N , " Applied Solar Energy ", 2 (5), 10, (1966).

The parameters of a Cassegrainian mirror system to be modified and used as a solar energy concentrator were mathematically and graphically analysed to clarify the basic laws making correct parameter selection possible. The computations resulted in determining the size of the solar image in the focal spot of the system as a function of the parameters of the elements comprising the system. The system showed a very high degree of energy concentration.

0755

Enhancement of the CPT Concentration with a CPC : Singh R N, Mathur S S and Kandpal T C, " Proceedings Indian National Solar Energy Convention ", Bombay (India), p.94, (1979).

The geometrical concentration characteristics of a two stage system employing a composite parabolic trough (CPT) as a primary and a compound parabolic concentrator (CPC) have been studied. The advantages of using such a system at low insolation levels are discussed.

0756

Sequential Use of a Composite Focussing Parabola and a CPC : Singh R N, Mathur S S and Kandpal T C , " Optik ", 55, 87, (1980).

The sequential use of a composite focussing parabola and a CPC is described. The utility of the sequence as a solar concentrator system for high temperature applications is brought out by calculating its geometric concentration characteristics.

0757

Investigation of Two Mirror System Containing a Paraboloid Mirror and an Inflated Film R e f l e c t o r : Umarov G Ya. and Alavutdinov D N, " Applied Solar Energy ", 4 (3), 42, (1968).

Optical and energetical characteristics of a two mirror system are discussed. The mirror system contains an inflatable reflector at various internal pressures.

0758

Parabolo – cylindric Concentrator with Reflector : Umarov G Ya. and Alavutdinov D N, " Applied S o l a r Energy ", 6 (3), 14, (1970).

Data on the design, optical and power characterisitcs of a parabolo – cylindrical concentrator w i t h supplemental reflector are given. This system transforms a linear local spot into a round spot. Such units may be used in the national economy for the thermodynamic transformation of solar energy to electricity.

0759

Two Mirror Solar Energy Concentrators Using Nickel Parabolic Reflectors : Umarov G Ya., Alimov A K , Alavutdinov D N and Zakirova V N, " Applied Solar Energy ", 6 (6), 53, (1970).

A double mirror system consisting of two concave mirrors significantly increases maximum flux density and the equilibrium temperature of an absolutely black body. One consisting of concave and convex parabolic mirrors has little decrease in maximum flux density and adjustment power, whereas the system's angle of opening is significantly decreased, improving efficiency of the device.

0760

Determination of the Geometric Parameters of Two Mirror Solar Energy Systems : Umarov G Ya., Zakhidov R A and Kirgizbaev D A, " Applied Solar Energy ", 9 (1), 10, (1973).

The paper solves the task of selecting parameters and mutual agreements of reflecting surfaces in a twin mirror concentrating system. The paper outlines the differential equation which makes it possible to estimate the secondary concentrator knowing one optical axis with a paraboloid and the concen – trating beam of rays within one focus.

0761

Calculation of the Radiant Energy Field or a Biparaboloidal Radiation Furnace with a Carbon Arc : Umarov G Ya., Zakhidov R A and Sokolova Yu. B, " Applied Solar Energy ", 11 (2), 86, (1975).

The formation of the radiant energy field in the secondary focus of a biparaboloidal furnace with a carbon arc is considered. The field is considered for an ideal system with a point source and the effect of source dimension and distance between mirrors on the formation of the field is investigated. Formulas are derived and an estimate is made for the variation of the field of radiant energy in the working zone of a biparaboloidal furnace with a source of realistic dimensions.

0762

Long Focus Solar Installation for the Illumination of Seeds : Umarov G Ya., Alimov A K and Alavutdinov D N, " Applied Solar Energy ", 12 (2), 51, (1976).

A two mirror system for the direct illumination of seeds and plants is described. The system consists of a main 1.5 m film mirror and a flat film reflector. The optical and energy parameters of the system are evaluated.

0763

Investigation of Optical Energy Characteristics of Multimirror Radiation Furnace : Umarov G Ya., Zakhidov R A, Vainer A A and Shamuzafarova G Sh., " Applied Solar Energy ", 15 (1), 21 (1979).

The process of radiation concentration by a biparaboloidal optical furnace with arc source is investigated by methods of mathematical and luminous modelling. The theoretical and experimental results are
compared. Recommendations are given for utilization of the proposed methods in studying the influence
of actual operational factors on the radiation field of this type of unit.

0764

Design of Non-imaging Concentrators as Second Stages in Tandem with Image-Forming First State Concentrators : Winston R and Welford W T, " Applied Optics ", 19, 347, (1980).

It has been shown that paraboloidal mirrors of short focal ratio and similar systems can have t h e i r
flux concentration enhanced to near the thermodynamic limit by the addition of non-imaging compound
elliptical concentrators.

0765

Integrated Precision Parameter for a Cassegrainian System : Zakhidov R A and Vainer A A," Applied Solar
Energy ", 11 (3 - 4), 107, (1975).

A method is devised by the authors to calculate the precision and temperature parameters of a Cassegrainian system with mirrors of different precision. It is established that the conditions imposed on
the precision of the second mirror are much less stringent.

0766

Secondary Reflectors for Economical Sun Tracking Energy Collecting Systems ; A Concept :Anonymous, " A
Concept Published by NASA ", Pasadena, California (USA), (1975).

Two secondary reflectors are proposed such that they track the reflected energy from stationary reflectors at ground level, thus reducing the weight and size of moving components of a sun tracking system.

Photovoltaic   Concentrator   Systems

0767

Solar Energy Conversion at High Solar Intensities : Backus C E, " Journal of Vacuum Science & Technology ", 12, 1032, (1975).

The concentration of sunlight offers distinct advantages for solar electrical generation either by thermal conversion or by photovoltaics. The present high cost of solar cell generated electricity can be
reduced considerably by using concentrators. Cells can be used with any of the concentrator designs,
and the major concern is keeping them at acceptable operating temperatures. Planer silicon cells,ver-
tical multi-junction, and gallium aluminium-arsenide cells all look attractive for concentrating systems.

0768

Photovoltaic III Concentrators : Backus C E, " IEEE Spectrum ", 17(2), 34, (1980).

General design considerations are outlined of photovoltaic solar power concentrating systems.

0769

The Effect of Concentrated Sunlight on Silicon Solar Cells : Backus C E, Rule T, Sanderson R & Chambers B, " Proceedings Silver Jubilee International Congress ", Georgia (USA), (1979).

This paper presents the latest data generated on silicon concentration cells that has been generated
at ASU (Arizona State University). Scores of cells from various manufacturers have been tested under
controlled conditions in concentrated sunlight upto concentrations of several suns. These cells have
shown gradual improvement of performance and currently can be expected to convert sunlight of upto 18%
efficiency for concentration ratios in the 10 - 300 sun range.

0770

Minimizing End Shadowing Effects on Parabolic Concentrator Arrays : Bardwell K, Lambarski T J, Turfler
R M and Rogers C B, " Proceedings IEEE Photovoltaic Spec. Conference ", San Diego (USA),p.765,(1980).

Collector and shadowing effects on energy output from a north/south oriented, tracking photovoltaic
(PV) parabolic collector are analysed. An optimum bypass diode protection design and PV cell placement along the receiver tube minimize the effects of collector end shadowing. This paper discusses
the methods and results of two circuit analyses used to study PV Cell placement and bypass diode configurations. Finally, the optimal design selected for bypass diode and PV cell placement is presented.

0771

Concentration Ratio and Efficiency in Thermophotovoltaics : Bell R L, " Solar Energy ", 23, 203, (1979).

Very high conversion efficiencies may be expected from solar cells for appropriately tailored Spectra.
In thermophotovoltaics, concentrated sunlight is used to heat a 'black body' cavity which re - emits
lower temperature radiation. Solar cells immersed in the cavity can absorb the higher energy photons

present and convert these to electric power with high efficiency. A generalization of this scheme is considered as a model for calculating the conversion efficiency expected for a real thermophotovoltaic system.

0772

Photovoltaic Engineering Services Pertinent to Solar Energy Conversion : Bell R O, Ho J C T, Kurth W and Surek T, " Argonne National Laboratory Report ANL - K - 75 - 3171 - ", (USA), (1975).

The application of the compound parabolic concentrator (CPC) for use with solar cells was investigated. Experiments with state of the art Si cells in CPC and under solar concentration were performed. A theoretical model for calculating the behaviour of Si solar cells with concentration was developed. Detailed calculations of the energy distribution in the CPC were made. Finally a cost effectiveness analysis shows that the CPC system will produce power at very much lower cost than will flat panel solar cell arrays.

0773

Consideration Generales Sur Les Concentrateurs Simples pour Cellules Photovoltaiques (General Considerations Concerning Simple Concentrators for Photovoltaic Cells) : Bellugue J, " Acta Electron ",20(2), 153, (1977).

Concentration of solar light which appears at first sight to be a simple means of decreasing photovoltaic watt cost is in fact a delicate process and demands a very intricate optical concentration system whose efficiency has the same function as that of illuminated cell. In fact this system should be considered as a veritable optical instrument. For simple concentration there are general relations between pointing precision concentration value and optical efficiency and the problem can thus be w e l l stated.   The different types of simple concentrator mirror or cylindrical lens, combination of conical mirrors, combination of parabolic mirrors, Fresnel lens and complementary combination of these devices among each other must be carefully studied or else, either their optical efficiency is diastrous or they are limited to very low concentrations or the light spot they supply endangers the photovoltaic cell.

0774

Cost Effective Concentration Ratio in High Concentration Photovoltaic Systems : Bendt P and Rabl A, " Proceedings ISES Silver Jubilee International Congress ", Georgia (USA), (1979).

In the high concentration approach to photovoltaic energy conversion one uses cells of high efficiency and high cost. With larger concentration ratio one reduces the combination of cell cost to system cost, but one also loses some diffuse and circumsolar radiation. There is an optimal geometric concentration ratio which maximizes the ratio of collected energy (long term average) over collector cost. The present paper describes the optimization of the concentration ratios; this step depends on circumsolar radiation and on optical errors. The spread of the solar image on the receiver is described by a Gaussian with angular standard deviation σ optics. Circumsolar effects are taken into a c c o u n t by developing an effective solar profile and its associated angular width σ Sun. A graphical method is derived which yields the optimal concentration ratio as function of σ optics, σ sun and cell/a p e r t u r e cost ratio.

0775

Photovoltaic Concentrators for On - Site Power : Boes E C, " Proceedings of the 4th National Conference", Albuquerque, New Mexico (USA), p.31. (1979).

The concept of using optical concentrators with photovoltaic cells for solar to electric power conversion has developed into a major solar electric technology. This paper describes photovoltaic concentrator technology. It discusses the current photovoltaic concentrator array design in t e r m s of their performance, durability, developmental status and cost.

0776

Economics of Photovoltaic Concentrators: Boes E C, Edenburn M W and Schueler D, " Proceedings I S E S (American Section) Conference ", Denver, Colorado (USA), 2.2, 21, (1978).

This paper is essentially a current status report on the economics of photovoltaic concentrators. It summarises the primary price objective for photovoltaic concentrators, the current price status a n d the latest projections for several specific photovoltaic concentrator arrays, and information on t h e improvement in the economies resulting from utilization of arrays for combined electric - thermal applications.

0777

Overview of the DOE's Photovoltaic Concentrator Project : Boes E C, Edenburn M W and Schueler D G , " Proceedings ISES (American Section) Conference ", Denver, Colorado (USA), 2.2, 25, (1978).

As the manager of DOE's Photovoltaic Concentrator Development Project, Sandia Laboratories is pursuing development of low cost photovoltaic power by taking advantage of sunlight concentrators w h i c h replace high cost photovoltaic cells with lower cost reflective or refractive optical surfaces. The project includes developing solar cells which are designed for high solar intensity applications. The paper presents a summary of DOE's Photovoltaic Concentrator Development Project, including a description of project goals, strategy, results and major activities.

0778

Concentrating Collectors for Thermophotovoltaic Convertors : Bracewell R N, Price K M and Swanson R M, " Proceedings ERDA Conference on Concentrating Solar Collectors ", Georgia (USA), p.4 - 17, (1977).

An assessment of the feasibility of a solar electric thermophotovoltaic (TPV) system is being carried out at Stanford University. In the TPV system, sunlight reflects from a paraboloidal concentrator and reaches the TPV convertor located at the focus. The concentrated sunlight enters the c o n v e r t o r through a window and receives slight additional concentration from a CPC secondary concentrator . The sunlight then passes through a throat into an evacuated cavity and heats a refractory radiater.   A photovoltaic cell faces the radiator, receives incandescent radiation from it and converts this radiation into electricity.

0779

Photovoltaic Energy Conversion Using Concentrated Sunlight : Burgess E L, " Optical Engineering",16(3), 305, (1977).

A development program is described in which sunlight concentration techniques are used to effect an immediate reduction in cost per unit power for photovoltaic systems in which cell cost dominates the total system cost. Current examples of concentrator solar cell technologies are single crystal silicon and gallium arsenide. Implementation of cost reduction by the use of sunlight concentration is not dependent on the development of low cost, mass production cell technologies,but emphasises high cell efficiency and low cost concentrator systems.

0780

Status of the Photovoltaic Concentrator Applications Experiments : Burgess E L, " Proceedings 13 th IEEE Photovoltaic Spec. Conference ", Washington (USA), p.35, (1978).

The Department of Energy has initiated a three phased plan which will result in the design, installation and operation of a number of different concentrating photovoltaic experiments. The systems, which will range in size from 20 to 500 KW peak electrical output, will be installed in a variety of onsite applications and operated for a period of 1 to 2 years to obtain both, technical performance data and data related to non technical issues. Seventeen projects entered into Phase I, System Design June 1, 1978. Brief descriptions of each of these projects are given.

0781

Photovoltaic Concentrator and Flat Panel Applications Experiments : Burgess E L, " Proceedings I S E S Silver Jubilee International Congress ", Georgia (USA), (1979).

The US Department of Energy has initiated a photovoltaic applications experiment program to gain on site operation experience with photovoltaic systems. The program plan calls for a series of experiments to be initiated in the 1978 - 86 time frame. These experiments are multiphased consisting of design, fabrication and system operation phases. The design phase for the first two sets of experiments has just been completed. One set consists of seventeen concentrator array experiments and the second set consists of twelve flat panel array experiments. The experiment sizes range from 20 to 500 KW peak electrical. This paper describes each application and selected design . The important technical and non-technical factors to be investigated in the system operation phase are outlined along with the data collection reduction and discrimination program initiated to support this program. Finally important technical information derived during the design phase of all twenty nine experiments is discussed.

0782

Photovoltaic Concentrator System Technology and Application Experiments : Burgess: E L, " Proceedings International Gas Turbine and Solar Energy Conference ", San Diego (USA), (1979).

One approach to reducing cost of photovoltaic systems producing electricity from sunlight is to concentrate sunlight on the photovoltaic cells, increasing their output and, hence decreasing the amount of cell area required for a given power output. This concept essentially trades high cost cell area for lower cost concentrating optics area. Photovoltaic concentrator technology is described and its future potential is discussed. In addition, brief summaries of several concentrator applications experiments which are in the system design phase are presented. The purpose of these experiments is to provide on site experience with photovoltaic concentrator systems.

0783

20 KW Gallium Arsenide Photovoltaic Dense Array for Central Receiver Concentrator Applications : Cape J A, Sahai R and Harris J S, " Proceedings 2nd International Photovoltaic Solar Energy Conference ", West Berlin (Germany), p. 1027, (1979).

The objective of this project is to explore the feasibility of operating photovoltaic subsystems on a Central Receiver Tower on which is focussed,highly concentrated(upto 1500 suns)sunlight by a large field of mirrors. Essential to the design of the photovoltaic subsystems, which is called a Dense Array, is achieving a nearly 100 % active area utilization of the incident solar flux.This is done in the present concept by overlapping a series of solar cell modules, rather like shingles, so that the current collector strips and interconnects, cool out lines, and other non active elements are shaded from the insolation. The array which will produce 20 KW presents a frontal area of 0.13 sqm and consists of four panels, electrically in parallel. Each panel consists of 16 modules of 16 GaAs cells each connected in series. This will result in a system output of $\approx$240 volts and 80 - 85 amperes at 1000 suns AM1. The design goal is to achieve greater than 16% overall conversion efficiency including area losses.

0784

Performance of a Solar Battery Using Quasi - Cylindrical Array of Plane Mirrors as a Concentrator : Deb S and Saha H, " Solar Energy ", 17, 67, (1975).

A battery using a quasi - cylindrical concentrator composed of an array of flat mirrors is first described. This is followed by a critical examination of the principal factors limiting the battery efficiency. The effect of concentration on cell performance is studied in detail,and a way in which this can be utilised to counteract to some extent, the effect of diurnal variation of insolation is indicated. An estimate is made of the overall battery efficiency and a case made for using the same for small/medium scale power generation for agro-irrigational purposes in arid/semi-arid regions.

0785

Perspective on Utility Central Station Photovoltaic Applications : DeMeo E A and Bos P B, "Electric Power Research Institute Report EPRI - ER - 589 - SR ", (USA), (1978).

This report develops nominal cost and performance goals for solar photovoltaic conversion devices intended for large scale electric utility applications. The objective is to provide an improved basis for establishing research and development priorities for photovoltaic devices and conversion concepts.

Comparisons are made among a number of generic power plant conceptual designs, with the aid of an array design parameter that is defined to include array area related costs, over all power plant efficiency and average available insolation.

0786

Photovoltaic Concentration Array : Donovan R L and Broadbent S, " Proceedings 13th IECE Conference ", San Diego (USA), 12, 1593, (1978).

A Photovoltaic Concentration Array (PCA) for terrestrial application is described. The PCA uses point focussing acrylic Fresnel lenses to concentrate sunlight on high intensity solar cells. The objective of the PCA development was to obtain economical photovoltaic power generation by replacing relatively high priced solar cells with lower cost lenses. The PCA operates at a concentration of 40, t r a c k s the sun in two axes, and is passively cooled. An inherent advantage of concentrating systems o v e r flat arrays is that they can afford high cost, high performance cells since the cost impact of t h e cells becomes secondary to total system cost.

0787

Photovoltaic Concentrator System for Roxborough Park : Donovan R l, Miller W D and Burleson J, " Proceedings 14th IECE Conference ", Boston (USA), p. 245, (1979).

The Roxborough State Park Visitors Center located 25 miles south west of Denver, Colorado is the planned site of a photovoltaic concentrator application experiment. The intent of the experiment is to demonstrate an energy independent application by combining the photovoltaic concentrator system w i t h existing active and passive solar heating and cooling systems. The experiment represents remote installations of moderate size where the true cost of utility power is high due to transmission l i n e construction in rough terrain. A description of the overall experiment concept is presented a l o n g with a description of the photovoltaic concentrator system and results of system studies.

0788

Overview of the ERDA Photovoltaic Concentrator Project : Edenburn M W, Schueler D G and Burgess E L, " Proceedings ERDA Conference on Concentrating Solar Collectors ", Georgia (USA), p. 1 - 25, (1977).

As the manager of ERDA's photovoltaic concentrator development project, Sandia Laboratories is pursuing development of low cost photovoltaic power by taking advantage of sunlight concentrators which replace high cost reflective or refractive optical surfaces. This paper describes Sandia's concentrator development project and discusses various concentrator designs being developed. It also discusses aspects of concentration which are peculiar to photovoltaic application and which must be accomodated in the concentrator design.

0789

Analytical Evaluation of a Thermophotovoltaic (TPV) Convertor : Edenburn M W, " Proceedings ISES Silver Jubilee International Congress ", Georgia (USA), (1979).

A solar thermophotovoltaic convertor uses concentrated sunlight to heat a cavity enclosed emitter that illuminates photovoltaic cells with a thermal radiation spectrum. Radiation incident on the c e l l s with energy below the cell's band - gap energy is reflected back to the emitter rather than being lost to the conversion process. The spherical cavity convertor was analysed by considering r e - radiation losses through the throat, the cell's spectral response to the cell's reflectance to below - band g a p radiation, emitter temperature, convertor geometry, sunlight concentrator, convertor geometry and sunlight concentration requirements. The emitter is assumed to be a uniform temperature, black body emitter. The convertor efficiency, the optical efficiency of the concentrator and the overall conversion efficiency for a TPV convertor, with and without a selective absorber, have been calculated and their variation with concentration ratio have been studied. It has been found that t h e conversion efficiency depends very strongly on concentrator accuracy. High conversion efficiency requires very accurate concentrators that will be expensive. The practicability of TPV conversion cannot be assessed without determining how accuracy affects concentrator cost.

0790

Photovoltaic Concentrator Design Comparison : Edenburn M W, Boes E and Shafer B, " Proceedings I S E S Silver Jubilee International Congress ", Georgia (USA), (1979).

In the past few years a variety of different designs for photovoltaic concentrators have been proposed. This paper provides a comparison of some of these designs. The comparison is given both in terms of descriptive design features such as geometric concentration ratio or type of optics used, and in terms of design quality expressed in terms of component efficiencies, array material costs and $/W. The selection of concentrator array designs for inclusion in this comparison was based primarily on the degree of development of designs and the desire for a representative variety.

0791

Cost Studies on Terrestrial Photovoltaic Power Systems with Sunlight Concentration : Evans D L a n d Florschuetz L W, " Solar Energy ", 19, 255, (1977).

A systems simulation program for comparing the energy costs associated with various alternative c o n centrating photovoltaic solar systems to energy cost expected with flat arrays is described. The a p lication to line focus and point focus concentrators is presented in a parameterized way. The results show that concentration offers a distinct cost advantage at high cell costs. However, they also show that concentration has the potential for being a viable alternative to the flat concentrated arrays for cell costs as low as 50/sqm. Also, for a given concentrator cost, cell cost and cell cooling effectiveness, there exists an optimum effective aperture area to cell area ratio. For reasonable projected cell costs, this optimum ratio is below 30 for passively cooled devices and below 60 for actively cooled systems.

0792

Parametric Cost Analysis of Photovoltaic Systems : Evtuhov V, " Solar Energy ", 22, 427, (1979).

A parametric cost analysis of photovoltaic systems based on different design philosophies has been

104

carried out. The analysis takes into account the fixed costs involved in a photovoltaic installation (such as site preparation, array foundations and structure, inversion equipment, back up capacity etc.), the cost of concentrators at different concentration ratios, ranging from unit (i.e. flat plate array) to 10^3, solar cell cost per unit area, system efficiency, etc. Resulting parametric curves allow the determination of the cost of electricity for a given set of system constraints. They also allow one to determine the optimum concentration ratio for a given solar cell cost per unit area. The analysis confirms many of the accepted conclusions, presents them in a concise form, and in addition, allows one to see the relationship between the various system design regimes and sensitivity of the total cost of electricity produced to the cost of the various system components.

0793

Concentrator Cost Bound for High Efficiency Photovoltaic Power Production : Faiman D, " Solar Energy ", 22, 397, (1979).

An upper bound on the cost per unit aperture area of the concentrator used in photovoltaic power production has been placed in a completely model independent manner.

0794

Concentrator Photovoltaic Systems for Economical Electricity and Heat : Furber J, " Proceedings ISES Conference ", New Delhi (India), p.720, (1978).

This paper describes how to choose components and how to design a small (several KWe) modular photovoltaic / concentrator system. Sources of components are listed, alongwith prices and an evaluation of products. Operation and maintenance procedures are described, and expected system costs are detailed.

0795

Optimization of Photovoltaic Receiver for a Parabolic Trough Concentrator : Garner C M and Biggs F , " Proceedings 14th IEEE Photovoltaic Spec. Conference ", San Diego (USA), p. 743, (1980).

The overall efficiency of a photovoltaic solar collector depends on design parameters such as the position and the shape of the receiver, the size of the solar cells and whether a one or two bus solar cell is used. A model has been devised (and a computer code written to implement it) to calculate the pattern of flux density concentrated on V - shaped receiver by a parabolic trough and to determine the corresponding I-V curve of the photo cell. The authors summarise the model and briefly describe the computer code.

0796

Combined Photovoltaic / Thermal Electrical Central Power Plant : Gluck D F, " Proceedings 14th I E C E Conference ", Boston (USA), p.255, (1979).

A preliminary design and performance evaluation of the subject power plant has been completed. DC electric energy is produced by boiling water cooled gallium arsenide solar cells covering a central receiver illuminated by a heliostat field with a concentration ratio of one thousand. The steam produced is used to power a low pressure turbine generator, with the exhaust steam directly condensed in a dry air heat exchanger. A rough comparison of this combined power plant with a photovoltaic plant with cooled solar cells showed the BBEC of the combined power plant to be 9% lower.

0797

Novel Versions of the Compound Parabolic Concentrator for Photovoltaic Power Generation : G o r s k i A, Graven R, McIntire W, Schertz W W, Winston R and Zwerdling S, " Proceedings 12th I E E E Photovoltaic Spec. Conference ", Baton Rouge (USA), p.764, (1976).

The Argonne National Laboratory is now engaged in the final stages of assembly and evaluation of two novel concentrator / photovoltaic panels based on the compound parabolic concentrator concept. In the first version of these terrestrial panels, the concentrator is in the form of a solid transparent dielectric bar, the (DCPC) with silicon solar cells bound to the exit apertures. In the second version the concentrator is in the form of hollow troughs of light weight plastic that has been metallized and protectively over coated (the CPC configuration). Tests for these panels will allow evaluation of a method that has the potential for significantly reducing the cost of photovoltaic power conversion through the use of low cost plastic concentrators.

0798

On the Design of CPC Photovoltaic Solar Collectors : Graven R M, Anthony J and McIntire W R," Proceed- ISES Conference ", New Delhi (India), p. 725, (1978).

Two new photovoltaic solar collectors have been designed, built and tested. A reflective style Compound Parabolic Concentrator (CPC) optical system was used for one collector and a dielectrical style (DCPC) optical system was used for the other collector to concentrate sunlight onto custom designed photovoltaic solar cells. For these two panels only periodic angular adjustments are required, which eliminate the need for two-axis tracking. A modular design individual sub units containing an array of series and parallel circuits allows a user to select a variety of operating conditions. The design considerations for these panels are summarized in this paper.

0799

Reduction of Intensity Variations on Photovoltaic Array with Compound Parabolic Concentrators : Greenman P, O'Gallagher J, Winston R and Costigue E, " Proceedings ISES Silver Jubilee International Congress ", Georgia (USA), (1979).

A method for reducing the intensity variations has been found in the context of the use of C P C enhanced photovoltaic arrays in space. The solution developed was the intentional introduction of irregularities into the reflecting walls of the CPC. These irregularities may be characterized by σ the root mean squared angular deviation of the reflector slope from that of a CPC with smooth, specular walls. If σ is a nonnegligible fraction of a CPC acceptance half angle, $\pm \theta_c$, the resulting intensity distributions are found to be much smoother than those for specular reflectors, although some light is scattered away from the absorber. By controlling σ, the trade off between uniformity of the intensity distribution and scattering loss can be optimized.

0800

Flat Plate vs Concentrator Solar Photovoltaic Cells ; A Manufacturing Cost Analysis : Grenon A and Coleman M G, " Proceedings 14th IEEE Photovoltaic Spec. Conference ", San Diego (USA), p.488, (1980).

A manufacturing cost analysis for both, flat plate and concentrator silicon solar cells has been performed. The cost of concentrator solar cells is shown to be a strong function of production volume as which, in turn, is a strong function of the concentration level. Manufacturing of concentrator solar cells in a stand-above factory would require enormous volumes of concentrator systems to be produced to allow major cost reductions to occur. The manufacturing of both, flat plate and concentrator cells in the same factory and utilizing similar processes will allow rapid cost reductions to occur for concentrator cells.

0801

Design and Operation of the Solarax Two - axis Tracking Linear Concentrating Collector System : Hamilton R C, Levine · A S, Wohlgemuth J H and Wrigley C Y, " Proceedings 14th IEEE Photovoltaic Spec. Conference ", San Diego (USA), p.772, (1980).

A two - axis tracking linear concentration system which is optimized specifically for efficient operation in photovoltaic systems was designed. Among the unique features of the design are a high efficiency, high reliability receiver design and a collector providing substantial stability and durability under environmental stress. Test data is presented for a prototype system designed t o deliver 1.2 KWe from each of two four collector arrays. Collector efficiency of 10.3% is reported alongwith selected temperature data.

0802

Array of Directable Mirrors as a Photovoltaic Solar Concentrator : Ittner W B, " Solar Energy ", 24, 221, (1980).

This paper describes a simplified method for calculating the images generated by a special type o f concentrator i.e. an array of independently steered mirrors on a single frame intended to direct the solar image onto a flat photovoltaic solar cell target. The overall system efficiency f o r various mirror configurations, characterized by such parameters as the maximum mirror angle, target - mirror plane separation and mirror aiming accuracy is discussed in terms of the specifications desirable i n an optical concentrator designed specifically to illuminate uniformly a photovoltaic solar cell target.

0803

Fresnel Optics for Solar Concentration on Photovoltaic Cells : James L W and Williams J K, " Proceedings 13th IEEE Photovoltaic Spec. Conference ", Washington (USA), p. 673, (1978).

The conversion efficiency of concentrator solar cells is shown, both theoretically and experimentally, to be reduced by illuminating the cells with a nonuniform flux density. The structure cost is reduced by increasing the allowable tracking error. Thus flux density uniformity, optical transmission allowable tracking error, cost per unit area and lifetime are important criteria for photovoltaic concentrating optics. Acrylic Fresnel lens designs are discussed in terms of these criteria.

0804

Characteristics of Axicon Concentrators for Use in Photovoltaic Energy Conversion : K u r z w e g U H , " Solar Energy ", 24, 411, (1980).

This note examines the concentration and cost characteristics of axicon concentrators when u s e d i n photovoltaic energy conversion. The required semi - vertex angle of the outer cone for minimum energy conversion costs is determined. It is shown that this angle depends on the relative cost of s o l a r cells compared to an equivalent area of reflector material and also on the concentration used. In general the higher the concentration of the axicon the lower the overall cost of electricity production upto a point where the cell efficiency decreases due to concomitant heating.

0805

On a Class of Axisymmetric Concentrators with Uniform Flux Concentration for Photovoltaic Applications: Kurzweg U H, " Solar Energy ", 24, 507, (1980).

A general differential expression for the concentration characteristics of all axisymmetric reflector absorber combinations capable of yielding a uniform flux density at the absorber surface is derived. Of the infinite number of combinations capable of such a uniform flux,an axicon and a new bell shaped reflector are analysed in greater detail.

0806

Quasi - Static Concentrated Array with Double Side Illuminated Solar Cells : Luque A, Ruiz J M, Sangrador J, Cuevas A, Gomez J M, Eguren J, Del Alamo J and Sala G, " Proceedings ISES Silver Jubilee International Congress ", Georgia (USA), (1979).

A two - stage optimal (in the sense of Rabl) quasistatic concentrator has been developed for double-sided illumination. It provides a practical gain of 8 with an acceptance angle of ±10.5°. The cells are immersed in water so that a passive cooling is provided by natural convection in water from the cell to the concentrator walls where there are refrigerating fins. A difference of temperature with the ambient of 12 deg C has been measured for irradiance on the cell of 750 MW / sqcm and wind speed of 1 m / sec. It is expected that this system will be able to work the whole year with 4 to 8 changes of elevation angle throughout the year.

0807

A Novel Mildly Concentrating Cone Collector/Photovoltaic Total Energy System : Manasse F K, Strong S , Manelas A and Brown J , " Proceedings ISES Silver Jubilee International Congress ", Georgia(USA),(1979).

The use of aluminium conical collectors, to mildly concentrate solar radiation onto photovoltaic cells, can result in enhanced electrical perforamance at reduced cost per watt while also providing thermal scavenging for space heating. A novel module design using single axis tracking, simple cell cooling and electrical integration is described. Each 4 ft X 4 ft, 33 cell module produces more than 65 watts with a solar concentration ratio of less than 5 : 1 and provides over 23,000 Btu / day of h e a t e d air. This total energy system has been tested in several residential and commercial installations. Numerous test results for residential air systems are given. Our experience shows that almost 60% of the electrical and space heating needs can be provided for an average 1500 sqft home in New England (USA), by a 10 module or 160 sqft array. A novel 8 modules two axis tracking scheme used in a Westchester (NY, USA) bank which produces over 0.5 Kw and more than 150,000 Btu / day is also described. This system provides all the electricity necessary to operate an automatic letter machine as well as partially lighting and heating the building. Recent test and performance data for a newly completed H U D supported installation are also given.

0808

Technology and Performance of a High Concentration Solar Cell Power Supply : Mandelkoru J & Yekutieli G, " Israel Journal of Technology ", 17(1), 12, (1979).

A prototype, 25 Watt output, high concentration solar cell power supply has been constructed using immediately available components (1977). The guidelines for the system were the use of available components. Concentration ratio of about 100 X was used to minimize cell cost without incurring high cost for critical optics, tracking and a minimum efficiency of 7 - 8%. The design called for simple structures and subsystems in order to reduce cost. The power supply consists of 36 X 1 cm cells in series, a corresponding number of 10 cm X 10 cm Fresnel lenses mounted above the cells, a sun position detector and tracker. The overall operating efficiency is 7.5%. The supply is described in terms of individual components and subsystem design, structure and fabrication. Performance problems a n d future expectations are discussed.

0809

Design and Construction of a One Kilowatt Concentrator Photovoltaic System : Marchi D L, " Sandia Laboratories Report No. SAND - 77 - 0909 ", (USA), (1977).

The construction detail of a system which uses Fresnel lenses to concentrate sunlight on silicon solar cells are described. The cells are cooled either passively by convection or actively using a pumped fluid coolant. Construction and operation of the array have disclosed several unique problems ; further modifications and improved future designs are being considered.

0810

Novel Low Cost Photovoltaic Concentrator : Masters W, Maraschin R and Kennedy W, " Proceedings 13 th IEEE Photovoltaic Spec. Conference ", Washington (USA), p.686, (1978).

An investigation of the optical quality of standard headlight reflectors designed for a photovoltaic concentrator based on headlight technology is made. Future development activities are also mentioned.

0811

Concentrator and Solar Photovoltaics : Merchant M C, " Sunworld ", 4(1), 21, (1980).

In reviewing the recent developments in solar photovoltaic electricity production using concentrator systems, this article describes the different types of solar cells available and their constraints & requirements. Concentration methods and tracking schemes are dealt with in detail. Present and planned applications and installations in the USA conclude the presentation. It is seen that more work is required by researchers and engineers to determine which of the concentration and tracking m e t h o d s will bring solar photovoltaic electricity into common use.

0812

Photothermal Conversion of Solar Energy into Electricity : Moragues J A, " Report Comm. Nac de Energia St ", Buenos Aires (Argentina),   (DOE 9-tr-159), (1978).

Brief description of various collectors that can be used for electricity production by solar p h o t o-thermal conversion are given including solar stills, shallow solar ponds, flat plate collectors, fixed collectors with moderate concentration, cylindrical parabolic concentrators, fixed concentrators with faceted mirrors, concentrations with concave rotating multitple mirrors, Fresnel lenses, paraboloidal rotating concentrators, hemispherical fixed concentrators, central receiver systems with linear focus and central receiver system with point focus. Other topics include thermodynamic cycles, energy transport, energy storage and economies. A review of some of the programs throughout the world in development of solar thermal power conversion is included.

0813

High Level Concentration of Sunlight on Silicon Solar Cells : Napoli L S, Swartz G A, Liu S G, K l e i n N, Fairbanks D and Tamutus D, " RCA Review", (USA), 38,76, (1977).

Silicon solar cells have achieved 15.5% conversion efficiency at 330 suns concentration level. Design rules regarding the fabrication and use of these solar cells at high concentration levels are derived. Finally, a system consisting of lens arrays and solar cells designed to generate 100 watts is described in detail. Preliminary data shows that the system efficiency is 8.3% during conditions of 74 mW / sqcm insolation level.

0814

Design and Development of a 300 Watt Solar Photovoltaic Concentrator ; Final Report, May 17, 1976 t o November 30, 1977 : Napoli L S and Swartz G A, " Sandia Laboratories Report SAND-78-7027", (USA), (1978).

The design, fabrication and performance of high level concentration (300 X) with refractive lenses on small silicon solar cells (o.4 sqcm) with passive cooling are described. Four systems were fabricated using polar axis (2 - axis) tracking. Two systems were rated at 100 W peak, and two were rated at 300 W peak. The maximum power measured on the 300 W unit was 236 W at 93 mW/sqcm direct insolation.   Solar cells were designed and fabricated using bulk high resistivity n - type silicon which showed a maximum efficiency of 17.5% independent of concentration level from 50 to 400 X. At 1000 X the efficiency dropped to 14%. Lens arrays made from vacuum formed sheets,and epoxy filled, showed good performance on a selected basis but non-uniformity and uV degradation dictated a change to Fresnel lenses. The Fresnel lens arrays are approximately 83% efficient but the imaging was quite non-uniform,causing degradation in cell fill factor and collection efficiency. Heat transfer calculations and simulations show that the cell temperature is about 25-35 deg C above ambient temperature. Thermal spreading of the heat from the cell area to an area equivalent to the lens area is the most important function of the heat sink. The addition of fins do not significantly reduce cell operating temperature. An optimization analysis was performed on passively cooled photovoltaic concentrators showing the various trade-offs that exist with lens size and concentration level. The conclusion is that the minimum manufacturing costs of $0.70/peak watt can be obtained at an optimum lens size of approximately 100 sqcm and concentration level of 100 suns. Deviation from this optimum either degrades performance, increases costs or both.

0815

A Study of Combined (Photovoltaic - Thermal) Solar Energy Systems : Nevile R C, " Proceedings 2nd International Conference on Alternative Energy Sources ", Miami (USA), (1979).

This paper is devoted to a theoretical analysis of a combined system. Incoming light is concentrated using lenses and allowed to fall on the solar cells which then generate electric power. The energy rejected by solar cells~80% of the insolation, is available to heat the thermal energy collection medium and delivered to the load requiring heat energy. Several levels of optical concentration are studied with an upper limit imposed by the rapid decrease in solar efficiency with increasing temperature and the increasing difficulty of accurate manufacture and operation with increasing concentration.

0816

Thermal Electric Performance Analysis for Actively Cooled Concentration Photovoltaic Systems : O'Leary M J and Clements L D, " Solar Energy ", 25, 401, (1980)

The photovoltaic cell within a composite stack of the receiver for a concentrating photovoltaic electrical power system is modelled as a combination of electricity and waste heat generator. The waste heat generator concept leads to analytical solutions describing temperature and power production profiles for the photovoltaic receiver. The analytical solutions considered the cases where convective losses to the surroundings may either be neglected or included. Examples of operating modes using the model are given.

0817

25 KW Fresnel Lens / Photovoltaic Concentrator Application Experiment at Dallas - Fort Worth Airport : O'Neil M J, " Proceedings 14th IEEE Photovoltaic Spec. Conference ", San Diego (USA), p.125, (1980)

The U S Department of Energy (DOE) has recently selected five different photovoltaic concentrator systems to be built and field tested at various locations around the nation. The only one of these systems which uses a linear refractive concentrator will soon be built and tested by E-systems at the Dallas-Fort Worth Airport, its mass production economies and prototype collector testing results are presented.

0818

Use of Concentrated Sunlight with Solar Cells for Terrestrial Applications : Ralph E L," Solar Energy", 10 (2), 67, (1966).

Methods for obtaining less expensive power generation from sunlight are described. The use of large area solar cells in conjunction with simple conical concentrators indicate that costs can be decreased by a factor of five compared to present day bare solar cells systems. The solar cells used in this experimental concentrator system were specifically designed to operate at high intensities. The concentrator system was designed to operate efficiently with no auxiliary cooling equipment and to produce one half watt per cell, or about three times the bare solar cell output. Experimental voltage current characteristics curves for the system were obtained under equilibrium temperature conditions. The effects of sun angle of incidence on system performance was also determined.

0819

Econometric Analysis of Concentrators for Solar Cells : Roy A S, " Proceedings ERDA Conference on Concentrating Solar Collectors ", Georgia (USA), p. 9-15, (1977).

A generalized model is formulated for correlating the electricity cost $[C(\cent/KwH)]$ to the system parameters : the collector annualized cost $Co(\cent m^{-2} yr^{-1})$; the cell annualized cost $C_e(\cent m^{-2} hr^{-1})$, the contration ratio (R), the orientation parameter (G), the insolation $[I_n (KWh\, m^{-2} yr^{-1})]$, the optical efficiency (η_0) and the cell efficiency (η_c) of concentrator solar cells systems. The model enables the economic assessment of various concentrator cell systems and the evaluation of the sensitivity of the cost of the produced electrical energy to various possible technological assessment. Optimization & the setting of spectrum of target goals for cells and collectors are thus made possible. An example is included showing minimum generated electricity costs for particular concentration ratios (between 4 and 400) and other parameters.

0820

Hybrid Silicone - Glass Fresnel Lens as Concentrator for Photovoltaic Applications : Sala G & Lorenzo E " Proceedings 2nd International Photovoltaic Solar Energy Conference ", West Berlin (Germany), p.1004, (1979).

A hybrid Fresnel lens made of glass and transparent silicone has been developed. Both materials are of high resistence to outdoor conditions. The objective is to design and manufacture 144 square hybrid Fresnel lenses of 40 X 40 cm^2 to act as concentrators in the 2 KW Ramon Areces photovoltaic array. The dispersive characteristics of the silicone resin used, Sylgard 182 of Dow Corning, has been measured. A lens providing a quasi - uniform current density generation on the receiver has been designed. The lens has 350 circular grooves but only 35 different tilt groove angles. Variation of optical and adherence properties of the lens are not significant after thermal and chemical aging cycles. The estimated cost of the hybrid lenses is $ 30/sqm which is competitive. The manufacturing process is very simple and needs a very small equipment investment.

0821

Static Concentrators for Two Sided Solar Cells : Sangrador J & Sala G," Solar Energy ",23,53,(1979).

The use of cylindrical mirrors to concentrate radiant energy onto a two sided photovoltaic absorber has been studied. Both theoretical and experimental results for the illumination intensity at the exit aperture were obtained for such a concentrator. A mathematical approach was devised to obtain the theoretical intensity distribution. Experimental intensity distribution was measured for a set of different mirrors. All the mirrors had the same profile but were made of different materials and had different reflecting surfaces.

0822

Chromatic Dispersion Concentrator Applied to Photovoltaic Systems : Sassi G, " Solar Energy ",   2 4, 451, (1980).

A chromatic dispersion concentrator which collects the different monochromatic components of the solar spectrum separately in subsequent concentric rings in the focal zone is proposed. It operates without an increase in the energetic losses compared to any other type of concentrator. If different photovoltaic elements with energy gaps equal to the photon energy falling on the focal zone are put in the latter, energy losses due to incomplete utilization of the solar spectrum and to incomplete utilization of the energy of a single photon can be drastically reduced. How the losses due to the voltage factor and the fill factor of the photovoltaic elements of the system can be reduced compared to the normal silicon cells is also demonstrated. The other contributions to losses in the conversion process have only been mentioned, forseeing their possible variation.

0823

The DOE Photovoltaic Concentrator Project : Schueler D G, Shafer B D, Boes E C and McDowell J F, " Proceedings ISES Silver Jubilee International Congress ", Georgia (USA), (1979).

The DOE Photovoltaic Concentrator Technology Development Project has a primary goal of developing low cost reliable photovoltaic arrays for use in major power system applications. The overall approach in this effort is to develop photovoltaic arrays which use refractive or reflective optics to concentrate sunlight onto high performance solar cells thereby replacing solar cells area with less expensive optical materials. A specific project long range goal is to achieve commercial readiness of concentrator arrays at a selling price of $0.50 per peak watt by 1986. Currently there are four major areas in which the efforts are being focussed. These are the developments of high efficiency c e l l s for use in concentrated sunlight, the development of reliable low cost novel concentrator concepts, the development of components to be used in concentrators such as lightweight parabolic toughs, and the fabrication and evaluation of complete arrays that have low cost potential. The paper summarizes the overall DOE photovoltaic concentrator project, and how developments such as those highlighted above will make it possible to achieve the long range DOE goals.

0824

Status of the US Department of Energy Photovoltaic Concentrator Development Project : Shafer B D, Boes E C, Edenburn M W and Schueler D G, " Proceedings International Photovoltaic Solar Energy Conference", West Berlin (Germany), p.524, (1979).

The overall objective of this project is to develop a photovoltaic development technology which results in low cost, long life photvoltaic arrays at a price of less than $0.50 per peak watt by 1986. The general approach is to identify and develop those concentrator concepts which have the highest potential for low cost and long life by supporting technology and increasing solar array production capability. The principle criterion is to develop concentrating arrays that provide the lowest life cycle energy cost. The work that the DOE is funding in the area of concentration cells; concentrator modules and complete arrays is surveyed. Highlights of work done to date in cell development include 18 - 20% efficient silicon cells and 28.5% efficient PV cell assemblies using Ga A1As Si Cells in conjunction with spectral splitting. In the area of concentrator modules a number of different types have been fabricated and evaluated.

0825

Low Cost High Performance Point Focus Concentrator Array Design : Shafer B D, Edenburn M W,Garner M & Togami H, " Proceedings 14th IEEE Photovoltaic Spec. Conference ", San Diego (USA), p.754, (1980).

Sandia Laboratories has initiated the design and fabrication of optimized photovoltaic concentrator arrays which are meant to demonstrate the technology readiness of actual hardware for the 1980 goal of $2.80/W$_p$. Two basic approaches are being pursued in parallel. These are, a point focus Fresnel lens concept and a linear focus parabolic trough. This study presents the point focus design.

0826

Concentrating Solar Collectors for Thermal and Photovoltaic Applications : Simon M, " Proceedings of AIChE Symposium ", San Francisco (USA), 76 (198), 27, (1980).

The author presents the development work for a 3 KWe concentrating silicon cell module with 4 h e l i o-static mounted cylindrical parabolic troughs and concentration factor of 20 to 40. A model calculation for the optimization of the concentrator factor is presented first. Depending on the future developments öf the specific cell costs and specific collector costs, a concentration between 10 to 50 will be optimal for cylindrical parabolic collectors. The preparation work for a prototype is presented with the modular 4 trough collector system which is optimized for maximum daily energy output, low series production, maintenance cost and long service time.

0827

Solar Cells with Concentrating Collectors and Integrated Heat Use System : Simon M, Pfeiffer H, Kohlmanusperger J and Gall S, " Proceedings 2nd International Photovoltaic Solar Energy Conferenc ", West Berlin (Germany), p.541, (1979).

The authors present the development work for a 4 KWe concentrating silicon cell module with 4 heliostatic mounted cylindrical parabolic troughs and concentration factor of 20 to 40. A model calculation

for the optimization of the concentrator factor is presented first. Depending upon the future development of the specific cell costs and specific collector costs a concentration between 10 to 50 will
be optimal. The preparation work for a prototype is presented with the modular 4 trough collector system which is optimised for maximum daily energy output, low series production and maintenance cost and
long service time.

0828

Hybrid System Consisting of Silicon Solar Cells with Concentrators and Heat Pump : Stojanovic M S and
Milinkovic S, " Proceedings 2nd International Photovoltaic Solar Energy Conference ", W e s t B e r l i n
(Germany), p.1021, (1979).

A discussion is presented of a solar system consisting of silicon photovoltaic cell with concentrators
and a heat pump. In addition to heat energy of low quality (temperature about 50 degC) which is obtained by cooling the solar cells, energy can be added by the heat pump driven by the electric power
generated in the same solar cells. The importance of the considered hybrid system lies in the f a c t
that it is autonomous and can be used as a refrigerator. Analysis of the suggested system based o n
technical characteristics of available commercial units (silicon solar cells, Fresnel's concentrators
and heat pump) shows that we are approaching an economical use of solar cells for the production o f
electric energy and that its conversion to heat energy via heat pump is justified.

0829

Proposed Thermophotovoltaic Solar Energy Conversion System : Swanson R M," Proceedings IEEE Conference ",
67, 446, (1979).

A solar electric system is proposed and discussed. This system uses concentrating mirrors focussing
on a thermophotovoltaic (TPV) convertor. Within the TPV convertor, the concentrated sunlight heats a
refractory radiator. A silicon photovoltaic cell faces the radiator, receives incandescent radiation
from it and converts this radiation into electricity.

0830

Concentrating Photovoltaic Array Testing : Watkins J L, " Proceedings ISES (American Section) Conference ", Denver, Colorado (USA), 2.2, 92, (1978).

A dsecription of the Photovoltaic Advance Systems Test Facility is given including the data a c q u i -
sition and control system. Several of the concentrating photovoltaic arrays are described along with
their characteristics. An example of accelerated aging for a photovoltaic module is given and the results are presented. Finally, plans for future expansion of the facility and future testing are detailed.

0831

Design and Analysis of Solar Cell Concentrating System Using a Flux Redistributor : Wennstrom J E and
Warner R E, (Unknown).

A concentrating system which focusses sunlight on an actively cooled array of silicon solar cells was
built and tested. A concentration ratio of 11.7 was provided by a parabolic cylinder with a secondary
mirror arrangement near the focal line to redistribute the solar flux. Relationship between optical,
electrical and thermal efficiencies are presented for two mounting positions. Solar cell voltage current characteristics and the effects of pointing error on cell output power were examined. Test results
show the flux redistributor has an equivalent reflectance of 0.75, but it provides convection suppression, a 1.0 intercept factor, flux uniformity across the cell surface and insensitivity to tracking or
mirror surface errors. The cells produced $0.05\,W\,cm^{-2}$ with photovoltaic efficiency of 6.5% and a system
efficiency of 4.1% for the conversion of direct beam sunlight to electricity at a cell temperature of 64°C.

0832

Evaluation of a Plastic Parabolic Concentrator for Terrestrial Photovoltaic Applications : Zimmerman D K
" Proceedings 13th IEEE Photovoltaic Spec.Conference ", Washington (USA), p.680, (1978).

The experimental evaluation of an air pressure stabilized plastic film parabolic concentrator is described. The study objectives were to determine the fabrication feasibility and optical performance of
the concentrator. The concentrator was made from gores of aluminized polyester film supported on a
light weight frame work. Shape of the reflector is maintained by a slight vacuum back of the reflective surface. The concentrator and receiver are enclosed by an inflated dome of transparent plastic
film, which provides wind and weather protection for the array and the concentrator. Three concentrators were fabricated and two were optically tested. A laser ray trace of concentrators was used to
determine the local deviations of the surface slope, the best optical axis and the best focal plane.
The concentration ratio, image quality and overall efficiency of the concentrators were determined
using a solar simulator. The results indicate that concentration ratios well above the 200 : 1 g o a l
can be achieved.

0833

Concentrating Photovoltaic Solar Array (CPSA) Conceptual Design Study ; Final Report : Zimmerman D K
and Bishop C J, " Sandia Laboratories Report No. SAND - 77 - 7019 ", (USA), (1977).

The objectives of the study, to assess the applicability of an air supported dome as a protective enclosure for a concentrated photovoltaic system array (CPSA) were : (i) develop a primary conceptual
structural design for a Fresnel lens concentrating photovoltaic system with and without protective
air supported enclosure ; and (ii) estimate preliminary capital and maintenance cost for each system
to assess which concept offers the greatest low cost potential. The study plan and r e s u l t s a r e
presented.

0834

Photovoltaic Concentrator Test and Evaluation Plan : Sandia Laboratories Report No. 78-0945,(USA),(1978).

This report presents the current planning for the test and evaluation of concentrating photovoltaic
arrays and array components. This planning is subject to change and evolution as more data is obtained. The photovoltaic concentrator test and evaluation task involves several related measurement acti-

vities including (i) determination of baseline collector performance and monitoring of effects of real time environmental exposure (ii) evaluation of severe environmental survivability of collector designs and materials,(iii) accelerated aging of collectors and their components and materials, to assist in estimating design lifetime . Data from these measurements is analyzed and used to compare collector designs by performance, operation and maintenance requirements, energy costs (environmental survivability and overall reliability). Potential material problems are identified and remedial actions suggested to the designer. Results are used to help set reliability, lifetime and performance goals for new collector designs. The information generated from this task is employed as input to systems modeling and design and to potential array users.

Tracking of Solar Concentrators

0835

<u>Heliostat Mount</u> : Applebaugh C R, " Final Invention Disclosure (McDonnell Douglas Corporation) No.PB – 242, 935 / 5GA / NSF – 75 – 72 – GI – 39456 ", (USA), (1975).

This device is designed to support and move a large reflector in such a way that t h e reflection of the sun or other moving objects is directed to a fixed target point located at any position in the hemisphere above the horizon. The device must have sufficient strength and rigidity to maintain the reflected image accurately on the target point under gusty wind conditions. This device provides a stronger more rigid design than present devices to maintain accurate positioning under wind loading.

0836

<u>Solar Collector Control System</u> : Beck E J, " US Patent Application 624, 148, Filed Oct. 20,(1975)".

The patent application discloses a system for controlling the movement in azimuth and elevation o f a large number of sun following solar energy collectors from a single central controller. The s y s t e m uses servo signal generators, a modulator and a demodulator for transmitting the servo signals and stepping motors for controlling remotely located solar collectors. The system allows precise tracking of the sun by a series of solar collectors without the necessity or expense of individualised solar trackers.

0837

<u>Thermal Heliotrope – A Passive Tracker</u> : Byxbee R C and Lott D R, " Proceedings 4th Aerospace Mechanical Symposium Jet Propulsion Laboratories ", (USA), p. 127, (1970).

By utilizing solar energy to activate a bimetal helix, a passive heliotrope has been developed. Initial model tests have proved the heliotrope concept and development work is being performed to establish feasibility. Three types of thermal heliotropes,namely a reset tracker, an incremental tracker and a seasonal adjuster are discussed.

0838

<u>Steering a Field of Mirrors Using a Shared Computer Based Controller</u> : Carden P O, " Solar Energy ", 20, 343, (1978).

A system is described for steering a multiplicity of two axis mirrors by means of a shared computer based controller. No local feedback is employed at the mirrors. The system is applicable t o power towers as well as to multi focus paraboloidal mirror systems. The operation and economic advantages of this system over alternatives is discussed and the results achieved with an actual system are given.

0839

<u>Solar Energy Powered Heliotrope</u> : Crawford R, " US Patent 3, 678, 685 / July 25, (1972)".

A solar energy powered heliotrope which functions in a passive, stored energy manner to orient a solar array towards the sun is discussed. A bimetallic motor element is activitated by solar energy to generate a work output. A constant torque spring assembly coupled to the motor element stores u s e f u l energy as a function of the work output. A bimetallic sensing element detects misorientation between the solar array and the sun. An escapement mechanism is actuated by the sensing element whenever misorientation between the solar array and the sun is detected : the escapement mechanism automatically and incrementally releases the stored energy to rotate the solar array in discrete increments until it is oriented towards the sun.

0840

<u>Suntrack Systems for Solar Concentrators</u> : De Gezzele A, " Proceedings 2nd International Solar Forum ", Hamburg (Germany), p.151, (1978).

Methods to track the incident angle of collectors are outlined. Different systems are examined both from a mechanical and electrical view point. The advantages and disadvantages are discussed and test results are given for various systems.

0841

<u>A Combined Digital – Analog Tracker for Terrestrial Applications</u> : Durbin R C, Boone J L and Kern F J, " Proceedings 2nd Annual UMR-MEC Conference on Energy ", Univ. of Missouri,Rolla (USA),p.263,(1976).

A combined digital – analog tracker is suggested to allow maximum efficiency in a solar electrical energy converter utilizing a twelve foot parabolic collector. The analog tracker compares solar beam radiation to ambient (diffuse) light to obtain optimum placement of the collector when the sun is visible. The digital portion of the tracker utilizes a wired program which derives information on solar position from a non-volatile random access semiconductor memory. This arrangement allows accurate mapping of the sun even when the sun is obscured by atmospheric phenomenon which would make mapping impossible.

0842

Synthesis of Four Bar Linkage for Solar Tracking : Dimarogonas A D and Mourikis A, " Solar Energy", 25, 195, (1980).

A method is presented for the synthesis of four bar linkage to provide adequate tracking of solar collectors. The design procedure starts with the selection of a number of accuracy points on the altitude versus hour function to yield a first form of a four bar linkage. An optimization algorithm improves on the original design for minimum error and optimum structural characteristics. Based on the method, mechanisms were designed for a certain locality with negligible tracking error, acceptable even for focussing collectors. The design procedure can yield inexpensive, yet accurate tracking with very simple seasonal adjustments.

0843

Sensors in the Automatic Control of a Solar Power Installation : Egorov A V, Elevich G V, Shavrin N V and Shulmeister L F, " Applied Solar Energy ", 4 (4), 52, (1968).

Photoelectric sensible elements in automatic control system of a solar energy installation a r e described. Connecting diagrams of these elements on load are given. Construction a n d recommendations for use are also described.

0844

A Novel System for Rapid Acquisition of the Sun in Solar Energy Conversion Systems : Eyman E D a n d Weakley R H, " International Journal of Control (First Series) ", 5, 23, (1967).

A method based on momentum exchange from rotating flywheels to a platform, to initially acquire t h e sun is described. The basic principle involved in the Momentum Exchange Acquisition System (M E A S) concept is that a large short term torque can be applied to the platform by breaking a small high speed inertia wheel. This can be accomplished with a very small expenditure of power by e m p l o y i n g two counter-rotating flywheels. One flywheel is braked to give the platform an impulse which will start rotating it about the corresponding axis; the second flywheel is braked to stop the platform. The acquisition sequence is described, level detector settings are discussed and the problem of c o m p u t e r simulation is discussed.

0845

Passive Solar Array Orientation Devices for Terrestrial Application : Fairbanks J W & Morse F H, (Unknown).

A passive solar array orientation device, called a thermal heliotrope is described and several terrestrial applications are illustrated. The thermal heliotrope consists of a bimetallic helical coil that serves as the motor element, producing torque and angular displacement. A control mechanism in t h e form of one or more shades completes the basic device. In comparison with electro-mechanical tracking systems, the thermal heliotrope is electrically passive,has relatively few parts, and is low cost.The principle of operation, several models built for space application and the design considerations, for several terrestrial thermal heliotrope units are presented. It is suggested that the use of thermal heliotrope for solar array orientation could significantly reduce array cost, thereby increasing the competitive economic posture of solar arrays for terrestrial applications. The thermal heliotrope modified for terrestrial use is readily acceptable to orient solar energy concentrators.

0846

Solar Powered Tracking Device for Solar Concentrators : Farber E A, Morrison C A, Ingley H A, Pytlinski J T and Wiggins D B, " Journal of Environmental Sciences ", 18, 11, (1975).

In support of a solar cooking project for the University of Florida Solar House, a project was initiated to design and construct a concentrating device with a tracking mechanism which would be completely solar powered and independent of conventional energy sources. This article discusses the techniques used in the development and construction of this solar powered, fluid mechanical tracking device.

0847

Dynamic Response Analysis of a Solar Powered Heliotropic Fluid – Mechanical Drive System : Farber E A, Ingley H A, Morrison C A and Cope N A, " Proceedings 2nd International Conference on Alternative Energy Sources ", Miami (USA), (1979).

The need for a tracking mechanism developed while researching methods of providing a 24 hour cooking capability with solar energy. Decisions were made to utilize a liquid heat transfer medium (cottonseed oil) and a sensible thermal storage system in association with a large parabolic concentrating collector to implement this cooking capability. In order to maximise the solar collection by t h i s collector, a tracking device was necessary to drive it from an east facing to a west facing orientation. Furthermore, it was decided to consider a feedback control instead of a clock motor. The results of this research produced not only a drive mechanism with feedback control but also one w h i c h required no external utilities source for its power. The unit, as presently designed utilizes a l o w boiling point fluid and two power cylinders connected to a drive mechanism for the collector.Pressure differences in the sensors due to fluctuations in collector alignment provide the driving potential to maintain the collector in proper aspect to the sun. The paper provides design data and operational data for this solar powered tracking device. Illustrations of its use and a discussion of its utilization for other solar energy projects will also be presented.

0848

A Clockwork Driven One Axis Tracking System : Gupta K C, Borania C H, Jain P K and Sathe A P," Proceed-

112

ings Indian National Solar Energy Convention ", Bombay (India), p.114, (1979).

Tracking is achieved by rotating the concentrator around an axis parallel to the axis of the earth. This suffices to account for the principal component of the apparent motion of the sun, the rest being taken care of by once‑a‑day manual adjustment. The concentrator assembly consists of the cages of total area 1 sqm. The assembly is mechanically balanced and mounted on pivot bearing so that the little power required to rotate it can be provided by a system of falling weights controlled by clock work. Further, it is shown that wind load should not hamper the operation of the system.

0849

Simple Solar Tracking System : Gupta K C & Sathe A P, " Proceedings ISES Conference ", New Delhi(India), p. 1336, (1978).

A falling weight controlled by a pendulum drives a shaft which is parallel to the axis of the earth at a rate of one revolution per day. The concentrator (convex lens) is mounted on the shaft and it thereby tracks the sun. Changes in the declination of the sun are accounted for correctly when the device is adjusted manually to point towards the sun each morning. A method is presented for correctly aligning the rotating shaft parallel to the axis of the earth by making observations of the focal spot of the sun.

0850

Passive Solar Tracking System for Steerable Fresnel Elements : Hitchock R D," US Patent 3, 986, 021/October 12, (1976)".

An angular tracking servo system is described for passive tracking of the sun. Solar radiation is used to control the altitude of a mirror element in an array of Fresnel reflectors. The array collects and focusses solar energy onto a high efficiency conversion device. The energy required to move the mirror element is supplied by a gear system which is attached through a pivot arm to a vertically moving float immersed in a chamber containing water.

0851

Optimal Control of Sun Tracking Solar Collectors : Hughes R O, " Proceedings ISES (American Section) Conference ", Denver, Colorado (USA), 2.1,913, (1978).

Using the disciplines of Modern Control Theory, an optimal tracking control for a point focussing solar concentrator is derived. By converting the tracking problem into a regular problem with a t r i m rate input a very low positioning error is achieved. A representative example with a corresponding computer simulation is presented.

0852

Sun Tracker Control System : Kerner T, " International Jet Propulsion Laboratory's Space Program Summary No. 37‑33", (USA), 4, 52, (1965).

A sun tracker control system designed to keep a parabolic mirror pointed at the sun to within a 30‑second error is discussed. The mirror drive has two degrees of freedom, one in azimuth and one in elevation. For these two axes the two central systems that govern the motions are identical. A block diagram of the mirror control system and also that of a breadboarded system in the laboratory with some modifications is presented. The linear sensor is substituted by a potentiometer generated voltage in the loop, and the gear train driving the mirror has been reduced to a typical one-stage gear t r a i n driving a potentiometer. The system performed satisfactorily in the closed loop configuration.

0853

A Guiding System for Solar Furnaces : Laszlo T S, deDufour W F and Erdell J, " Solar Energy ", 2 (1), 18, (1958).

An electronic system was developed to guide solar furnaces and instruments of similar axial arrangement. The driving motors are supplied with a basic current. To this is added algebraically a correction current developed by a photosensing and amplifier system. This arrangement permits precise even following of the sun's apparent motion. The system is exceptionally free from "hunting".

0854

Development of a Heliotropic Orientation Device for High Concentration Ratio Solar Power Generation Systems : Leovic W J, " NASA Report No. NASA‑CR‑57020 ", (USA), (1964).

In this paper the design and evaluation of a passive orientation device for use in alleviating t h e stringent sun‑vector alignment requirements associated with lightweight, high quality, paraboloidal concentrators is described.

0855

Solar Concentrator with Polyester Film for Reflecting Surface and Pendulum Arrangement for Tracking Movement : Marathe C R, " Proceedings Indian National Solar Energy Convention ", Bhavanagar (India), 638, (1978).

In this paper a low cost design for a solar concentrator of parabolic cylindrical shape with pendulum arrangement to control its daily rotation is described.

0856

A Guiding System for a Solar Furnace : Moore J G & Amand P St.," Solar Energy ", 1(4),27, (1957).

A guidance system for solar furnace is described. This system will automatically track the sun about any two orthogonal axes. A single photosensitive detector is used in conjunction with a rotating shutter to generate an A‑C error signal. The amplitude of the error signal is dependent upon t h e radial displacement of the solar image from the optic axis. The phase of the error signal is dependent upon the angular position of the image. The error signal is amplified and used to operate t w o motors which direct the seeking system so that the optical axis is pointed towards the sun.

0857

Solar Powered Tracking Device : Morrison C A, Farber E A, Ingley H A and Wiggins D B, " Building Systems Design ", 73(1), 7, (1975 - 76).

The paper discusses the design, construction and testing of a completely solar powered tracking de - vice. Two opposed power cylinders respond to heat sensitive sensors so as to create a mechanical ba - lance of the system when the surface of the concentrating device is normal to the incident solar rays. While temperatures in excess of 200 deg F are possible through the use of flat plate solar collectors, the mass flow rate of effluent under such conditions is diminished to such an extent that it becomes impractical to use this type of device when it is required to generate high temperature fluids.

0858

Response Characteristics of a Thermal - Heliotrope Solar Array Orientation Device : Morse F H, " Pro - ceedings 5th Aerospace Mechanical Symposium ", p.33. (1971).

The thermal heliotrope is a passive solar array orientation device containing a bimetallic helix that rotates when activated by solar energy. The rate and extent of the rotation depends upon the proper - ties of the two metals and the temperature of the helix. An energy balance analysis is performed to determine the temperature distribution in the helix. By initially restricting the analysis, a simpli - fied equation governing the response of the heliotrope is obtained. In order to gain insight into the response of the heliotrope, a series of experiments were performed. The results of these tests and the implications are presented.

0859

Sun Direction Detection System : Schmidt L F and Pace Jr. J D, "US Patent Application 616 472/GA, (1975) .

A sun sensor detection system is described which includes an illumination detector and a sun angle de - tector. The illumination detector provides a low resistance output wherever the sun is within a selec - ted field of view and a high resistance output whenever the sun is outside the field of view. The sun angle detector provides an output voltage related to the direction of the sun with respect to the nor - mal direction. The output of the sun angle detector is fed to the altitude control circultory to con - trol the vehicle altitude with respect to the sun as a function of the output of the sun angle detector.

0860

Tracking and Shadowing Models for Solar Collector Systems : Schimmel Jr. W P, Hickox C E and Lee D O , " ASME Paper No. 77 WA/Sol - 8 ", (USA), (1977).

A unified, systematic method, based on vector algebra , is presented from which numerous geometric re - sults, important to the effective collection of solar energy, can be deduced. Results were obtained in a straightforward manner from the first principles. Since the results are general in nature, local weather bureau data can be used effectively when comparing various proposed systems. Furthermore, the generality of the results allows the influence of system parameters to be investigated with a minimum difficulty . In addition to dealing effectively with the tracking requirements of various solar col - lectors and heliostats, the analysis yields convenient solutions to several shading problems of some importance in solar energy collection.

0861

Solar Tracking Control System(Sun Chaser): Scott D R & White P R, " Report DOE/NASA/TM-78225 ",(1979).

The solar tracking control system (sun chaser) is believed to be an improved method of tracking the sun in all types of weather conditions. The sun chaser will follow the sun from east to west in clear or cloudy weather and reset itself to east position after sun down in readiness for the next sunrise. A description of the sun chaser hardware and its operation together with results is presented.

0862

On Automatic Solar Tracking Unit : Sharafi K Sh., " Applied Solar Energy ", 1(1), (1965).

Two solar tracking mechanisms designed to maintain the orientation of a solar energy collector are described. Both systems operate on the thermal effect of the sun. The first system consists of two thin walled metal tubes attached to a U - shaped glass tube containing mercury. Electric contacts are located in the glass tube at either end and form open circuits with a motor which can rotate the col - lector. When the unit becomes disoriented,the sun's rays fall on one of the metal tubes. The gas in - side becomes heated and expands, and displaces the mercury towards a contact which closes the circuit and operates the motor to reorient the collector. The second system consists of two blackened bime - tallic plates parallel to each other and fastened on one end. An insulated fork is attached to the free end of one plate with the free end of the other plate inside of the fork. When heated by the sun the plates drop their free ends closing a contact in the fork which operates the motor and reorients the collector.

0863

A Low Cost Electromechanical Heliostat Sensor and Control System : Thornton J P and Clevett M L , " Proceedings ISES Conference ", Los Angeles (USA), (1975).

Studies have indicated that the heliostat sensor and control assembly, including supporting structure, may represent upto 45% of the initial installation costs. Calculations show that an overall accuracy, including heliostat and receiver tower deflection, in the order of ±1 degree is desirable during oper - ational periods for maximum system efficiency. To keep the system performance within this limit, the sensor and control system must have a pointing accuracy in the order of ±5 arc minutes. This paper describes a sensor and control system which employs a unique solid state electronics circuit using silicon solar cells as sensors in conjunction with modified off the shelf electromechanical drive sys - tems. A breadboard model system, including heliostat, was designed, built and tested at the Martin Marietta Solar Test Facility. Initial tests with a small heliostat indicate a tracking accuracy with - in ±2 minutes of arc.

0864

<u>Solar Tracking System</u> :   White P R & Scott DR, " US Patent Application 6060435, Filed Feb.25, (1979)."

A solar collector is angularly oriented by motor wherein the outputs of two side by side photodetectors are discriminated as to three ranges, corresponding to a low light or darkness condition by east control circuit to light intensity lying in an intermediate range, direct sunlight by differential tracking circuit. The first output drives the motor to a selected maximum easterly angular position to await sunrise ; the second enables the motor to be driven westerly to the earth's rotational rate when clouds are present which prevent accurate tracking ; and the third output, differentially controls the the direction of rotation of motor through east drive relay and west drive relay to effect tracking of the sun.

0865

<u>Computer Control in Sun Tracking Mechanism for Solar Thermal System</u> ; Yamada T, Noguchi T and Tsukada K, " Proceedings ISES Silver Jubilee International Congress ", Georgia (USA), (1979).

In order to design the highly accurate suntracking mechanism in the heliostat of solar thermal p r o cesses such as solar furnace, solar power plants etc, the programmed control by a digital computer & automatic suntracking mechanism with the photosensor have been coupled. The tracking accuracy so defined is within 1/120 degrees.

0866

<u>Solar Tracking Device</u> : Zerlaut G A & Heiskell R F," US Patent 4031385/January 21, (1977)."

A solar tracking system is disclosed, utilizing a photovoltaic device, for following the changing solar position. When the sun ceases to be the brightest object in the sky, such as during periods of partial cloud cover, the system seeks to acquire the brightest in preference to the sun. The photovoltaic device is automatically overridden. The override system is clock driven and moves the tracking system to approximate the solar position so that the sun can be immediately reacquired by t h e photovoltaic device when the sun reemerges as the brightest object in the sky.

0867

<u>Automatic Solar Tracker</u> : " Automatic Solar Tracker, published by NASA ", Pasedena, California (USA), (1975).

No servopower is required by a device which tracks the sun to focus energy on a collector.   Differential pressure of a considerable fluid such as freon against a fixed position equalizes the radiant energy on a pair of black body elements which heat the fluid to drive the mechanism.

Materials Problems in Solar Concentrators

0868

<u>Study of Metallized Reflecting Surface Aging</u> : Alimov A K, Novikova I A and Sutyagine V M, " A p p l i e d Solar Energy ", 14 (4), 45, (1978).

The effect of sunlight and weather on the metallic (aluminium) coating of metallised polymer films is studied. Recommendations are given on protection of the metallic layer.

0869

<u>A Study of the Reflection Efficiencies of a Water White Silver Backed Glass Mirror at Different Angles of Incidence</u> : Allison F and Steele K, " Air Force Missile Development Center Report AFMDC – TN –59–27 ", Holloman Air Force Base, New Mexico (USA), (1959).

The reflection efficiencies of the silver backed water white glass mirror contemplated for the Holloman solar furnace have been determined at various angles of incidence, by both, computational and experimental methods. The results indicate practically negligible differences in the mirror efficiencies at all angles at which the solar radiation will impinge on the heliostat and the condensor mirrors of the presently designed furnace. The mirror efficiency experimentally determined approximates 92% of the incident angles considered.

0870

<u>Environmental Effects on Solar Reflector Structures</u> : Allred R E, " Proceedings ISES Silver J u b i l e e International Congress ", Georgia (USA), (1979).

Long term stability in the outdoor weathering environment is a primary requirement to the economical use of solar energy. Environmental threats to solar reflector structures include thermal and mechanical fatigue, moisture, ultraviolet radiation, corrosion, hail and reaction with atmospheric pollutants. A reflector structure is defined here as the structural backing to the reflective surface of a solar collector. The purpose of the reflector structure is to provide and maintain the desired geometry of the collector and to protect the reflective surface from hail, wind and corrosion. A program is being conducted at Sandia Laboratories to determine the response of candidate solar reflector structure materials to those threats. Candidate materials include paper and aluminium core honeycomb with steel, aluminium and plywood face sheets, sheet moulding compounds (SMC), foamed glass, plywood, fibreglass and mycarta face sheets and isocynurate foam core with steel face sheets.   Specific results presented include observations of accelerated and real time aging of reflector structure materials, hail impact on sandwhich panels as a function of the diameter and observations on the degradation of silvered glass reflective surface. Experimental methods, developed for these studies have also been presented.

0871

<u>Glass for Solar Concentrator Application</u> : Bouquet F L, " Jet Propulsion Laboratory Report No. JPL-5102-105 ", (USA), (1978).

Materials for highly reflective surfaces for applications to solar thermal power systems are treated in this report. Their primary emphasis is on the use of second surface glass mirrors with comparison to alternate candidates. Flat, parabolic and Fresnel lens systems are contenders for solar thermal concentrators of various power requirements. The emphasis in this report is on glass for parabolic reflective surfaces. The technology status and experimental data on glass, metallic and polymeric concentrators are presented.

0872

<u>Selection of Materials and Development of Processes for the Fabrication of Solar Reflector P a n e l s</u>: Boyner M R and Laur H K, " Proceedings National 11th SAMPE Technical Conference ", Boston (USA), page 424, (1979).

The MDAC Central Receiver solar thermal power system is based on the concept of multiple solar energy reflectors (solar reflector panels) directing the sun's energy on a receiver unit, where the concentrated heat from the sun is used to convert water into steam. This study presents the trade studies and tests performed to select the core material, the back face sheets and the adhesives and sealants used to bond and seal the reflector panels. The processes for adhesive application and panel assembly are discussed.

0873

<u>Solar Reflectance of Unprotected and Protected Aluminium Front - Surface Mirrors</u> : Bradford A P & Hass G, " Solar Energy ", 9, 32, (1965).

Solar reflection and absorption characteristics of mirror coatings were studied. High vacuum evaporation method for producing highly reflecting coatings, and film combinations with any desired values of solar reflectivity and thermal emissivity are reviewed. It is shown that films of uniform thickness over large areas can be easily obtained by use of suitable shutters and rotating sector wheels & by well controlled movements of mirror substrates during vacuum deposition.

0874

<u>Electroforming Aluminium Composites for Solar Energy Concentrators ; Final Report</u> : Buschow A G, H e s s I J and Schmidt F J, " NASA Report No. NASA - CR - 66322 ", (USA), (1967).

The tests, experiments, developments and studies relating to the optimization of an aluminium electrodeposition process are described. The efforts towards increasing the strength of the aluminium deposits by codeposition of glass fibers and hollow silica microspheres, as well as a preliminary design of a plating cell, scaled by a factor of 15, which would be suitable for electroforming upto 10 ft diameter solar concentrators are also discussed. Significant increase in electroform s t r e n g t h were achieved.

0875

<u>Composition and Method for Electroforming Substrates</u> : Carlson J A and Lui K, " US Patent Application No. SN - 112365 / February 3, (1971)".

An electrolytic bath and technique for electroforming large area foil structures and particularly a hollow core aluminium substrates for a light weight solar panel are discussed. The electrolytic bath includes an amount of at least 0.3 molar to about 1.0 molar and a mixture of an aromatic and an aliphatic ether preferably containing at least 10% to about 60% aromatic ether, suitably about one third aromatic ether. Improved plating is obtained by recirculating and filtering the plating bath, controlling the temperature of the bath between 10 deg C and about 30 deg C and controlling the c u r r e n t density between about 10 to 25 amperes per square foot.

0876

<u>Development Status of Aluminium Solar Concentrators</u> : Castle C H and Kovalcik E S, " Paper at A I A A/ ASME 3rd Biennial Aerospace Power Systems Conference ", Philadelphia (USA), (1964).

Review of the stretch formed aluminium method of fabrication of solar concentrators, as well as t h e thick film deposition of aluminium by vacuum evaporation is presented. Stowage problems are discussed along with problems of structure. Temperature problems resulting from the launch and heat input from the sun, earth and payload are considered. The results of several programs are discussed : Sunflower, binary concentrator - radiator, 60 inch thermonic concentrator. The performance o f t h e concentrator with respect to surface errors is based upon data obtained from aluminium concentrators. A l i s t o f technological advances and a consideration of further areas of research are mentioned.

0877

<u>Research and Development in Solar Mirror Quality Assurance Performance; Annual Report</u> : Coleman W J, Lind M A, Hampton H L and Gordon N R, " Pacific North West Laboratory Report PNL-2427",(USA), (1977).

A survey of the current technological status of solar reflectors has been initiated for the Energy Research & Development Administration (ERDA). The survey is directed towards the areas of quality assurance and performance of solar mirrors. This report is intended to present a program overview and give the initial results of some preliminary studies currently under way. Specific topics that a r e covered in this report include : (i) An extensive literature survey of the weatherability, durability & life expectancy of solar materials has been initiated. Some preliminary results of that s u r v e y are discussed. (ii) General optical degradation effects on solar materials are discussed:(iii) Preliminary results of recently initiated studies on the aging of solar materials are presented. The need for careful materials preparation is emphasised. (iv) A study which addresses the problem of cleaning of solar reflectors has been initiated. Some preliminary results on cleaning agents are presented. (v) The need for careful material preparation is emphasised. (iv) A study which addresses the problem of the cleaning of solar reflectors has been initiated. Some preliminary results on cleaning agents are presented. (v) The need for higher accuracy of laboratory instrumentation for aging studies is emphasised. (vi) The development activities associated with a low cost easy to use instrument to measure specularities are presented.

0878

Effect of Outdoor Exposure on the Solar Reflectance Properties of Silvered Glass Mirrors : Freese J M, " Sandia Laboratories Report No. SAND 78 - 1649 ", (USA), (1978).

Results of the experiments performed to obtain a broad general understanding of the effects of dust accumulation and weather conditions on glass mirror reflectivity have been presented. The specular reflectance loss of silvered glass heliostat mirrors has been studied over a nine months period during which the mirrors were exposed to outdoor weather conditions. A bidirectional reflectometer was used to measure the reflected beam profile of approximately 9 cm square samples exposed to the environment on outdoor racks.

0879

Development of Portable Reflectometer for Field Measurements of the Specular Reflectance of Solar Mirrors : Freese J M," Proceedings ISES Silver Jubilee International Congress ", Georgia (USA), (1979).

A portable reflectometer designed for in the field measurements of solar mirror materials has been developed. This instrument is of a convenient size for use by one operator in an outdoor environment. Instrument design, operation and calibration is discussed.

0880

Material Degradation and Corrosion Problems of Solar Concentrators : Frohardt M W and Fryberger T, " Proceedings ISES Silver Jubilee International Congress ", Georgia (USA), (1979).

To achieve the high temperatures necessary for electric power generation, solar concentrators are used to reflect and direct solar energy to a receiver (absorber). Large reflector surface areas must endure outdoor exposure and maintain high optical efficiency for the projected life of the system (30 years). This paper gives a brief overview of solar reflector technology and the materials degradation and corrosion problems associated with it.

0881

Development of Defects in Reflective Coatings Exposed to High Temperatures : Gaziyev U K, Koltun M M & Trukhov V S, " Applied Solar Energy ", 8 (4), 109, (1972)..

The effect of high temperature on aluminium mirror coatings with a metal oxide surface protective film is described. High temperature treatment results mainly in the formation of openings in the reflective coatings defect and rupture concentrations in the reflective surface smoothness have also been observed on the surface of the samples. These alterations result from the interaction of the aluminium film with the glass and the protective coatings and also from oxidation process.

0882

Thermal Stability of Mirror Reflecting Coatings with Various Protective Backings : Gaziyev U K, Koltun M M and Trukhov V S, " Applied Solar Energy ", 9 (1), 26, (1973).

The paper describes an attempt to prevent the interaction of aluminium film with the substrate under the impact of high temperature. This is done through the application of intermediate layers between the glass and aluminium reflecting surfaces.

0883

Studying the Effect of Increased Moisture on Mirror Coatings of Solar Energy Concentrators : Gaziyev U K, Koltun M M and Trukhov V S, " Applied Solar Energy ", 10 (2), 71, (1974).

To consider the influence of moisture on reflection properties, reflecting coatings of several compositions are tested and the results of the elevated moisture on coatings is tabulated.

0884

Weather Testing of Solar Utilization Materials : Giligan J E, Brzuskiewicz J and Gaumev S J, " Proceedings ERDA Conference on Concentrating Solar Collectors ", Georgia (USA), p.5 - 15, (1977).

A program of research and experimental testing is described in which the optical and mechanical performance of materials for use in solar energy utilization devices is determined before and after exposure to outdoor weathering tests. Materials which are currently in use and others which are being considered or developed for these applications are being exposed to outdoor weathering in P h o e n i x (Arizona), Miami (Florida) and Chicago (Illinois). The results of these tests, primarily the effects of outdoor exposure on optical and physical properties, will ultimately be compiled in a handbook, along with cost, availability and other pertinent information.

0885

Heliostat Reflective Surface Substrate Characterization : Gordon N R, " Pacific North West Laboratory Report No. PNL - 2810 ", Washington (USA), (November 1978).

The purpose of the survey and analysis is to provide background information and procurement, specifications for materials which have potential application in heliostat construction. The work performed under the reflective s u r f a c e s u b s t r a t e s characterization and specifications task is covered. The properties of foam core sandwich constructions are emphasised. The research studies reported were directed towards determining the relationship between viscoelastic deformation of the adhesive and deflection of the mirror module.

0886

Evaluation Techniques for Determining the Reflectivity, Specularity and Figure of Solar Mirrors :

Griffin J W, Lind M A and Philipp L D,"Solar Energy Research Institute Report SERI / TR - 98366 - 1, (1980).

This report is intended as a primer for those persons interested in the optical evaluation of heliostat mirrors. A cursory review of the current options for measuring the reflectance, specularity,and figure of mirrors is presented. The extension of both traditional and novel optical laboratory techniques to field applications is also discussed.

0887

The Evaluation of Solar Mirror Figure by Moire Contouring : Griffin J W and Lind M A, " Pacific Northwest Laboratory Report PNL - 3286 ", (USA), (1980).

This report details the theoretical and experimental considerations necessary to fully implement Moire topography on mirror surfaces. A procedure to de-specularize the mirror is demonstrated which conserves the surface morphology without damaging the reflective surface. The Moire fringe patterns observed for the actual mirror facets are compared with theoretical contours generated for representative dish facets using a computer simulation algorithm. A method for evaluating the figure error of the real facet is presented in which the error parameter takes the form of an average absolute deviation of the surface slope from theoretical. The experimental measurement system used for this study employs a 200 line / inch Ronchi transmission grating. The mirror surface is illuminated by a collimated beam at 60°. The fringe observation is performed normal to the grating. These parameters yield contour intervals for the fringe patterns of 0.073 mm. The practical considerations for extending the techniques to higher resolution are discussed.

0888

Specularity Measurements by Fourier Transform Examination : Hampton H L, Hartmon J S and Lind M A, " Paper at the Institute of Environmental Science Seminar on 'Testing Solar Energy Materials and Systems'"Gaithersburg (USA), (1979).

This paper describes how the Fourier transform technique can be used to examine the scattering properties or specularity of solar reflecting materials and also discusses some instrumentation which is being developed to measure the effect.

0889

Use of Flexible Reflective Surfaces for Solar Energy Concentration : Jacobi W I, " Journal of Vacuum Sciences and Technology ", 12, 169, (1974).

A flexible metallized film, stretched drum-head fashion, can form a mirror that can be oriented to direct sunlight to a central receiver. The use of taut membranes is being explored as part of a study of the feasibility of a large scale solar thermal power plant. The evaluation of any reflector requires, measurement of the spectral response, specularity and overall flatness. Integrating the spectral response to the energy distribution of sunlight gives the total reflected energy. Of equal importance is the specularity or divergence of the reflected beam. The combination of these factors will determine a candidate reflector's suitability . A device for measuring the bidirectional reflectance distribution function is described and data are presented. A discussion of membrance reflector development and data from live tests are presented.

0890

Reflection Coefficient for a Black Surface Glass Mirror : Kent T and Vant Hull L L," Proceedings ISES Conference ", Winnipeg (Canada), (1976).

Many systems, the solar tower in particular, use back silvered mirrors to reflect solar energy to a collector while the variation with angle of incidence of the transmission of light through glass is well documented, mirrors have not previously been considered. Studies in the variation of reflectivity with angle of incidence are presented considering multiple reflection, polarization, spectral thickness and absorptivity of the glass. In most cases of interest, the reflectivity varies only a few per cent for angles of incidence between ±80°, and of course goes to unity at grazing incidence(90°).

0891

Specularity Measurements for Solar Materials : Lind M A, Hartmon J S and Hampton H L, " Proceedings SPIE Symposium ", 161, 98, (1978).

A technique using Fourier transform analysis which is suitable for measuring the specularity of Solar glass components in the m rad and sub-m rad is discussed and demonstrated. A brief mathematical background as well as illustrative examples are included. A number of methods for image analysis are discussed with particular emphasis given to electronic integrating detectors. Typical Fourier plane image distributions are given for a few common solar utilization materials and details of the instrument used to produce the images are considered. The limitations and capabilities of various instruments are outlined alongwith methods for further enhancing the utility and sensitivity of the technique.

0892

Heliostat Glass Survey and Analysis : Lind M A and Rusin J M, " Pacific Northwest Laboratory Report PNL - 2868 ", (1978).

A comprehensive survey of both foreign and domestic sources of low distortion, high transmission flat glass with a nominal thickness of 3 mm was undertaken. The purpose of the survey was to determine the characteristics, availability and cost of glass for use in second surface heliostat mirrors for the Barstow pilot plant and future commercial central receiver plants. Information obtained from the manufacturer and the results of the investigation performed at Sandia Laboratories at Albuquerque and Livermore were compiled with the PNL characterization data to generate the specifications for the GFE glass to be used in the Barstow pilot plant. During the course of the survey, nine of the major glass manufacturers were contacted for information and assistance. These manufacturers were PPG, Corning, Ford, LOF, CE, ASG, Fourco, Sahold & Guardian. Eleven different flat glass samples from seven of the domestic sources and one foreign source were characterised for solar transmittance, flatness and durability. The samples were representative of four different manufacturing processes : float, fusion, rolled and twin ground. The results and implications of these glass characterization studies and a brief summary of the manufacturer survey are presented in this report.

118

0893

Summary Report of the Solar Reflective Materials Technology Workshop : Lind M A and Ault L E, " Pacific
Northwest Laboratory Report PNL - 2763 ", (1978).

This document provides an overview of the Solar Reflective Materials Workshop which was held in Denver
(Colorado), on March 28th - 30th, 1978. No attempt has been made to reproduce the complete proceedings
of the conference : but the authors of this summary have compiled the highlights of the presented papers and workshop sessions. An executive summary is included in the first part of the document which
outlines the basic purpose and organization of the conference and summarises the presentations and the
conclusions of the participants.

0894

Heliostat Mirror Survey and Analysis : Lind M A, Buckwatter C Q, Daniel J L, Hartmon J S, Thomas M T and
Pederson L R, " Pacific Northwest Laboratory Report No. PNL - 3194 ", (USA), (1979).

The purpose of this work is to provide background information and procurement specifications for materials and processes which have potential applications in heliostat construction. This report covers
the work performed under three tasks: the mirror industry survey, mirror degradation analysis and advanced concepts for mirror protection. The first three chapters of this report are devoted to this
purpose. While the results presented here are by no means conclusive, they are a solid first step in
what should be a much more comprehensive investigation of the solar mirror lifetime problems.The concepts fall into two general categories : encapsulation and modification of the deposition process,
excluding it, may extend lifetime. Several more techniques currently being investigated for encapsulation are discussed. In addition, some exciting new work on modifying the chemistry at the glass/
silver interface is discussed.

0895

Research and Development of Magnesium / Aluminium Electroforming Process for Solar Concentrators Final
Report Mar 13, 1966 - May 10, 1967 : Lui K, Guidotti R and Klein M, " NASA Report No.NASA-CR-66427 ",
(USA), (1967).

Studies to optimize organic solvent plating baths for the deposition of aluminium and magnesium were
conducted. The program was divided into electrotechnology studies, process development and electro -
forming studies. High purity aluminium was electroformed from a mixed ether bath of aluminium chloride
and lithium aluminium hydride which showed potentially useful structural characteristics. The effects
of current density, bath impurites and various solvent and salt additions were studied. The aluminium
electroforming process was scaled up and an aluminium paraboloid mirror, 30.5 cm was electroformed to
demonstrate the bath's practicability.

0896

Experimental Investigation into Reflectances of Mirror Surfaces : Mavashev Yu. Z, Rudshtein V L, Sali-
kova F S, Khudaeva G B and Valleva I A, " Applied Solar Energy ", 13 (4), 20 , (1977).

It has been established that when mirror elements are used in solar installations and they must operate
in vacuum, at low temperatures, and in the presence of vapours and gases preference should be given
to electropolished reflecting surfaces.

0897

Device for Measuring the Spectral Sensitivity of High Temperature Materials : Mavashev Yu. Z, Nechaeva
L V, Arifov P Y, Kulagin A I and Khodzaeva T N, " Applied Solar Energy ", 15(2), 41, (1979).

The article describes a device for measuring the spectral emissivity and true temperature of materials
at high temperature, the device is based on a solar furnace. Values of the spectral emissivity of
some materials as measured using the device in question are presented.

0898

Specular Reflectance properties of Mirror Materials : Pettit R B," Proceedings ISES Conference",Winnipeg
(Canada), (1976).

The absolute specular reflectance of flat mirrors at discrete wavelengths over the range 400 mm   to
900 mm is obtained as a function of the reflected beam width from 10 m rad (0.057°) to 16.9 m rad
(0.97°). For some of these material, the reflected beam profile could be described by a normal distribution. In this case the reflectance properties are completely characterized by solar reflectance 'Rs', and a standard deviation. However for other materials, the reflected beam profile is described by the sum of the two normal distributions. Data for selected material and examples of the
curve fitting procedure are presented.

0899

Characterization of the Reflected Beam Profile of Solar Mirror Materials : Pettit R B, " Solar Energy", 19, 733, (1977).

The absolute specular reflectance of flat mirrors was obtained as a function of the reflected beam
width from 1 m rad to 17 m rad using a bidirectional reflectometer . The characteristics, calibration &
advantages of this reflectometer are discussed. The mirror materials studied include samples of silvered glass, metallized plastic films and polished, bulk aluminium obtained from a variety of commercial supplies. For some of these materials, the reflected beam profile could be described by a single
normal distribution. However, for the other materials, the reflected beam profile was described by the
sum of the two normal distributions. Examples of the curve fitting procedure and data for selected
materials are presented.

0900

Laser Ray Trace Bi-directional Reflectometry Measurements of Various Solar Concentrators : Pettit R B
and Butler B L, " Proceedings ERDA Conference on Concentrating Solar Collectors ", Georgia (U S A),
p.6 - 31, (1977).

In order to characterize the optical quality of reflecting parabolic solar concentrators, two measurement techniques were developed. Laser ray trace measurements were used to determine the focal length and focussed beam size of the parabolic structures, while bi-directional reflectometry was developed to measure the specular reflectance properties of mirror materials. The overall optical quality of a concentrator is calculated by combining results from these two measurements. Model parabolic structures were fabricated by commercial supplier from a variety of materials including forest, products metals ceramics and plastics. Depending upon the materials and manufacturing processes, average focal lengths were within ±4% of their design value while slope errors ranged from ~1.5 m rad to ~8 m rad. Mirror materials studied included silvered glass, metallized plastic films and polished aluminium sheeting. For some mirrors, the specular beam profile was characterized by a normal distribution. Thus the reflectance properties are completely described by a reflectance and a standard deviation. However, for the remaining mirrors, the specular base profile was characterized by the sum of the two normal distributions. Measured solar reflectance values ranged from 0.83 to 0.96 while standard deviation values ranged from ~0.2 to over 30 m rad.

0901

<u>Structural and Material Optimization for Solar Collectors</u> : Reuter Jr. R C and Allred R E," Proceedings ISES Silver Jubilee International Congress ", Georgia (USA), (1979).

Sandia Laboratories has been involved in a program which has a strong emphasis on the definition and selection of viable, line focussing collectors for large scale solar energy applications. Specifically these collectors take the form of 90 degree rim angle parabolic troughs with silvered glass reflective surfaces. This power deals with the structural and material considerations which have been brought to bear on the problem of eliminating unqualified materials and structural designs from the myriad of candidates, and to optimize those which demonstrate technical merit for this application. The paper covers detailed descriptions of the performance requirements placed upon the reflector support structure, analytical methods applied for structural evaluation and material considerations and tests utilized for candidate material appraisal. Results are presented in such a way that the collector system designer can select a material and construction category for his reflector support that will be technically adequate and yet have a potential for low cost.

0902

<u>Development and Application of the Moire Method in Solar Engineering Problems</u> : Sucherpakov Yu. K, & Tairova L P, " Applied Solar Energy ", 13(5), 6, (1977).

The capabilities of the Moire method in measuring the rotation angle of the normal and the curvature of reflecting surfaces used in solar engineering are examined. An approach is proposed that makes it possible to study the deformation of surfaces of arbitrary form. Techniques for visualizing the Moire pattern are discussed. Results are presented of studies of the formation of reflecting surfaces of circular and hexagonal film type facets.

0903

<u>Use of Electrostatic Spraying to Produce Solar Energy Concentrators</u> : Sobirov O Yu, Gafurov A M, Vilkova S N, Zakhidov R A and Umarov G Ya, " Applied Solar Energy ", 10(5), 13, (1974).

An electrostatic spraying method of distributing the frozen granules of polymer resin uniformly over the surface of a metallized polyethylane terephthalate film for manufacturing solar concentrators is discussed.

0904

<u>Solar Absorptance of Second Surface Mirrors for High Angles of Incidence</u> : Stultz J W," Journal of Spacecraft and Rockets ", 13(10, 57, (1976).

Tests and calculations indicate an increase rather than a decrease in solar absorptance of second surface mirrors at angles of incidence greater than 80°. In practical applications, this effect will be insignificant because, although the solar absorptance is high, the projected area is compensatingly small, resulting in an insignificant increase in the solar or albedo load. However, a spin stabilized spacecraft can expose second surface mirrors to all angles of incidence. Also reflected solar energy from a planet or other spacecraft surface could impinge predominantly in the region of high angles of incidence. If a large area is involved the total load could be significant.

0905

<u>Solar Reflector Foaming Technology Development</u> : Swanson P, "Air Force Air Propulsion Laboratory Report No. AFAPL - TER - 64 - 128 ", (USA), (1964).

Detailed study, test and fabrication of a 2 ft and 10 ft diameter solar collectors have demonstrated that it is not practical to use a pressure distributed foam system to form a large sized aluminised Mylar solar reflector in space. The report discusses design of the mixer unit and fabrication equipment, gives results of reflectance, absorptances and emissivity testing, and describes the fabrication of 2 ft and 10 ft diameter inflatable mirrors. Details of the testing program to determine the physical properties of mechanically mixed foam are included.

0906

<u>Mirrors for Solar Energy Application</u> : Taketani H and Arden W M, " McDonnell Douglas Astronautics Co. Report MDCG - 7213 ", (USA), (1977).

Silvered mirrors with highly specular reflectance properties, within 4 milliradians, were studied using a variety of instruments. Pure silver, when applied by commercial deposition methods, was found to have a total reflectance efficiency (specular plus scattered) of 98.3% when chemical deposition technique was used. In contrast, the theoretical total efficiency of bare vapour deposited silver is 98.7%. The specular reflectance efficiency at 4 m rad was 92.8% for chemically deposited silver and 94% for vapour deposited silver. No appreciable differences in reflectance efficiency were found among the different commercial chemical silvering processes nor for varying silver thickness for the mirrolab process. To protect the silver, second surface mirrors using sheet and float glass were investigated. Reflectance efficiency in excess of 90% was achieved for both mirrors. However, subsequent tests

indicated that first surface mirrors required more formulation development to enhance abrasion properties and weather resistance and second surface mirrors required low iron glass to achieve high reflectance efficiency.

0907

Specular Mirrors for Solar Energy Applications : Taketani H & Arden W M, "SAMPE " 10(1), 54, (1978).

The reflectance efficiency of heliostat mirror surface is a critical parameter in the central tower
concept for conversion of solar energy. Reflectance efficiency was measured on silver coated surfaces
and on second surface mirrors of float and sheet glass for comparison in a solar heliostat. Abrasion
tests were conducted to determine the effect of impingement of sand particles on reflectance. Washing
studies were performed to identify the most effective solution and cleaning procedure for the surfaces.

0908

Studies on Thermal Effects on Solar Furnace Mirrors : Troong T Y, " U S Air Force Missile Development
Centre Report No. AFMDC - TR - 59 - 15, Solar Furnace Support Studies ", Holloman Air Force Base, New
Mexico (USA), 2, 133, (1959).

Two studies are presented in this report. The first is concerned with the warping of a back silvered
glass mirror when heated by solar rays incident to either the face or the back of the mirror. The
second study is concerned with the mounting of concave mirror segments forming the paraboloidal concentrator of a solar furnace.

0909

Film Tension in Film Concentrators : Umarov G Ya. and Gafurov A M, "Applied Solar Energy", 4(3),36,(1968).

The film tension dependence on the angle of opening was experimentally determined. The data has shown
that with increasing the angle of opening in a film concentrator, the film tension also increases, but
the film tension has a maximum; when the angle continues increasing the tension of the film decreases.
The explanation of this phenomenon is that at such tensions the film begins to extend and its corresponding limit of tension diminishes.

0910

The Use of Polyurethane Foam in Making Solar Energy Collectors : Vilkova S N, Novikova I A and Alavutdinov D N, " Applied Solar Energy ", 1(4), (1965).

A method for making solar energy collectors from rigid foamed polyurethanes lined with   mirror - like
plastic films is described.

0911

Polystyrene Foam Reflectors :Vilkova S N, Novikova I A, Gafurov A and Umarov G Ya., " Applied Solar Energy ", 2(5), 18, (1966).

Specific impact and compression strength data on compositions of polystyrene foams used as a substrate
for a solar energy collector mirror surface are presented. Various fillers, asbestos, glass f i b r e ,
capron fibre and capron interlayers were added to the polystyrene to raise its physical strength and
to increase its heat resistance. The collector is to be used under the conditions of a hot   Central
Asian summer. The specimens were subjected to heat treatments, irradiation exposure, and light a n d
weather exposures to determine how well the compositions stand up with respect to strength.

0912

Two Stage Method of Mirror Manufacturing on a Self - Hardening Base : Vilkova S N, Guner E A, Gafurov A M,
and Umarov G Ya., " Applied Solar Energy ", 4(3), 38, (1968).

The method consists of two stages. First is a thin aluminium mirror on epoxy resin, which does not get
turbid with time. The epoxy resin was freely poured on the mirror surface after being polymerized at
room temperature. It was warmed thoroughly at 195 - 200 deg C. By means of this treatment, all air dis-
solved in the resin comes out. Mirrors manufactured by this method have good optical properties.

0913

Cleaning of Mirror Face Coatings by Triacetylcellulose Films : Zakhidov R A, Levashova T M, Altman Ts.M
and Dubrovskii L A, " Applied Solar Energy " 15(2), 65, (1979).

It is shown that mirror face coatings can be successfully cleaned by triacetylcellulose films, particularly in the case of strong contamination.

0914

Geometry of Film Concentrators with Radial Seams : Zarubin U S, Umarov G Ya., Gafurov A M and Solody -
nikov Ya. A, " Applied Solar Eenrgy ", 4(1), 25, (1968).

Comparison of experimental and theoretical data is made. If two radial joints are present, the effective tension of any point in the concentrator increases twice as compared with a concentrator without
joints. If the concentrator has radial joints, the tension at point of the concentrator i n c r e a s e s
three times.

Miscellaneous Concentrator Designs / Patents Etc.

0915

Description and Test Results for a Low Temperature 3 KWe Rankine Cycle Energy Conversion S y s t e m :
Abbin Jr. J P, " Sandia Laboratories Report No. SAND – 77 – 1538 ", (USA), (1978).

This report describes a 3 KWe Rankine cycle system operating with 155 deg F water as the heat s o u r c e
and 45 deg F water as the cooling source that has been assembled and tested at Sandia Laboratories. This
report describes the unit and presents test data on the system. Gross electrical generating efficiency was approximately 3.5%.

0916

The Temperature of Cavity – Type Solar Absorbers with a Circulating Fluid : Abrams M, " Sandia Laboratories Report No. SLL – 74 – 5209 ", (USA), (1975).

An analytical expression is obtained for the temperature of a cavity – type absorber of solar radiation
which has a circulating heat transfer fluid.   This expression relates effective cavity temperature to
the temperature of the incoming fluid, cavity geometry, emissivity, coolant flow rate and the incident
radiant flux.

0917

A Photo – electric Probe Measuring the Density of Concentrated Solar Radiation : Afyan V V, Shakhparonyan V V and Vartanyan A V, " Applied Solar Energy ", 11(3 – 4), 111, (1975).

A photo – electric probe for concentrated solar radiation is described. It has a receiving a r e a o f
0.8 sqmm and a calibration stand. A calibration graph showing the probe current as a function of the
incident intensity in the range of 0–5.10^4 W/sqm is given. Results obtained with the p r o b e in t h e
field of a solar concentrator are reported.

0918

Combination of Radiant – Flux Radiator for High Temperature Solar Plant : Afyan V V and Vartanyan A V,
" Applied Solar Energy ", 15(1), 160, (1979).

A combination radiant flux regulator for a high temperature solar plant with paraboloidal concentrator
is considered. It takes the form of low shielding cyclinders operating simultaneously, an outer and
inner shield, it permits regulation of two parameters, the radiant flux density at the focus and the
power of the solar plant. An experimental model of the controller has been tested.

0919

Development and Investigation of Solar Thermoelectric Generator Using a Tubular Module : Agabaev Ch.,
Guhhman G A, Kolomoets N V, Markman M A, Milevskaya N G & Rozyev N," Applied Solar Energy ", 15(5),77,
(1979).

Experimental expressions are given for the capacity of a solar thermoelectric generator as a function
of incident solar radiation density, wind speed, ambient temperature.   The variation i n generator
capacity with duration of operation is also given.

0920

Properties of a Solar Concentrator with Hexagonal Glass Facets : Alimov A K, Alavutdinov D N and Abduazizov A, " Applied Solar Energy ", 11(3 – 4), 14, (1975).

The energy parameters of a faceted concentrator 5 m in diameter have been investigated. The facets were
i n the form of hexagonal paraboloids mounted on a paraboloidal base. This gives an effective working-
ing area that is 10% greater as compared with a concentrator using circular facets. The power generated in the focal plane of the concentrator was 9.0 Kw for an incident radiation of 784 W/sqm.

0921

Heliotherapy Unit : Alimov A K, Alavutdinov D N, Ovechkin N F, Sadykov A S, Mashampin A M & Suyarov Kh.,
" Applied Solar Energy ", 14(4), 53, (1978).

A heliotherapy unit consisting of a portable concentrator, shutter and treatment booth is developed
and fabricated. The portable concentrator uses a paraboloidal long focus glass concentrator   w i t h
aluminium mirror coating, lacquer protection and an orienting mechanism. Some practical experiments
were found to be successful.

0922

Suggested Methods of Aligning the Planes of the Solar Furnace Heliostat Mirrors into Parallelism :
Allison F and Hughes G, " Solar Energy ", 2(3 – 4), 46, (1958).

Three possible methods for measuring the alignment of the planes of the individual mirrors of the heliostats are proposed. Two of the methods are optical and one is mechanical. The optical methods r e-
quire, in one case, a theodolite of high precision and large aperture, and, in the other case, a specially built instrument. The mechanical method would use high precision spirit levels with the heliostat set in the horizontal plane.

0923

Research and Development on Solar Energy for Electrical and Mechanical Power Generation : Almanza R,
Garibay J, Lopez S, Munoz F and Zarate R, " Proceedings ISES Silver Jubilee International Congress",
Georgia (USA), p. 1095, (1979).

The first thermal mechanical conversion system investigated and still under study is a device of 0.5 KW made of focussing collectors which follows the sun during the day. It consists of 24 sqm of collectors made of parabolic cylinders, covered with aluminized acrylic. In the focal line there is a copper tube with a selective surface of CuO which in turn is situated inside glass tubes which are evacuated to reduce convection losses. The working fluid is steam reaching temperatures higher than 150 deg C and pressure of 3.5 atm. This system is connected to a piston engine with an efficiency of 20%; the whole system is capable of operating for 4 hours during clear days.

0924

A Concentrating Collector System to Supply Industrial Process Hot Water : Amedure G, Rost D F, Alexander C F & Schuler H F, " Proceedings ERDA Conference on Concentrating Solar Collectors ", Georgia (USA), p. 8 - 15, (1977).

Design and performance characteristics of a concentrating collector system to supply industrial process hot water are discussed in this paper. The concentrating collector is the General Solar Systems tracking concentrating solar collector. The system is capable of producing over 146 Kw at 88 deg C during a sunny noon.

0925

A Line Focus Segmented Mirror Concentrator : Anderson D E and Stickley R A, " Proceedings of SPIE Conference ", 85, 24, (1976).

A solar concentrator system is described which is based on the use of a long array of east-west mounted focussing reflectors. Each of the reflective segments projects a line image of the sun into the entrance aperture of a stationary receiver through which a working fluid flows. Typically, ten segments in parallel are initially aligned to focus on the same receiver aperture ; the mean solar flux across the aperture can be in the order of 40 times that falling on the mirrors. The mirrors are steered by linked rotation about their individual centers to track the sun in elevation. Simple logic is also incorporated to rotate the mirrors into a face - down position on lack of sun or lack of demand. This provides for protection during storms and at night, and for defocussing in case the system overheats. The system is simple, well suited to installation and maintenance operation and offers the potential for high efficiency in the intermediate temperature range at a low life cycle cost.

0926

A Pyrometer for Solar Furnaces : Andors S P, Grigorenko L F, Kozak V A, Mavashev Yu.Z, Mazaev G A, Mosin L M, Nechaeva L V and Profatslova I I, " Applied Solar Energy ", 15 (4), 26, (1979).

A new pyrometer for measurement of temperature of bodies heated in solar furnaces is described. Its effective wavelength lies in the 1.38 μm water - vapour absorption band. Technical data are given for the pyrometer.

0927

Comparison of Linear Focussed and Flat Plate Solar Collectors in Livermore, California : Angvick G L, Brune J M, Eicker P J, Green R M and Haseltine E H, " Sandia Laboratories Report No. SAND - 74 - 824 ", (USA), (1975).

The study had several purposes. First the study developed and implemented methodologies and computational models for the prediction and evaluation of the performance of solar energy collectors. In particular, solar energy and collector heat balance models were developed which would allow comparison of flat plate and linear focussed collectors. The models were used in a sensitivity study comparing the annual performance of hypothetical flat plate and linear focussed collecting systems operating in Livermore valley in California.

0928

A Radiation - Cavity Solar Collector to High Temperature Applications : Antoniak Z I and Palmer H B, " Proceedings ERDA Conference on Concentrating Solar Collectors ", Georgia (USA), p. 2-79, (1977).

A 1.5 m long experimental model of a previously proposed high temperature solar concentrator - collector in which argon is employed as a working fluid has been studied. The effect of using a selective absorber in place of graphite absorber reported earlier was investigated. No measurable gain in efficiency was observed. A computer model of this system which takes into account most of the influential variables has been formulated. It yields temperature profiles that normally agree with the experimental data at all axial positions within 10 deg C. This good agreement permits the formulation of a second computer model, of the full scale device, with confidence. The effect of various parameters has been investigated in an optimization study of the full scale collector.

0929

Mirror and Lens Solar Energy Concentrator : Aparisi R R, Kolos Ya. G and Shatov N J, in " Semiconductor Solar Energy Convertors (Ed. V A Baum) ", p. 101, (1969).

A composite concentrator, consisting of a truncated conical mirror (made of sheets, strips or sections) and a torroidal lens, is considered. For small units, the lens is made by cutting from one large piece of glass ; but for larger units, the lens is made of the composite Fresnel type. Design formulas are given and a prototype mirror and lens concentrator is described. Experimental results obtained under natural conditions during illumination of the concentrator with direct solar readiation, are given. For a radiation density of 600 Kcal/sqm/hr the maximum concentration of the radiation density in the focal region is 86270 Kcal/sqm/hr. This concentration can be used to illuminate cylindrical thermoelectric generators, or lasers pumped with solar or other radiation.

0930

Lightweight Reflector Assembly and Method : Argoud J J and Walker W L, " U S Patent Application 617202, Filed September 24, (1975) ".

An invention for a lightweight solar reflector assembly having a glass cellular substrate and a method of forming the reflector assembly is described. The novelty of the invention appears to reside in the method of forming a large low cost reflective surface for use in a solar concentrator or antenna. The invention also includes the reflective surface and a light weight cellular glass substrate having the same coefficient of thermal expansion to provide a high quality optical reflective surface.

0931

Design of Field Tests and Analysis of Experimental Results for LSE 7 Meter Collector Foundations Shenandoah, Georgia : Auld H E, " Sandia Laboratories Report No. SAND - 79 - 7075 ", (USA), (1979).

Four reinforced concrete cylindrical piers, typical of foundations utilized for double axis - tracking solar collector systems, were tested to determine eccentric horizontal and vertical failure loads.The results from these tests were found to compare favourably with the results from a previous experimental program and with theoretical calculations which incorporate the geotechnical parameters of the test site. Recommendations are made for the design of the foundations for the solar collector system to be constructed at this site.

0932

Study of Low Cost Foundation / Anchor Designs for Single - Axis - Tracking Solar Collector S y s t e m s : Auld H E and Lodde P F, " Sandia Laboratories Report No. SAND - 78 - 7048 ", (USA), (1978).

The requirements for the foundations of single axis tracking solar collector systems are defined. Ten preliminary foundation systems capable of meeting these requirements are evaluated in relation to deployability and cost effectiveness. A detailed design is presented for the optimal system with two alternative designs.

0933

Study of Foundation Design for Single Axis Tracking Solar Collector Systems Under Reduced L o a d i n g Conditions : Auld H E and Lodde P F, " Sandia Laboratories Report No. SAND - 79 - 7016 ",(USA), (1979).

The objective of this study was to assess the impact of reduced wind loads, caused by fencing and multiple row shielding effects, on the design of foundations for single axis tracking solar collector systems. The cost and dimensions of both cylindrical piers and rectangular footings for s t a n d a r d, h a l f standard and one quarter standard wind loading conditions are presented.

0934

Analysis of Field Test Results for Single Axis Tracking Solar Collector Foundations : Auld H E,"Sandia Laboratories Report No. SAND - 79 - 7023 ", (USA), (1979).

Five reinforced concrete cylindrical piers, typical of foundations utilized for single axis tracking solar collector systems, were tested to determine eccentric horizontal and vertical failure loads. The results from these tests were found to compare favourably with the results from theoretical calculations which incorporate the geotechnical parameters of the test site. Recommendations are made for the incorporation of these results into the design of foundations for future solar collector systems.

0935

Calculating the Radiant Field of Concentrators with Artificial Sources : Azimov S A, Adylov G T, Mallaeva Kh. M, Pirmatov I I and Riskiev T T, "Applied Solar Energy", 14(5), 19, (1978).

A procedure for calculating the radiant field of concentrators with arbitrary sources is proposed. Account is taken of the reflector inaccuracies which is sufficient when the source angular dimensions are comparable with the mean square deviations of the reflector surface normal from the ideal direction. The radiation field of the " Uran " radiative heating unit is considered as an example.

0936

The ANSALDO 35 Kw Solar Power System : Bado G, Tome G, Angelino G G and Macchi E, " Proceedings I S E S Silver Jubilee International Congress ", Georgia (USA), p.1090, (1979).

The conceptual design of a 35 Kw solar power system based on the use of linear concentrating collectors is presented. Each collector module consists of two parabolic troughs capable of b o t h, elevational and azimuthal alignment. The thermal engine of the system is an organic fluid (perfluorocarbon) multistage turbine prime mover with an efficiency of 25 % at 300 deg C collectors cooling l o o p e x i t temperature. The system is designed for stand alone and for grid connected operation and, if requested, for the combined generation of power and heat.

0937

Enhancement of Solar Arrays in Space Through the Use of Concentrators : Baker I, " Proceedings 1 4 th IECE Conference ", Boston (USA), p. 1278, (1979).

Results are presented of a conceptual design study and comparative evaluation of solar concentrators for enhancing the performance of photovoltaic arrays in space.

0938

Turning Collectors - A Compromise Between Stationary and Heliostatic Ones : Barak A Z, " Proceedings ISES Conference ", Los Angeles (USA), (1975).

A third type of solar collectors, namely the turning collector is described and discussed. It has some properties of both stationary collectors (including periodically adjustable ones) and heliostatic collectors with diurnal suntracking, minimizing their disadvantages and offering performances which might be better economically for quite a few possible applications.

0939

Solar Collectors with Improved Thermal Concentration : Barak A Z, "U S Patent No.3994279/Nov.30th,(1976).

Reduced heat losses from the absorbing surface of the energy receiver of a cylindrical radiant energy collector is achieved by providing individual, insulated cooling tubes for adjacent parallel longitudinal segment of the receiver. Control means allow fluid for removing heat absorbed by the tubes to flow

124

only in those tubes upon which energy is being directed by the reflective wall of the collector.

0940

Radiant Energy Reception Regulation Device : Baranov V K, "Applied Solar Energy", 3(2), (1967).

A device which uses the property of the triple reflector to reflect rays falling on its surface is described. When this surface contacts with a liquid, the radiation flux passes through the surface.

0941

Methods of Designing Profiles of Focussing Concentrators and Focussing Wedges for Parabolic Solar Concentrator : Baranov V K, " Applied Solar Energy ", 12(6), 17, (1978).

Techniques are described for rapid and practically exact construction of profiles of parabolic-torroi-
dal focussing concentrators and parabolic - cylindrical focussing wedges. Values of the radius vectors
needed for constructing the profile of a focussing concentrator and a focussing wedge with a parametric
angle of 20 are given.

0942

Equipment for Concentrating and Collecting Solar Energy with Solar Tracking System & Suitable Process:
Barr I R, " German Patent No. 2550194 / A1 / 26th May, (1976)".

The solar collector consists of a concave semi - cylindrical reflector concentrator with a longitudinal
axis of the concentrator. The system has arrangement for movement, which permits movement in two directions. The movement is controlled by a solar sensor. The strength of solar radiation is measured
simultaneously. The tracking system is only actual if this radiation is above a certain threshold level.

0943

A Prototype of Solar Linear Concentrators Plant Installed to Supply Industrial Heat: Barra O A, Barutti A,
and Carratelli E P, " Proceedings ISES Silver Jubilee International Congress ",Georgia (USA), p.1046, (1979).

A 50 sqm solar linear concentrator plant has been realized in the " Birra Peroni " Italian brewery to
verify the problems arising in a real application of solar energy exploitation to supply   industrial
heat in the range 120 - 300 deg C before designing a large size plant. The peculiar aims of this experimental plant are : (a) to test a mathematical computer model, previously carried out, by means of experimental measurements in order to obtain a general computer code useful to design these plants; (b) to
measure the plant conversion efficiency ; (c) to solve the problem deriving from jointing the s o l a r
plant with the pre-existing brewery water-steam network. In the present paper, the plant is described,
the arguments (a), (b) and (c) are introduced and the preliminary results obtained are shown.

0944

Shadow's Effect in a Large Scale Solar Power Plant : Barra O, Conti M, Santamata E,Scarmozzino R and
Visentin R, " Solar Energy ", 19, 759, (1977).

An analytical - numerical method for the evaluation of the effect of shading in a typical solar p o w e r
plant made of 'N' lines of concentrating cylindrical parabolic collectors which track the sun by running around an East - West directed horizontal axis is presented.

0945

Adjustment of Flat Orientators for Solar Furnaces : Bashnyk A Ya., Mavashev Yu. Z, Nechaeva L V a n d
Rudshtein V L, " Applied Solar Energy ", 11 (3 - 4), 17, (1975).

A method of adjusting flat faceted orientators is described. It is accurate, fast and requires little
additional equipment.

0946

Radiant Energy Distribution on Reflection from Mirrors : Baum I V," Applied Solar Energy ",10(1),27,(1

General integral relationships representing a number of solutions of theoretical problems which a r e
accurate (with the exception of minimally simplifying assumptions) and are valid for a broad range of
variation of the variables are reported. The results can be processed on a computer when it is desirable to obtain a family of distributions, depending on the mirror parameters.

0947

Optical Furnace with an Artificial Source : Baum I V, "Applied Solar Energy", 10(4), 111, (1974).

The theoretical principles that can be used in the design of optical furnaces are reviewed. The possibility of producing a uniformly illuminated zone with relatively low flux density is estimated.Special
formulas are given for the parameters of optical systems.

0948

Comparative Analysis of Models of Irradiance Field Formation in Wide Aperture - Optical System: Baum I V
and Mamedniyazov S O, " Applied Solar Energy ", 13(5), 19, (1977).

A comparison is made of the energy flux density distribution on the receiver of a paraboloidal c o n-
centrator obtained in the multifactor wide aperture optics model with the approximate distributions
for " exact " and " non exact " concentrators classified on the basis of the value of the accuracy criterion 'Q' and various particular models used in solar engineering (including the Aparisi approximation formula).

0949

Equalization of Irradiation Field on Receiver Surface : Baum I V and Mamedniyazov S O, " Applied Solar
Energy ", 13(5), 27, (1977).

The form of a reflector that creates a uniform irradiance field on the surface of a flat receiver is
determined and the edge nonuniformity losses due to finite angular dimension of the source are calculated.

0950

Mass Transfer Through Aperture in Chamber Within Turbulent Flow : Baum V A and Sakr I , " I z v e s t i y a Otdelenie Tekhrucheskikh Nauk, Energetika Aotomatika ", No. 3, p.114, (1962).

Mass transfer studies made in connection with construction of solar furnace having inlet for beam reflected from concentrator are presented in this paper.

0951

Basic Optical Considerations in the Choice of a Design for a Solar Furnace : Baum W A and Strong J D, " Solar Energy ", 2(3-4), 37, (1958).

Simple relationships exist between the aperture of a solar furnace, the target size, the angle of convergence, the maximum attainable concentration ratio, the overall efficiency and the amount of spill light surrounding the target for optical systems having continuous unobstructed surfaces and no spherical aberration, the concentration ratio is determined solely by the angle of convergence. V a r i o u s systems differ considerably, however in the amount by which their aperture exceeds that of an i d e a l system of equal performance, in the resulting efficiency at which they operate ; in t h e   associated amount of waste light spilled outside the target ; in the distribution of this spill light; and in the relative convenience with which they can be fabricated and adjusted. A paraboloidal mirror with heliostat does not seem to be an optimum choice. The optical properties afforded by other two mirror systems are discussed and are illustrated by example. Also presented is a structural design for a t w o mirror system which appears to possess some practical advantages over the heliostat paraboloid combination.

0952

Novel Design of Evacuated Concentrating Spherical Collector : Bayazitoglu Y, " Proceedings   International Conference on Alternative Sources of Energy ", Miami (USA), p. 353, (1977).

The proposed design is that of a spherical collector consisting of a fixed evacuated glass sphere internally reflecting out the lower hemisphere and with a fixed receiver-absorber plate contained inside it. Convection suppression is provided simply by evacuating the sphere. Concentration is achieved by reflecting the solar rays from the lower hemisphere on to the stationary, plane and receiver absorber. The receiver-absorber plate is fixed in such an orientation that during the sun's virtual diurnal motion, the paraxial line of the hemispherical reflector always falls on a portion of it. Thus the receiver-absorber can be made to receive the maximum amount of radiation of different daily positionings of the sun. A properly chosen receiver-absorber thickness may also make the collector less sensitive to seasonal variations and allows it to collect any imperfectly reflected energy. The complete concept is that of a collection consisting of an array of such glass spheres attached to the main c o n d u i t carrying the heat transfer fluid.

0953

Production of Foam Plastic Concentrators and their Characteristics : Bazarov B and Baum V, " Proceedings International Conference on Solar Electricity ", Toulouse (France), p.957, (1976).

Solar concentrators were constructed of rigid polyurethane foam. The physical and mechanical properties were determined and the results tabulated. The properties and service life tests performed during a period of one year indicate that pulverized rigid polyurethane foam can be successfully u s e d for the fabrication of solar concentrators.

0954

Investigating the Geometric Quality of Cast Polyurethane Foam Duplicates for Solar Energy Concentrators : Bazarov B A, Kapelyushuikov V M & Kalinin B A, " Applied Solar Energy ", 13(2), 17, (1977).

An experimental method is considered that makes it possible to investigate the quality of solar concentrators based on rigid polyurethane foams. Fundamental computational relationships are given for evaluating the correspondence of a geometric shape of the fabricated duplicates to the reference die.

0955

Optical Analysis and Optimization of Line Focus Solar Collectors : Bendt P, Rabl A, Gaul H W & Reed K A, " Solar Energy Research Institute, Report No. SERI / TR - 34092 ", (USA), (1979).

This paper describes a macroscopic approach that yields all the parameters needed for the optical design of line focus parabolic troughs in closed analytical form, requiring only a minimal computation. The goal of the optical analysis developed in this report is to determine the flux at the receiver as a function of concentrator configuration, receiver size, width of sun and optical errors (e.g. tracking, reflector contour).

0956

Determining the Optical Quality of Focussing Without Laser Ray Tracing : Bendt P, Gaul H and Rabl A, " Journal of Solar Energy Engineering (Trans. ASME) ", 102, (20, 128, (1980).

This paper describes a novel alternative to the laser ray trace technique for evaluating the optical quality of focussing solar collectors. The new method does not require any equipment beyond that which is used for measuring collector efficiency ; it could therefore become part of routine collector testing. An angular scan is performed ; i.e. the collector output is measured as a function of misalignment angle over the entire range of angles for which there is measurable output (typically of a few degrees).

0957

Interaction Entre le Champ d'Heliostats et le Systeme Thermodynamique d'une Centrale Solaire a-tour : Bessus B, Mersier C, Pharabod F and Abatut J L, " Proceedings International Conference on Solar Electricity ", Toulouse (France), p. 559, (1976).

The geometrical concept of a mirror field, the individual size of each mirror and the size of the receiver are the primary elements of an optimal choice of the thermodynamic conversion system. In this study, keeping a criterion the optimal thermoptical ratio, the influence of the maximal energy density at the receiver focus as well as the average energy concentration ratio is analysed. Nature of h e a t

126

transfer fluids, temperature of cycle, and thermic exchange systems are discussed. Some total ratios are given as examples.

0958

Mathematical Modeling of Solar Concentrators : Biggs F and Vittitoe C N, " Proceedings ISES Conference", Winnipeg (Canada), 2, 220, (1976).

A computational capability that models the operation of any solar energy collector that uses flux concentrators is a valuable aid in planning, design, construction, calibration, safety analysis and operation of the system. In addition to the usual optical considerations, the model should treat such imperfections as reflecting surface slope errors, sun tracking and alignment errors, and mirror focussing errors. It should properly account for the angular distribution of the incoming sun rays and the effects of atmospheric transmission on this distribution. A model with these capabilities is described and two computer programs for implementing it are illustrated.

0959

Concentrator Quality Evaluation : Biggs F and Vittitoe C N, " Proceedings ISES (American Section) Conference ", Denver, Colorado (USA), 2.1, 850, (1978).

The performance of a reflecting solar concentrator depends, of course, on its surface reflectance, but there are other important factors. Among these are the sun tracking errors, surface slope errors and surface irregularities. It is appropriate to use statistics and describe and analyse these non-deterministic factors. A scheme for specifying the quality of a solar concentrator that includes all these effects is described and illustrated.

0960

Modelling the Beam Characterisation System : Biggs F, Vittitoe C N and King D L, " Advanced Instrumentation, 34 (Pt. 2) (Proceedings of ISA) ", p.535, (1979).

The beam characterization system (BCS) recently developed for heliostat evaluation at the central receiver test facility at Sandia Laboratories, measures, digitizes, records and analyses a flux density pattern in a beam of reflected sunlight. Since the BCS collects data with a given set of conditions to determine optical specifications which can predict heliostat behaviour under other sets of conditions, it is necessary to use a theoretical model of the system to interpret results. A statistical method is used to handle stochastic variables such as sun tracking errors and surface slope errors. A cone-optics technique is used to incorporate the statistics into a consistent model of the optical behaviour of a heliostat. An overview of this model is given.

0961

A Self-Supporting Solar Power System for an Isolated Community : Bignon M J, " Proceedings ISES Silver Jubilee International Conference ", Georgia (USA), (1979).

There are many remote and sunny places in the world where electric power cannot be supplied by a distribution network. Solar energy is well suited for middle sized local generation but it must be suplemented to achieve the required reliability. This paper describes a self supporting combined solar thermal power production system developed with the help of a grant from the French National Scientific and Technical Development Fund.

0962

Solar Radiation Heat Transfer to High Temperature Heat Carriers : Bilgen E," ASME Paper 74-WA/HT-14,(1974) .

In this paper, the fundamental aspect of propagation of heat flux due to solar radiation to a high temperature heat carrier through a metallic wall of a collector is studied and an exact solution of the problem is obtained. Theoretical results are compared to experimental measurements made with a focussing type collector. The surface temperature of a flat plate collector is also predicted based on the exact method of analysis. A good agreement is obtained between theories and experiment. Overall heat transfer coefficient to a high temperature heat carrier is also determined.

0963

Estimating Hourly Solar Radiation for One Axis Tracking Focussing Collectors : Bingham C E, Posner D M and Bradly J O, " Proceedings 4th Annual UMR-DNR Conference on Energy ", (1978).

A method for converting total horizontal radiation to a direct radiation on the surface of a tracking collector is developed. A technique for estimating direct hourly radiation for a north-south axis, horizontally mounted parabolic trough is outlined & further adapted to an east-west tracking orientation.

0964

Horizontally Mounted Solar Collector : Black D H, " U S Patent Application No. 823061 ".

A horizontally mounted solar collector for collecting solar energy is described. It uses a vertical reflector assembly, a stationary reflector and a collector. The deflector assembly contains a plurality of vanes which change the direction of solar energy to the vertical while constantly keeping the same side of the deflector facing the sun. The vertical rays are then reflected off the stationary reflector and are then absorbed by a collector.

0965

Calibration of Solar Concentrator for Use in Power System Research : Blake F A, " Paper at American Rocket Society - Space Power Systems Conference ", California (USA), (1962).

The results of an experimental calibration program performed on a rigid solar collector are discussed. Evaluation of the geometry by detailed optical inspection (Modified Hartmann Test), of energy collection by calorimeter, and of energy density by radiometer tests is covered. The effect of the tracking system, vacuum chamber window and focal plane mounting hardware, on the concentrator performance is shown. Flux distribution patterns in planes perpendicular to the optical axis and on cylinders around the optical axis are discussed. Two methods of flux-distribution determination are reported comparatively. These are incremental-energy method using a water calorimeter, and point by point method using a radiomicrometer.

0966

Analytical and Experimental Determination of Radiation and Temperature Distribution Inside Solar Receivers : Blay D, Bernard J & Haziza C, " Proceedings International Silver Jubilee Congress ", Georgia (USA), (1979).

An analytical model for the determination of radiation and temperature distributions inside solar receivers of solar power plants is presented. Experimental investigation has also been made a n d t h e results are compared with the analytical findings.

0967

Economic Aspects of Solar Thermal Electric Power Generation in Small Utilities : Bluhm S A, Ferber R R and Mayo L G, " Proceedings ISES Silver Jubilee International Congress,Georgia (USA), p.1072, (1979)

A study was performed to examine the economic aspects of solar thermal power plants in synthetic small utilities. Insolation typical of Albuquerque, New Mexico, was used to simulate siting in the South Western United States. The results indicate the solar thermal plants may be competent in oil dependent municipal utilities of the Southwest in the 1985 to 2000 time period.

0968

Availability of Direct, Total and Diffuse Solar Radiation to Fixed and Tracking Collectors in the USA: Boes E C, Anderson H E, Hall I J, Prairie R R and Stromberg R T, " Sandia Laboratories Report No. SAND – 77 – 0805 ", (USA), (1977).

This report describes the results of a thorough study of solar radiation availability for the USA. The availabilities of both direct and total solar radiation to various types of suntracking collectore are described. Availabilities of both direct and total solar radiation to a wide variety of tilted surfaces are also given, including surface with various azimuth angles. Finally the distribution of diffuse radiation on a horizontal surface is given. All of these availabilities are described on a seasonal basis, and some availabilities are presented by months.

0969

Average Solar Radiation Available to Various Collector Types : Boes E, Anderson H, Hall I, Prairie R R and Stromberg R T, " Proceedings ERDA Conference on Concentrating Solar Collectors ", Georgia (U S A), p. 9 – 7, (1977).

This paper describes the results of a recently completed analysis of the seasonal and geographic distributions of solar radiation available to various types of solar collectors in the USA. The results are presented in the form of tables of seasonal mean daily amounts of solar radiation available t o various configurations for tracking surfaces and to a comprehensive set of fixed surface orientations of 26 US sites. All solar radiation availabilities described are separated into direct and diffuse components so that they are applicable for both concentration and flat plate collectors. The p a p e r also presents some comparison of the relative amount of solar energy available to different solar collector schemes. A primary conclusion of the study is that there is generally more direct solar radiation available to tracking concentrating collectors than there is to fixed flat plate collectors.

0970

An Optimization Procedure for Solar Thermal Electric Power Plant Design : Bohon W M and Levy S L, " Proceedings ISES Silver Jubilee International Conference ", Georgia (USA), (1979).

The design of a solar thermal electric power plant to produce electricity at the least cost involves a great number of performance / cost evaluations. A practical approach to this optimization t a s k i s through sub–optimization e.g. maximise the performance (integrated energy output) of each plant design considered and then select the best design via differential cost estimates (minimize cost of electricity). A procedure is presented for designing a solar thermal electric power plant such that performance is maximized. The development of the optimization procedure and the accuracies of the employed mathematical models and assumptions are discussed. The practical application of this procedure to solar thermal electric power plant design is illustrated by specific examples.

0971

Analysis of Optical Behaviour and Collector Performance of a Solar Collector : Bordoloi K C, Murray T M and Chaves E J, "Proceedings Southeastern Conference ", (1978).

This paper is concerned with the study of the optical behaviour and the collector performance o f a non tracking half circular concentrator. This study pertains to geometrical considerations in relationship to the incident sunlight and the optical parameters such as the concentration ratio and the average number of reflections. Also a brief look at the thermodynamical parameters suggests that the concentrator performs satisfactorily under nontracking mode.

0972

Automatic Determination of the Spectral Emissivity and True Temperature of Bodies Heated in the Focus of a Solar Furnace : Borukhov M Yu., Zakrytnyi V F, Kovalenko A V and Mavashev Yu. Z, " Applied Solar Energy ", 10(4), 109, (1974).

The reflection method based on Kirchhoff's law has been used to determine the spectral emissivity and true temperature of materials heated in solar furnaces.

0973

Microwave Properties of Dielectrics Studies in Optical Furnaces : Borukhov M Yu., Zakrytnyi V F, Kovalenko A V and Mavashev Yu.Z," Applied Solar Energy ", 10(3), 59, (1974).

A method is described for the determination of the dielectric properties of materials in the m i c r o wave region at high temperatures, characteristic for solar furnaces.

0974

Solar Collectors as Energy Converters : Bosnjakovic F, " Studies in Heat Transfer ",[A Festschrift for E.R.G. Eckert](Hemisphere Pub. Corp. - Washington), p.331, (1979).

The thermodynamics of the solar collector and its energetic efficiency are reduced to several characteristic magnitudes. These include, in addition to the solar radiation available, operating temper-
ture, useful heat obtained, possible focussing arrangements, angle of incidence for the collector, as
well as the meteorologically determined extinction coefficient of the atmosphere. The energetic efficiency is understood as the ratio of the energy (work potential) of the useful heat obtained to the
energy of the available solar radiation. To determine it, the temperature of terrestrial solar radiation must be known. Therefore, in this paper the dependence of the efficiency of the collector on
its type and many of its operations as well as on atmospheric factors are examined and formulated in
general terms.

0975

Optimisation du Dimensionnement d'une Centrale Solaire Pour L'Obtentio d'un Prix de KWh Minimal: Boy -
Marcotte J and Potiron A, " Proceedings International Conference on Solar Electricity ", (France), p.
851, (1976).

The sizing of a thermoelectric power plant was achieved by modelling the basic elements of the system;
collector, piping, eventual storage, turbogenerator. Reasonable and consistent options have been made
prior to modelling the plant, the hot temperature was varied from 200 deg C to 300 deg C. The digital
code developed for the study made it possible to determine the size of the basic elements that lead to
the minimum cost of the KWh for a given cost per area of collector.

0976

Graphic Method of Calculating Conical Focons : Braslavskaya M V and Baranov V K," Applied Solar Energy",
4(4), 64, (1968).

The work deals with parameters calculation of focons used for concentrating the density of light fluxes
from different light sources including the sun.

0977

Technical and Economic Analysis of On Site Solar Electric Power Generation : Breadley J O, Costello D
and Rizzi A, " Proceedings ISES Conference ", Los Angeles (USA), (1975).

A computer model has been developed to synthesize several solar thermal and photovoltaic systems to
supply the electrical and space conditioning requirements of loads ranging from a single family resi-
dence to a community of 100,000 people. The model performs a present value economic analysis of the
effective life cycle cost of each system, taking into consideration tax status of the system owner
(corporate, public or private) and future trends in capital equipment and fossil fuel costs. Those
costs, alongwith costs associated with conventional fossil fueled systems are projected into the future to provide an economic comparison, over time, of solar and fossil fired systems.

0978

Hybrid Concentrating Mechanical Tracking Collectors in Mainframes Typical of Flat Plate Collectors —
Best of Several Technologies Aimed at Solar Cooling Reliability : Brent C R, " Proceedings ERDA Conference on Concentrating Solar Collectors ", Georgia (USA), p.2 - 85, (1977).

Random direct beam solar radiation measurements made over the past three years confirm the distinctly
flatter power curve available to tracking collectors compared to direct beam radiation falling on a
fixed collector. Knowing that an economic breakthrough in solar cooling would have to be made in order
for maximum solar utilization to occur in the south, the author combined the protective and insulating
properties of mainframes typical of flat plate collectors with tracking ($\pm 4°$) miniature concentrating
troughs in an attempt to optimize a high temperature solar collector capable of powering an absorption
chiller at 99 deg C. Prototype testing provided an efficiency near 20% at this temperature. Upgrading
of reflectors, glazing and absorber surface should provide higher efficiencies. Collector costs
are somewhat higher at $ 178/sqm but capital recovery calculations show that a system cost of $ 9500
can be amortized against fuel savings of $ 600 per year in 1978 within a pay back period of ten years.
The solar collector consists of 31-one dimensional tracking, reflecting trough of a depth, length and
critical wall angle (20°) to fit well into a rectangular 4 ft X 8 ft, 5 to 8 in deep mainframe and provide optimum concentration for their size (1.67 to 3.0 fold).

0979

Aerodynamic Stability of a Heliostat Structure : Brosens P J, " U S Air Force Missile Development Center, Report AFMDC-TR-59-15 Solar Furnace Support Studies ", Holloman Air Force Base, New Mexico (USA),2,
99, (1959).

A flat rigid mirror structure supported electrically and without damping is studied for stability in
the wind. The wind is assumed to have a constant velocity normal to the elevation axis of the mirror.
The stability of any translational motion of the mirror, perpendicular to the elevation axis is then
analysed. Regardless of the intensity of the wind the mass of the heliostat mirror, and the compli-
ance of its supporting structure such motions are unstable if, at a given angle of attack, the derivative of the aerodynamic normal force coefficient with respect to the angle of attack is negative. In
the case of this mirror, the available data on normal force coefficients is not sufficient to permit
drawing a definite conclusion as to whether the present structure is stable or not under the assumed
conditions.

0980

Oscillation of a Rigid Heliostat Mirror Caused by Fluctuating Wind : Brosens P J, " U S Air Force Missile Development Center Report AFMDC-TR-59-15, Solar Furnace Support Studies ", Holloman Air   Force
Base, New Mexico (USA), 2, 117, (1959).

The response characteristics of an elastically supported heliostat in random vibration are studied.
For wind fluctuation of importance,the horizontal and vertical translational motions are uncoupled.
Their excitation may be considered separately. Certain assumptions are made pertaining to the random

characteristics of the wind. It is found that the damping required to prevent undesirable   oscilla-
tions of the heliostat is small.

0981

Actual Case Studies of the Economics of Heating Boiler Feedwater Using Linear Focussing Solar Collectors : Brown J E, Carito J A, Guild P H, Noga M W and Silverman C G, " Proceedings ISES (American S e c -
tion) Conference ", Denver, Colorado (USA), 2.1, 828, (1978).

The technical and economic feasibility was investigated of augmenting the heating of boiler feed water in two south western steam-electric generating plants with state of the art solar equipment. T h e
plants studied were El Paso Electric Company's Newman Station Unit No. 1 and Sierra Pacific Power Company's Fort Churchill Station Unit No. 2. It was determined to be technically feasible to use state
of the art linear focussing solar collectors to heat boiler feed water. However, for each case approximately 20 per cent annual fuel cost escalation is required for the next 25 years t o economically
justify the capital investment.

0982

Flux Density for Ray Propagation in Geometrical Optics : Burkhard D G and Shealy D L, " Journal of Optical Society of America ", 63, 299, (1973).

A general formula is derived that specifies the illumination (flux density) over an arbitrary receiver surface when light rays are reflected by or refracted through a curved surface. The direction of
the deflected ray and its intersection with the receiving surface used with the equation for the surfaces, leads to a transformation that maps an element of deflecting area onto the receiving area, by
means of the jacobean determinant. A formula for the flux density along a ray path follows as a special case. An equation for the caustic surface is obtained from the latter. As an example radiation
flux density contours are calculated for a plane wave reflected from a sphere. Flux density and the
caustic surface are calculated for a plane wave reflected onto a plane from a concave spherical l e n s
and also for a plane wave refracted onto a plane through a hemisphere.

0983

Absorptance, Emittance and Thermal Efficiencies of Surfaces for Solar Spacecraft Power : Butler C P,
Jenkins R J and Parker W J, " Solar Energy ", 8(1), (1964).

An expression for the thermal efficiency of a surface used as receiver for spacecraft solar power generation is derived in terms of the solar absorptance, hemispherical emittance, and concentration ratio. Experimental equipment is described in which direct measurments may be made upto 1000 deg C by
using a carbon-arc-image furnace to simulate the sun. Emittance and absorptance data are presented on
six polished metals and four plasma flame sprayed coatings from 200 to 800 deg C. It is shown t h a t
thermal conversion efficiencies as high as 40% at 800 deg C can be realised with refractory coatings &
with a solar radiation concentration ratio of 50 or 6.9 Watts/sqcm.

0984

Relation Between the Surface Shapes of a Heater Receiver and the Concentration Ratio for a Parallel-
Ray Flux : Buzin E I, " Applied Solar Energy ", 2(4), 11, (1966).

Derivation of formulas relating the surface shapes of the reflector and heat collector and the intensity of the geometric concentration of parallel rays reflected (from the reflector) normal to the heat
collector surface. It is shown that solutions of ordinary differential equations can be used to determine the various concentrator systems that meet the requirements for any specific case. The results
a r e seen to be of interest for the use of solar energy in space missions (particularly when t h e
concentrator systems consist of foldable or inflatable elements that do not have a paraboloid shape).

0985

Reducing Heat Loss from the Energy Absorber of a Solar Collector : Chao B T and Rabl A, " U S Patent 3,
986, 490, 19th October (1976) ".

A device is provided for reducing convection heat losses in a cylindrical radiant energy collector. It
includes a curved reflective wall in the shape of the arc of a circle positioned on the opposite side
of the exit aperture from the reflective side walls of the collector. Radiant energy exiting the exit
aperture is directed by the curved wall onto an energy absorber such that the portion of the absorber
upon which the energy is directed faces downward to reduce convective heat loss from the absorber.

0986

Solar Energy Concentration with Liquid Lenses : Chauhan R S, " Solar Energy ", 18,587, (1976).

In this paper, an inexpensive method of producing liquid lenses for solar energy concentration has been
described. Both circular and cylindrical configurations have been discussed. Fabrication details of
absorber design for circular and cylindrical configurations have also been discussed.

0987

Ray Tracing Through Funnel Concentrator Optics : Chen C W & Hopkins G W, "Applied Optics ",17,1466,(1978).

A procedure and equations are presented for tracing meridional and skew rays through some funnel concentrators proposed as solar energy collectors. These concentrators are general t o r o i d s generated
from parabolic or elliptical surfaces. The ray tracing procedure uses an iterative technique w i t h
exact tracing to conical surfaces tangent to the concentrator.

0988

The Use of a Kaleidoscope to Obtain Uniform Flux over a Large Area in a Solar or Arc Imaging Furnace:
Chen M M, Berkowitz-Mattuck J B and Glaser P E, " Applied Optics ", 2, 265, (1963).

The use of properly designed light pipes to redistribute the energy of a solar furnace or an arc imaging furnace is discussed. Theoretical analyses concerning the principle of operation, as well as
formulae for estimating the flux uniformity and reflection losses are discussed.

130

0989

Circular Flat Plate Heat Exchanger for Solar Concentrator : Cobble M H, " Solar Energy ", 6, 164, (1962).

An analysis is given of a heat exchanger of the flat plate type, having circular boundaries that could
be used just out of the focus of a concentrator of the paraboloid of revolution type, to heat liquids.
The heat exchanger consists of two parallel plants, through which an incompressible ideal fluid is flowing radially while being heated.

0990

Heat Exchanger for Solar Concentrators : Cobble M H, " Solar Energy ", 7, 18,(1963).

The theoretical analysis of three types of heat exchangers for use with solar concentrators is developed. Under the assumptions of an incompressible inviscid fluid, constant energy rate input per unit
area, losses to the surroundings, an expression for the temperature field, an average exit temperature
is given. An expression for a determination of a fluid exit temperature, is derived and plotted for
the three types of heat exchangers.

0991

Analysis of a Conical Solar Concentrator : Cobble M H, " Solar Energy ", 7, 75, (1963).

The concentration that a conical mirror can theoretically attain, is developed for two types of targets ; a circular cylindrical target and a conical target. Under the assumption of a one-dimensional
sun, the optimum mirror cone half angle is determined for both types of targets. At the optimum mirror cone half angle, the concentration, assuming a two dimensional sun, is determined, and the optimum
concentration in terms of mirror radii is found for both types of mirrors.

0992

Temperature Fields of Solids Heated by Solar Concentrators : Cobble M H, " Solar Energy ", 7,134, (1963).

The steady – state temperature distribution is solved for four target solids under the assumptions that
the energy rate per unit area at the target surface is coordinate dependent, and that there are losses
to the surroundings at the exposed target surfaces. The analysis includes a cylindrical slab and a
sphere heated at the focus of a paraboloidal mirror, and a rectangular slab and a cylinder heated at
the focus of a parabolic mirror. The general solution for each target solid is given for any type of
distribution of target energy rate per unit area. In addition, an experimental target energy rate per
unit area is assumed and the exact temperature field is shown. A dimensionless external temperature
module is plotted versus a dimensionless coordinate for various Nusselt numbers, using the assumed
target energy rate per unit area.

0993

Efficiency of a Series of Thermoelectric Generators in a Solar Wedge Concentrator : Cobble M H, "Proceedings 2nd International Conference on Thermoelectric Energy Conversion ", Arlington (USA), p.100,
(1978).

A system for generating electrical power from solar energy utilizing a nontracking wedge to concentrate sunlight onto a series of thermoelectric generators, cooled by a heat exchanger, is analysed.
Expressions are developed for the temperature distribution in the 'p' and 'n' legs, and for the temperatures of the hot and cold junctions, and the temperature distribution in the fluid,the electrical
work efficiency and the efficiency of heating the cooling fluid. The wedge construction is given as
a function of the wedge half angle and the inclu ded angle of incidence. Utilizing the theory in an
example, the electrical work efficiency and fluid heating efficiency are shown plotted against concentration and mass flow rate in separate figures.

0994

Solar Thermal Electric Power Generation in Egypt : Cobble M H and Smith P R, " Proceedings ISES Conference ", Los Angeles (USA), (1975).

Progress report of a project under way to develop systems for producing electricity using solar c on-
centrators and conventional steam turbines is described. A primary objective of the project is   to
produce electricity for small groups of homes with equipment that is reliable and which uses a readily obtainable working fluid. The project consists of two phases. The first phase is to design,build
and to operate a small prototype power generation station. It consists of a 9 sqm segmented cylindrical parabolic mirror with a concentration of about 30 : 1. Mounted integral with this concentrator is
a single pass, multiple tube cavity boiler. The second phase of the project will be to use the information gained from the operation of the prototype system to design and build a 100 Kw thermal system capable of supplying electricity to small groups of homes.

0995

Verification of Wedge Concentration Using a Helium Neon Laser : Cobble M H and Thacher E F, " Proceed -
ings 24th Annual Technical Meeting, Institute of Environmental Sciences ", Texas (USA),p.287, (1978).

The concentration of a wedge utilizing multiple reflections is developed and the theoretical expres -
sion for the concentration ratio is given as a function of the wedge half angle, and the included angle
of incidence. A plot and a table of the resulting concentration ratios are given for various combinations of wedge half angles, and the included angle of incidence. The theoretical predictions of approach angles, points of contact and concentration ratios were compared to an experiment using a helium neon laser as a collimated beam, and the results, theoretical and experimental are presented.
The wedge is capable of high concentration for small wedge angles, and for a given included angle of
incidence there is no need for tracking.

0996

Simple Procedure for Predicting Long Term Average Performance of Non-tracking and of Tracking Solar
Collectors : Collares Pereira M and Rabl A, " Solar Energy ", 23, 235, (1979).

The Liu and Jordan method of calculating long term average energy collection of flat plate collectors is simplified (by about a factor of 4) and generalized to all collectors, tracking and nontracking . The only meteorological input needed is the long term average daily total insolation 'H'. To illustrate the method a flat plate, a 1.5 X compound parabolic concentrator, and a tracking line focus parabolic reflector are evaluated and compared.

0997

Collector Comparison Based on Yearly Average Performance : Collares Pereira M and Rabl A, "   S o l a r Energy Research Institute Report SERI/ TR-34-304 ", (USA), (1979).

The utilizability method is used for calculating the yearly average energy output (Q year)of the principal solar collector types and tracking models. The function dependence of 'Q'year on climate and on geographic and collector variables is analysed. It is shown that to an excellent approximation 'Q'year can be corelated in terms of two or three suitably chosen parameters. The corelations are presented in graphical form and as analytical curve fits. From these graphs one obtains 'Q'year by scaling abscissa and ordinate by the collector parameters (slope and intercept of instaneous efficiency curve and all day average incidence angle modifier). The method is illustrated by comparing flat plate,   compound parabolic concentrators (CPC), parabolic trough, linear Fresnel lens in polar mount and parabolic dish as a function of operating temperature.

0998

Accurate Temperature Measurements in Work with Solar Furnaces : Conn W M, " American Journal of Physics", 24, 581, (1956).

Measurement of temperatures above 1500 deg C in a solar furnace is discussed. If the sample is heated without melting, the true temperature can be obtained from the brightness temperature, determined with an optical pyrometer or phototube. If the sample is studied at or above the melting range, it is necessary to separate the radiation emitted by the sample from the incident or reflected radiation while the temperature is measured. This was accomplished by shutting off the incident radiation with a high speed rotating shutter or sector.

0999

Une Evaluation de L'aive Efficace sun Champ de Mirroire on Termes de Repartition du Verre Installe : Courrege Ph., " Proceedings International Conference on Solar Electricity ",Toulouse (France,p.831(1976) .

This paper introduces a formula which provides an upper estimate of the effective area of a plane regularly networked mirror field. That is a continuous type estimate by a double integral which depends on the sizes of mirrors and on their arrangement only via the distribution of set up glass.

1000

Anticlastic Reflecting Concentrators : Cowman C D and Shepherd Jr.P G, " Proceedings ERDA Conference on Concentrating Solar Collectors ", Georgia (USA), p. 5 - 45, (1977).

A low cost reflecting concentrator, with potential applications for solar energy collection, has been developed (U S Patent No. 4035'064). The optics of reflecting anticlastic (e.g. parabolic hyperboloid) surfaces are discussed with reference to two practical examples ; a plane mirror subjected to a twisting deformation and an edge defined minimum area reflecting surface. Results from a computer ray tracing study of some minimum area surfaces are presented with preliminary experimental data.

1001

Solar Collector : Coxon De W & Bregg G N, " U S Patent 4, 131, 109 / December (1978) ".

A solar collector assembly is disclosed having within its housing a series of radiant energy entrapping collector elements each having an energy absorbing base, a generally semicylindrical upwardly open reflector with concave, upper, receptor - reflector surfaces and convex lower receptor surfaces, a collector and radiation entrapping receptor projecting up in said reflector into the radiation reflection pattern of the concave surfaces. These elements form elongated fluid flow heat exchange passages.

1002

Solar Concentrator Flux Distributions Using Backward Ray Tracing : Daly J C, " Applied Optics ", 18 , 2696, (1979).

Flux distributions produced by parabolic and circular cylinder solar concentrators subject to surface slope errors and defocussing are determined. The technique presented traces a set of rays from a point on the absorber back through the concentrator optics to the sun. The solar flux at the a b s o r b e r point is the sum of the flux associated with each ray. Various models of the solar disk are i n t r o - duced by weighting the flux associated with each ray as a function of where it strikes the solar disk.

1003

Conceptual Design of a Combined Cycle Solar Hybrid Power System : Darnell J R, Westsik J H & Lam E Y, " ASME Paper No. 80 - C2 / Sol - 11 ", (1980).

This paper summarises the results of a DOE funded study of a Solar Hybrid Power System. The concept consisting of an air cooled solar central receiver which is used to displace a portion of the f o s s i l fuel requirement of a first of its kind, commercial size 112.6 MWe power plant,is described. Operation, performance and cost estimates are presented. The results of an assessment of the conceptual design show significant potential for utility application by the year 2000. Developmental needs a r e   also discussed.

1004

SLATS Line Focus Solar Collector : Davison J H and Wendt A J, " Proceedings ERDA Conference on Concentrating Solar Collecors ", Georgia (USA), p.2 - 29, (1977).

SLATS (Solar Linear Array Thermal System) is a high concentrattion (40 : 1) solar collector with single axis tracking. It is composed of planar modules containing 10 one-foot wide reflectors in which all reflectors are focussed on a stationary, linear overhead receiver. Each reflector concentrates reflected energy by a factor of 4 : 1 or more. These modules, depending upon the quantity of heat required and the site, may be connected in series forming rows 200 ft and more in length. SLATS is controlled electronically by an automatic closed loop servo system. The electronic control will automatically start operation each morning and end operation each evening. It also provides for shut-down in the event of cloud cover, over temperature and power failure. SLATS can deliver thermal energy at high temperature, with efficiencies exceeding 60 %. This is obtained by use of high quality mirrors, an insulated receiver of heat absorber and controlled real time sun tracking.

1005

Non Tracking Concentrating Solar Collector : DeGeus A M, " U S Patent 4, 222, 370 Filed 17th May,(1978)".

This device is a non tracking concentrating solar collector comprising a series of parallel parabolic reflector segments which are reflective on both surfaces and which have focal points lying in a plane which is approximately at a right angle to the direction of incident solar radiation. The incident radiation is received in the space between the upper ends of the reflector segments, the lengths of which are selected so that the incident radiation is reflected back and forth between the segments until the fourth and sixth reflections emerge downwardly from between the reflector segments in a beam which is much narrower than the space between the upper ends of the reflector segments. At least one collector for solar radiation is positioned beneath the reflector segments, other than the first segment, and parallel to these segments so as to receive the narrowed beam.

1006

Concentrator for Solar Air Heater : Demichelis F and Minetti-Mezzetti E, "Applied Physics ", 19, 265, (1979).

A focussing collector for air heater, the concentrator being Fresnel lenses, is examined here.The sun light is focalised on black body metallic sheets, contained in an insulated box, through which the energy is transferred to the fluid (dry air). The sheets, having the same light, are arranged at a distance from one another increasing from the center to the edges in order to trap all the incident energy. The energy reflected by the sheets is also trapped. A thermotechnical analysis of the system in order to determine the optimal conditions for the heat transfer is reported.

1007

Construction and Evaluation of Linear Segmented Solar Concentrators : Desai P, Williams J R, Rollins J and Lindsey A M, " ASME Paper No. 75 - WA/Sol - 6 ", (1975).

Two types of linear faceted collectors suitable to satisfy energy needs in the intermediate temperature range between 100 and 300 deg C are examined. These are the segmented plane solar concentrators (SPSC) and the faceted fixed mirror solar concentrator (FFMC). A comparative performance evaluation of the SPSC for several receiver configurations is presented. Calculated heat flux onto the heat exchanger is compared with measurements obtained by scanning a board spectral response detector across the focal plane in terms of the concentrator efficiency of the FFMC.

1008

Capteurs a Concentration du Rayonement Solaire Estimation du Rendement : Desautel J and Peri G (and--Desautel J & Imbart B, "Influence des Imperfection des Surfaces Collectrices "), "Proceedings International Conference on Solar Electricity ", Toulouse (France), p.397, (1976).

In this paper are presented two aspects of the problems in the field of solar radiation focussing collectors. The first is concerned with a method to determine solar collector's performance in view of the design and comparison of these collectors while the second aspect takes into account significant parameters due to the manufacture of the reflecting surface and such as defects of curvature, orientation, stiffness and also due to the pointing errors of the mounting.

1009

Inexpensive High Temperature Solar Collector : Dorman J and Lansing F L, " NASA Technical Briefs 2(2), 213, " (USA), (1977).

A high temperature solar concentrator was devised at California Institute of Technology. It consists of a series of parallel cylindrical lenses made of glass or plastic. The curvature of lenses is designed to give the maximum concentration ratio and short focal length. Experiments show that the collector can heat a working fluid to around 200 deg C. The liquid lens concentrates the solar radiation by a factor of 100 : 1. A typical lens is 10 cm wide and mounted to move with the rest of the collector on a transitional tracking unit. Seasonal adjustments can be made to the tracking unit.

1010

Feasibility Study of Pulse-Shaping for Solar Furnace : Drew G G, " Solar Energy ", 9, 217, (1965).

Basic characteristics necessary in thermal pulse for experimental work, which are changes in intensity and spectral distribution with time ; changes required in solar spectrum to match this pulse are described in detail. A possible mechanism for shutter, of The Defence Chemical Biological and Radiation Laboratory's solar furnace to provide required characteristics, is suggested.

1011

Performance Testing of the Suntec SLATS Solar Collector : Dudley V E and Workhoven R M, " Sandia Laboratories Report SAND - 78 - 0623 ", (USA), (1978).

This report presents the results of the tests made on Suntec SLATS solar collector Mid Temperature Solar Systems Test Facility at Sandia Laboratories in Albuquerque (New Mexico). The test objective is stated, the collector system and the test facility are both described, a definition of the performance test variables is given and the results of the test are given. The results are in the form of plots and tables of data. A summary of results and conclusions end the report.

1012

Concentrating Solar Collector Test Results, Collector Module Test Facility (Summary Report):Dudley V E and Workhoven R M, " Sandia Laboratories Report No. SAND - 78 - 0815 ", (USA), (1978).

This report summarises the results of tests on five concentrating solar collectors. Parabolic trough, fully tracking Fresnel lens and Fresnel reflector designs were tested. Efficiency and thermal losses were measured at temperatures about 100 deg C to 300 deg C. Performance curves are given for each collector.

1013

Concentrating Solar Collector Test Results - Collector Module Test Facility (CMTF), Summary Report (January - December 1978) : Dudley V E and Workhoven R N, " Sandia Laboratories Report No. SAND-78-0977, (1979).

This report summarises the results of tests on four concentrating solar collectors at output fluid temperatures from 100 deg C to 300 deg C. Collectors tested were manufactured by Scientific Atlanta Inc., General Electric, Del Manufacturing and Suntec Inc.

1014

Performance Testing of the Del Concentrating Solar Collector : Dudley V E and Workhoven R M, " Sandia Laboratories Report No. SAND - 79 - 0514 ", (1979).

This report summarises the results of tests performed on the Del Solar Collector at the Mid Temperature Solar Systems Test Facility. Test objectives are defined, test procedures are described, test results and conclusions are given.

1015

A Sequential Algorithmic Approach to Synthesizing Minimum Cost Solar Collector Fields : Duff W S," Proceedings ISES Conference ", Los Angeles (USA), (1975).

A methodology is presented in this article for finding the collectors, heat transport fluids, piping arrangments, field geometries, pipe diameters and insulation thickness that make up a solar collector field delivering thermal power with a specified mass flow rate, inlet temperature and outlet temperature at minimum cost.

1016

Optical and Thermal Performance Analysis of Three Line Focus Collectors : Duff W S, " ASME Paper No. 76 - WA / HT - 15 ", (1976).

An analytic approach developed earlier for two axis tracking collectors is extended to one axis tracking collectors. The parabolic trough, the linear array of fixed slots with a movable absorber are specifically addressed. The two measures, effective aperture and image spread, are developed for the three line focus concentrators. A derived measure, effective aperture, is suggested as an alternative to concentration ratio to express more adequately, the potential collector performance.

1017

Performance Comparison Method of Solar Concentrator : Duff W F and Lameiro G F, " ASME Paper No.74-WA/ Sol - 4 ", (1974).

A method for comparing the performance of various types of solar concentrators has been developed. This method is completely independent of all target characteristics with the exception of target shape. The method makes it possible to identify the characteristics of such diverse concentrators as a Fresnel reflector and a paraboloid reflector that delivers the same optical performance. This method can be applied directly to many end uses of the heat from concentrated radiation.

1018

Parametric Performance and Cost Models for Solar Concentrators : Duff W F, Lameiro G F and Lof G O G, " Solar Energy ", 17, 47, (1975).

The development of parametric performance and cost models for various solar concentrators is discussed. The equations are derived in the context of an optimization scheme which can be applied to many different problems which arise when heat is generated by means of solar concentrators. Thus, while the results presented were developed for finding a minimum cost solar electric energy power plant, the method employed has been found to have general applicability. Sensitive analysis of the subsystems is also discussed. Finally, numerous illustrative examples are presented.

1019

Minimum Cost Solar Thermal Electric Power Systems (STEPS) : Duff W S and Karaki S, " Proceedings ISES Conference ", Los Angeles (USA), (1975).

This paper reports on an eighteen month National Science Foundation sponsored study to determine the minimum cost system, or systems, for generating electric power by solar thermal means. This effort, begun in May 1973, has consisted primarily of estimating large volume manufacturing costs of candidate components, formulating required performance models and developing procedures whereby these components are synthesized into minimum cost of electric energy systems.

1020

Parabolic Reflector for Concentration of Solar Radiation Used for Energy Yield : Dvering G, " German Patent 2, 506, 905/A/August 28, (1976) ".

A parabolic mirror is described for concentrating sunlight for energy production. The mirror consists of a plastic sheet silvered on one side, which is stiffened by suitable construction. The equipment is positioned according to the sun's position by an electronic tracking arrangement. Various examples of construction are discussed.

1021

Development and Application of High Power Light Pipes : Dvernyakov V S, Pasichnyi V V, Kasich-Pilipenko I E, Ermina T V, Sattarov D C, Muraveva M I and Galant I E, " Applied Solar Energy" 15(4), 32, (1979).

It is shown that it is possible to develop glass-fiber light pipes designed to transmit high density

134

radiant fluxes from the focal zone of a concentrator and to manipulate these fluxes so as to realize technological processes. Standard optical materials were investigated ; it was established that certain heavy crown-glasses, light crown-glasses, quartz lightpipes and certain crystals are resistant to the action of concentrated solar radiation.

1022

The Cost and Value of Washing Heliostats : Eason E D, " Proceedings ISES Silver Jubilee International Conference ", Georgia (USA), (1979).

The equation is derived for the washing frequency that minimizes cost per unit energy for solar collectors of any type. Central receiver heliostats of two designs are used in an example, & the results for Albuquerque soiling rates show that washing 15 – 30 times per year is worthwhile. The optimal frequency is directly or inversely proportional to the square root of the relevant parameters, but it is not sensitive to number of collectors or plant size. The frequency for minimum cost per unit energy corresponds to the strategy of washing whenever the value of the extra energy from cleaner mirrors will pay for the cost of the wash.

1023

Collector Cost and Performance Trade-off Studies : Eason E D, " Proceedings DOE Workshop on System Studies for Central Solar Thermal Electricity ", Houston, Texas (USA), (1978).

The cost of solar central thermal electric power plant can be reduced by determining the most cost effective level of heliostat performance. This paper describes one method of carrying out the necessary trade-off studies and cost minimization. Example using preliminary 10 MW pilot plant, cost and performance figures serve to illustrate the methodology, type of data needed and potential results. The emphasis is on understanding the nature of the compromise between cost and performance, rather than establishing the best collector design for specific plant.

1024

Radiation Characteristics in Optimization of Solar Heat – Power Conversion Systems : Edwards D K and Nelson K E, " ASME Paper No. 61 – WA – 158 ", (1961).

Criteria are developed for comparison and selection of thermal radiation characteristics of concentrators, collectors and radiators ; solar heat power systems may prove feasible for auxiliary power suplies in space or in other regions remote from inexpensive fossil fuels.

1025

A Ray Tracing Approach to Solar Radiant Energy Exchange in a Specular Axisymmetric Spinning Cavity : Engel R K, " Proceedings Thermodynamics and Thermophysics of Space Flight Symposium ", Palo Alto, California (USA), p.15, (1970).

The evaluation of solar heating to a radiator surface recessed within a specular cylindrical cavity is determined. The analysis is based on deterministic ray tracing schemes. That is, surface intersections are treated deterministically as opposed to a Monte Carlo random selection approach. For open cavity configurations this procedure enhances computational accuracy and reduces computer time.

1026

Initial Tension Determination for a Shell of a Solar Concentrator : Faizullayev A, " Applied Solar Energy ", 9(1), 21, (1973).

The paper outlines the algorithm for solving the third order differential equation with variable coefficients and provides the digital results obtained. The Runge – Kutta method and Vegstein iterative method for clarifying Y_{n+1} have been applied. The same approach may be used for solving a number of non-linear differential equations.

1027

The FES Delta Focussing Solar Collector : Falbel G, " Proceedings SPIE Conference ", 68, 112, (1975).

This paper describes the design and measured performance characteristics of a focussing solar collector that requires neither single nor dual axis tracking of the sun, but achieves solar concentration gains exceeding 2 : 1 for both direct and diffuse solar energy. Both sides of a conventional flat plate collector absorbing plate are used to collect solar energy using a compound cylindrical reflector and no thermal insulation is required in the collector, thus effecting significant material and fabrication cost economies. The focussing capability allows improved high temperature performance as compared to conventional flat plate collectors without requiring a selective absorber coating on the collector plate. Ray traces defining the collector's acceptance field of view for both direct & diffuse solar energy, are presented and the theoretical and experimentally obtained efficiency are compared. Measured results obtained from an instrumented prototype collector, operating since February 1975 are presented, showing that a single collector collects approximately 50,000 Btu's per average sunny day throughout the year. Applications of the collector as a transparent window or curtain wall replacement in buildings are discussed and large production quantity cost estimates are presented.

1028

Design and Performance of a Concentrating Solar Collector : Farag I H, " Proceedings AIChE Symposium ", San Francisco (USA), 76, 6, (1980).

The paper describes the design consideration and the performance evaluation of a parabolic concentrating collector. The collector's surface is approximately 1.4 sqm of reflective material focussing on a 38.1 mm diameter blackened copper pipe. The heat transfer medium was chosen to be air. Preliminary results indicate a 43% efficiency of collection based on average daily insolation.

1029

Application of ASHRAE Standard 93 – 77 Concentrating Collectors for the Purpose of Predicting All Day Performance : Fiore P J and Wood B D, " Proceedings ISES Silver Jubilee International Congress ", Georgia (USA), (1979).

This paper reviews and evaluates the performance equations which can be used to corelate the data from the ASHRAE 93 - 77 instantaneous quasi-steady state thermal performance tests of solar concentrators. The application of the incident angle modifier for correcting the near normal incident tests for other incident angles is discussed for each of the generic concentrator types. A generalized calculating procedure is proposed for taking into account ambient conditions, effects of the incident angle changes tracking errors, and thermal time response in order to arrive at an all day performance for the given concentrating collector. Test data from a linear parabolic trough are used to illustrate the calculation procedures for calculating the all day performance.

1030

Measurements of Temperatures of Products Treated in a Solar Furnace by Means of IR Pyrometers : Foex M, Coutures J P and Benezech G, " COMPLES Bulletin No. 18, 67, (1970) ",

This paper presents the study of the advantages and drawbacks concerning the use of IR pyrometers for measuring temperatures of products treated in a solar furnace. It is shown that by employing certain interference filters, the temperatures ranging from 2000 to 3500 deg C can be measured by means of optical pyrometers if the parasitic effects due to reflection of the solar radiation are eliminated. Problems concerning the determination of radiation factors when the black body cannot be used are discussed.

1031

Solar Thermal Energy Collection System : Frank M W, " U S Patent 3, 875, 926 / April 8, (1975) ".

An application and a method for receiving and transmitting thermal solar energy are described. A collector of a solar energy reflects solar energy to a heat pipe which transmits vapour produced therein to a condenser to release and utilize the thermal energy carried by the vapour either to perform heating functions or to be dissipated as the vapour is condensed.

1032

Concentrating Collector Testing at the Solar Engineering Test Module : French R L, Mooney L G , Mc - Dowell J H and Uselton R B, " Proceedings ERDA Conference on Concentrating Solar Collectors ", Georgia (USA), p.6 - 19, (1977).

The Solar Enegineering Test Module (SETM), a facility designed for testing concentrating collectors at intermediate heat transfer fluid temperatures (≤ 530 F) and pressures (≤ 1050 psi) is operated by American Technological University (ATU) in support of the ATU / Fort Hood Solar Total Energy Military Large Scale Experiment, LSE No.1. This paper gives a summary description of SETM, its development and operation and includes preliminary test results.

1033

Comparative Evaluation of Distributed - Collector Solar Thermal Electric Power Plants : Fujita T, El Gabalawa N, Caputo R S and Herrera G G, " Proceedings 13th IECE Conference ", San Diego (USA), 2, 1600 (1978).

Distributed - collector solar thermal electric power plants are compared by projecting power plant economies of selected systems to the 1990 - 2000 time frame. The approach taken is to evaluate the performance of selected systems under the same weather conditions. Capital and operational costs are estimated for each system. Energy costs are calculated for different plant sizes based on the plant performance and the corresponding capital and maintenance costs. Optimum systems are then determined as the systems with the minimum energy costs for a given load factor. The optimum system comprises of the best combination of subsystems which give the minimum energy cost for every plant size. Sensitive analysis is done around the optimum point for various plant parameters.

1034

Projected Techno - Economic Improvements for Advanced Solar Thermal Power Plants : Fujita T, Manvi R & Roschke E J, " Proceedings IECE Conference ", Boston (USA), p. 59, (1979).

The projected characteristics of solar thermal power plants (< 10 MWe) employing promising advanced technology subsystems components are compared to current (or pre 1985) steam-Rankine systems. Improvements according to advanced technology developments options are delineated. The improvements derived from advanced systems result primarily from achieving high efficiencies via solar collector systems which capture a large portion of the available insolation and concentrate this captured solar flux to obtain high temperatures required for high heat engine / energy conversion performance. Advanced technologies including conversion systems such as Stirling Engines, Brayton / Rankine combined cycles and storage / transport concepts encompassing liquid metals and reversible reaction chemical systems, are considered. In addition to techno - economic aspects, technologies are also judged in terms of factors such as developmental risk, relative reliability and probability of success.

1035

Vacuum Solar Thermal Collector with Optimal Concentration : Garrison J D, " Proceedings SPIE Conference ", 114, (1977).

A fixed solar thermal collector tube giving approximately optimum performance has been designed by proceeding through a series of optimization steps, which almost uniquely fix the form of this design. While seeking optimum performance, features have been included which should lead to low cost mass production of these tubes.

1036

A Comparison of Solar Thermal Energy Collection Using Fixed and Tracking Collectors : Garrison J D, Craig G T and Morgan C, " Proceedings ISES (American Section) Conference ", Denver, Colorado (USA), 2.1, 919, (1978).

Solar thermal energy collection at seven sites is calculated for air and vacuum flat plate collectors and a vaccum collector using cylindrical and Winston concentration. The hourly intensity and angular distribution of solar radiation is predicted for the calculations using insolation measurements and an insolation model. Tracking configurations include two axis, vertical axis, polar axis and east - west axis tracking. Fixed collector arrays are tilted towards the equator at the latitude angle. These calculations indicate energy collection which varies linearly with cloudiness index K_T. Diffuse radiation is $\sim$15% of energy collected by concentrating collector arrays at low operating temperatures where large acceptance angles maximize energy collection. Smaller acceptance angles, required to maximize energy collection at higher operating temperatures, eliminate the advantage of Winston concentration for tracking arrays. Absorbers with high selectivity are necessary for high efficiency of fixed evacuated collectors at higher operating temperatures.

1037

Incidence Angle Modifier and Long Term Performance Prediction for Solar Collectors : Gaul H and Rabl A, " Proceedings ISES Silver Jubilee International Congress ", Georgia (USA), (1979).

For flat plate collectors, the measurment of the incidence angle modifier is straight forward and well known. In line focus collectors without end reflectors the incidence angle modifier is complicated by spillover of radiation at non-normal incidence. Since the spill over effect is dependent on the length of the collector module, the results from tests on (usually short) collector modules cannot be used directly to describe the behaviour of (usually long) collector arrays in practical installations. Analysis of incidence angle modifier must, therefore, separate the spillover of collector length. The formalism for this analysis is developed and applied to test results for several parabolic trough collectors, published by Sandia Laboratories. Tests results obtained from a Hexcel parabolic trough collector at SERI are also analyzed. The results are presented in two forms, first as a table or graph of optical efficiency versus incidence angle, and secondly, as a single number, the all day average optical efficiency (for typical operating times).

1038

Optical Design and Performance of an Inverted Segmented Mirror Solar Concentrator : Gharmalkar D R and Anderson E E, " ASME Paper No. 78 – HT – 32 ", (1978).

Expressions for the position and orientation of the mirror to prevent shading of adjustment mirror segments are developed from geometric optics. Optical efficiency, concentration ratio and receiver intercept factors are presented as functions of the design parameters. These parameters include the mirror size, number of mirror segments and the concentrating array, the focal distance between mirror array and the receiver and the receiver size. The influence of the improper orientation of the collector with respect to solar beam radiation and mirror thickness is also presented.

1039

Solar Collector Calibration : Ghosh S K, " Photogrammetric Engineering ", 32, 320, (1966).

A purely photogrammetric procedure was successfully applied for the analysis of structural deviations of a 44 ft parabolic solar reflector with a standard error of only 0.5 mm. Conventional photogrammetric equipment was employed ; a 'Wild' Phototheodolite camera and a 'Wild' A7 stereoscopic mapping instrument. The data was reduced with an IBM 7094 computer. Careful attention, the important details of the operation – photography, control, point identification base height ratio, plotter operation, a sound plan helped assure rigorous and accurate results.

1040

Multi-element Silicon Photoelectric Convertors for Investigating the Optical Energy Characteristics of Solar & Radiant Energy Concentrators : Gilberman A Ya., Kovalev I I, Krasilovskii V J and Madvedeva L M, " Applied Solar Energy ", 15(3), 5, (1979).

The article considers the possibility of employing multi-element PEC with a mess structure for investigating the OEC of solar and radiant energy concentrators in radiator / concentrator / receiver system. The sensitivity characteristics of matrix PEC and the value of the interconnection factor are given.

1041

Light Weight Solar Concentrator Development : Gillette R, Snyder H E & Timar R, " Solar Energy ", 5, 24, (1961).

The concentrator was designed to supply heat at 3140 deg F to a thermionic generator. The generator has a 0.465 sqm heat absorber area. It requires 440 watts of heat flux to strike the absorber. Area concentration ratio 1500/1 is possible.

1042

A New Collector of Solar Radiant Energy ; Theory and Experimental Verification : Giovanni F, " Proceedings U N Conference on New Sources of Energy ", Rome (Italy), (1961).

A new type of collector capable of operating at high efficiencies at temperatures of 4000 – 5000 deg C and economical enough for industrial production is proposed. Experiments on three different collectors of this type and the first results from a collector for a solar turbine engine, which was designed to supply 5 to 7 kg / hr superheated steam at 450 deg C and 150 atm, verify the theoretical predictions.

1043

Manufacture of Curved Glass Mirrors for Linear Concentrators : Giutronich J E & Mills D R, " Proceedings ISES Conference ", New Delhi (India), (1978).

This paper describes a technique for producing a glass cylindrical parabolic reflector of good specularity, improved figure and a long life expectancy. The method described is one which is easily adaptable to industrial application and is applicable to cylindrical shapes generally.

1044

Industrial Applications – The Challenge to Solar Furnace Research : Glaser P E, " Proceedings U N Conference on New Sources of Energy ", Rome (Italy), 6, 360, (1964).

The potential applications of solar furnaces to industrial processes that require high temperatures under controlled conditions has stimulated research in the field of solar furnace design and construction, developments of instruments, experimental investigations and process development. This paper discusses recent developments in the construction of reflecting surfaces, the measurement of the most important variables, experimental research already accomplished and the potential of industrial applications of solar furnaces.

1045

Specular Reflectance Properties of Solar Mirrors as a Function of Incident Angle : Glidden D N & Pettit R B, " Proceedings ISES Silver Jubilee International Congress ", Georgia (USA), (1979).

The specular reflectance properties of several solar mirror materials have been measured as a function of wave length, polarization and incident angle. Results of a 3 mm thick, silvered flat glass mirror, measured specular and hemispherical reflectance properties are in excellent agreement with reflectance values calculated from a multiple beam reflectance model. At 65° from normal, the average solar reflectance is decreased only ∼4% from the value of normal incidence. For the aluminized acrylic mirror, measured specular reflectance value at 15 m rad were in agreement with hemispherical reflectance values, while calculated reflectance values are as much as 2% at an incident angle of 65° from normal while the width of the specular beam increases slightly with incident angle.

1046

Fluorescent Solar Energy Collectors ; Operating Conditions with Diffused Light : Goetzberger A, " Applied Physics ", 16(4), 399, (1978).

The fluorescent energy conversion principle using several sheets of transparent material doped with fluorescent molecules to concentrate radiation, is extended to include diffused radiation. Two cases are treated here, only diffused radiation and secondly a composite spectrum consisting of 40% direct and 60% diffused radiation simulating the average illumination of a flat exposure in central Europe. In both cases, photovoltaic conversion efficiency is significantly higher than that with the AMI spectrum. This is due to the blue shift and narrow shape of the diffuse spectral distributions. With realistic boundary conditions the theoretical conversion efficiency is 1.56 times higher than for the AMI case. The highest theoretical conversion efficiency is now 38%.

1047

Thermal Energy Conversion with Fluorescent Collector – Concentrators : Goetzberger A, " Solar Energy ", 22, 435, (1979).

The potential of fluorescent collectors for thermal conversion is evaluated theoretically and its possible advantages merit attention. For thermal conversion heat losses are only of importance at the cooling pipe which can be well insulated. Equations are derived describing heat losses for various configurations and operating temperatures. Then conversion efficiencies are calculated. Results show that fluorescent collectors although having a lower initial efficiency than flat plate collectors, retain this efficiency at high operating temperatures and at low solar flux intensity.

1048

Fluorescent Planar Concentrators ; Performance and Experimental Results : Goetzberger A, Heidler K, Wittwer V, Zastrow A, Bauer G and Sah E, " Proceedings 2nd International Photovoltaic Solar Energy Conference ", West Berlin (Germany), p. 513, (1977).

Fluorescent concentrators operate on the principle of fluorescent light conversion and guidance by total internal reflection. An advantage of this type of concentrator is the possibility of also collecting diffused radiation and division of the incoming solar spectrum into different wavelength fractions. Each part can be converted by solar cells optimized for a narrow photoenergy range. The first experimental results of collector efficiency and stability are presented.

1049

Solar Energy Conversion with Fluorescent Collectors : Goetzberger A and Greubel W, " Applied Physics ", 14(2), 123, (1977).

A new principle for solar energy conversion is proposed and evaluated theoretically. Collection and concentration of direct and diffused radiation is possible by the use of a stack of transparent sheets of material doped with fluorescent dyes. High efficiency of light collection can be achieved by light guiding and special design of collectors. The optical path length in a triangular collector is computed. In combination with solar cells this type of collector offers the advantage of separating the various fractions of light and converting them with solar cells of different band gaps. Theoretical conversion efficiency under optimum conditions is 32% for a system with 4 semi-conductors. Thermal energy conversion offers several advantages over conventional collectors ; high temperature and efficiency even under weak illumination ; separation of heat transport and radiation collection ; low thermal mass. Thermal efficiency is computed to be between 42% and 60%.

1050

Solarenenergieumwandlung Mit Fluoreszentzkollektoren / (Solar Energy Conversion with Fluorescent Collectors) : Goetzberger A, Bauer G and Wittwer V, " Elektronikpraxis ", 14(3), 31, (1979).

The theory and some considerations associated with the design of fluorescent collectors are presented. These collectors offer several advantages compared with common concentrators and photovoltaic convertors. The most important feature appears to be the decrease of cost of electric energy generation which is realised by concentrating the solar radiation including the diffused portion, in inexpensive plastic plates. Concentration factors of the order of 100 are possible. It is also possible to separate the spectrum into several spectral regions and to match it optimally to semiconductors with diverse energy gaps. Fluorescent collectors can also be used in thermal transducers or hybrid systems.

1051

Mathematical Model for the Design of an Optimum Stationary Solar Energy Collector : Gonzales C T and Arnas O A, " Proceedings SPIE Conference ", 68, 103, (1975).

This is a general development of an equation to describe the optimum cavity – type stationary solar energy collector in two dimensional space, and then the extension into three dimensional space. The intent is to establish a starting point for those who wish to do research in the area of concentrating collectors for solar energy. The final design of the collector is dimensionally optimized to collect as much concentrated energy from the sun as is physically and economically feasible.

1052

Solar Power Collector Breadboard Test Program : Goralnick NS, " Proceedings SPIE Conference ", 168, 120, (1975).

Itek's objective in the field of high temperature solar energy has been to develop an efficient and inexpensive method of collecting the sun's energy ultimately for use in a solar thermal e l e c t r i c power plant. A unique and simple concept has emerged in the development process. On the basis of that concept, Itek received a grant from the National Science Foundation to conduct a solar power collector breadboard test program. This paper describes Itek's collector approach and discusses the test program. A summary is given including the program objectives hardware problems encountered, corrective actions taken, performance measurements and evaluation,and results of a preliminary operational system cost analysis.

1053

Solar Power Collector Breadboard Test ; Final Report : Goralnick N S, " Itek Optical Systems Report No. Itek – 75 – 82501 ", Lexington (USA), (1975).

The first section describes the Itek concept for solar energy power conversion and identifies the related patent issued to Itek. The second section describes the $1\,m^2$ collector breadboard manufactured and tested to verify the concept. The collector efficiency, calculated from the test data is compared with the analytical value, to verify the performance of the collector. Raw test data, mirror reflectance and absorber point reflectance measurements are available on request. The third section describes the trade-offs leading to Itek's segmented collector mirror approach, and then a preliminary operational system cost analysis is performed. The cost elements considered are discussed and summarized. The specific recommendations are contained in the fourth section.

1054

Seasonal Daily Thermal Energy Output of the ITEK Solar Collector : Goralnick N S, " Proceedings ERDA Conference on Concentrating Solar Collectors", Georgia (USA), p. 2 – 51, (1977).

Measurement o f overall efficiency as a function of temperature has been made on a demonstration collector system. The results are given based on local noon operation with peak solar insolation available. This paper describes the approach to determine the seasonal daily thermal energy output of the Itek Solar Collector utilizing insolation data from Sandia Laboratories.

1055

Solar Concentrators ; New Energy Alternative : Greene A M, " Iron Age ", 214 (6), 65, (1974).

Compound parabolic concentrators, used to collect solar energy, can heat water to about 140 degrees F; adequate for heating and cooking. One metal and glass design ; which utilizes the vacuum principle to reduce heat losses can heat water to over 200 deg F. Various designs using copper, silicon, etc., are described and potential usage and advantages for year 2000 are projected.

1056

Application of Circumsolar Measurements to Concentrating Collectors : Grether D F, Hunt A and Wahlig M, " Proceedings ISES Silver Jubilee International Congress ", (USA), (1979).

A systematic measurement program is being carried out to assess the effects of circumsolar radiation (the solar Aureole) on the operation of solar conversion systems using concentrating collectors . Graphs are presented that illustrate the variety of observed solar plus circumsolar profiles. The results of statistical studies of the circumsolar radiation are also presented. Monthly summaries of the angular flux distributions for each site are discussed. These summaries present the data in several ways t o aid in determining the energy-weighted effects of circumsolar radiation for a particular solar design. For this analysis, the concentrating solar energy plants are characterized by two simplified parameters ; an effective field of view and an operating threshold. The summaries include the overestimate that a pyreheliometer would make as a function of the field of view and threshold of the solar collector, the gain in intercepted energy made by increasing the receiver size, and the average time – of – day variation of circumsolar levels.

1057

Methods of Quality Control for Solar Energy Concentrators : Girilikhes V A , " Applied Solar Energy ",8(4), 71, (1972).

The paper examines the basic tests for solar energy concentrators which are conducted during manufacture and operation. The paper classifies methods for determining the optical, precision and optical-power characteristics of the reflecting surfaces and also describes the instruments and devices used in the Soviet Union. The paper assesses the advantages and disadvantages of the different methods and provides recommendations on their application in studying the quality of reflecting surface.

1058

Universal Power Characteristic for a High Temperature Solar Heating Source : Girilikhes V A & Matveev V M, "Applied Solar Energy ", 8(3), 51, (1972).

The universal energetic characteristics of the solar heat source, relating in dimensionless form all the defining parameters of the 'Sun–Concentrator–Receiver' system with the criterion of its energetic effectiveness are obtained . On the basis of the universal characteristic the relationship b e t w e e n the concentrator accuracy and the operating temperature is also studied.

1059

Comparison of Compound Parabolic and Simple Parabolic Concentrating Solar Collectors : G r i m m e r D P, " Solar Energy ", 22, 21, (1979).

The compound parabolic and simple parabolic collectors are analysed and compared for their ability to accept nondirect radiation and for their respective reflector arc lengths. The simple parabolic concentrator (SPC) can make use of some nondirect solar radiation if the absorber tube is intentionally enlarged so as to intercept defocussed radiation. A principal advantage of collecting nondirect radiation with a SPC rather than with a compound parabolic concentrator (CPC) is the reduced use of materials in the construction of the reflector.

1060

Development and Performance of the DEL Concentrating Solar Energy Collectors : Goranson G G, Riise H N and Eldridge B G, "Proceedings ERDA Conference on Concentrating Solar Collectors", Georgia (USA), p. 2 - 15, (1977).

The DEL collectors have been designed for high optical and thermodynamic performance and low cost in manufacturing, installation, operation and maintenance. The evolution of the unique DEL collector design is described relative to system considerations in space heating and cooling, industrial process heat, agriculture and power generation.

1061

Thermal Testing of Expandable Solar Collectors ; Final Report : Gunderson A W, " Air Force Systems Command , Flight Dynamics Laboratories Report No. AFFDL - TR - 66 - 132 ", Wright Patterson Air Force Base, Ohio (USA), (1966).

The solar collector test program was initiated to determine the structural response of expandable and unfurlable solar collectors to a simulated space thermal environment. The heating of the collectors was by T - 3 infrared heat lamps, and cooling was obtained by liquid nitrogen boil-off. The collector structures in both instances performed as expected and no unusual deflections were measured.

1062

Hyperbolic Cone - Channel Condenser : Gush H P, " Optics Letters ", 2,22, (1978).

An optical condensing system is described that consists of a lens and the polished interior of a cone generated by rotating a hyperbola about an appropriate axis. An optimum concentration of l i g h t is achieved in a relatively compact system. In certain circumstances the hyperbola reduces to a straight line simplifying the construction of the cone.

1063

Mid - temperature Solar Systems Test Facility (MSSTF) Project Test Results : Harrison T D & McCulloch W H, " Proceedings ISES Silver Jubilee International Conference ", Georgia (USA), (1979).

This report describes the results of three years of testing at the DOE / Sandia Mid-temperature Solar Systems Test Facility (MSSTF) at Albuquerque, New Mexico.   The MSSTF consists of line focussing a n d point focussing collector field subsystems, energy transfer equipment, high temperature thermal energy storage subsystems, a liquid boiler, an organic Rankine cycle turbine/generator and central data acquisition equipment. The solar energy impacting the aperture of the collector and the energy collected have been measured at intervals during the testing period. When degradation of collectors was observed, the causes were traced and identified. As a consequence, development activities have been initiated and ways to correct deficiencies were found.

Losses in the energy transfer equipment are defined as to quantity, location and cause. Proposals are made for minimizing losses in the energy transfer system and plans for their incorporation in the next generation equipment are described. Two high temperature thermal energy storage subsystems, a thermocline tank and a multitank system, both utilizing the sensible heat of a fluid as the energy storage mechanism have been tested. The demonstrated performance of these systems and design information for such systems are presented and discussed. From the test data, a thermal energy inventory is created to show the disposition of all energy which strikes the aperture of the collectors. From this inventory, the efficiencies of various elements of the solar total energy system are calculated.

1064

Status of Solar Energy Collector Technology : Heath Jr. A R, " Progress in Astronautics & Aeronautics (Power Systems for Space Flights) ", 2, 655, (1963).

Survey of the state of the art of solar energy collector development is presented. Discussed are the various types of collector concepts, methods and materials for construction. The concentrating ability and efficiency are presented for several collectors which have been constructed to date. For the various collectors analyses are made with regard to unit weight, specific power, launch package volume and operating temperature of the energy conversion device.

1065

Solar Collector Development : Heath Jr. A R and Manwell P T, " Astronautics and Aerospace Engineering ", 1, 58, (1963).

Discussions of the state of the art of solar collectors which concentrate the relatively low level solar energy to a usable density for the particular energy conversion method. Six different concepts of light weight solar energy collectors, which are developed to a point where quantitative data exist are considered, namely, the Fresnel, inflatable, one piece, petal, umbrella and inflatable r i g i d i s e d. These are sketched and a table lists some pertinent characteristics. The collectors are compared in terms of efficiency, specific power, package volume, unit weight and combined collector absorber efficiency.

1066

NASA Solar Concentrator Development : Heath Jr. A R and Hoffman E L, " Paper at 3rd Biennial Aerospace Power Systems Conference ", Philadelphia (USA), (1964).

A description of the various NASA solar energy concentrator development projects is given. Investigations were made on several of the concentrators and some results are indicated. Concentrators studied include (i) the one piece 5 ft diameter stretch-formed aluminium, and electroformed nickel concentrators (ii) expandable concentrators based on whirling membranes concept and a ground demonstration

140

model of the cone-and-column concept. Finally, studies in progress on inflatable-rigidized concentrators have recently shown that this concept has promise.

1067

Review of Solar Concentrator Technology : Heath Jr. A R and Hoffman E L, " Paper presented at the IECE Conference ", Los Angeles (USA), (1966).

Descriptions of fabrication techniques as well as a brief discussion of recent results from investigations made on concentrator are presented. In the area of one piece concentrators, the stretch formed aluminium process has been developed to the point where concentrator accuracy compares favourably with the high quality formerly obtained only by electroforming nickel. The aluminium electroforming processes have been scaled upto the point where 0.75 meter-diameter concentrators have been fabricated. In the area of expandable concentrators, a modified model of the whirling membrane concept has given improved concentration of energy, however, the designed parabolic cross section has not been attained.

1068

The Application of ASHRAE Standard 93 - 77 to Concentrating Collectors : Hill J E and Jenkins J P, " Proceedings ERDA Conference on Concentrating Solar Collectors ", Georgia (USA), p.6 - 11, (1977).

The American Society of Heating, Refrigeration and Air Conditioning Engineers (ASHRAE), has recently adopted ASHRAE Standard 93 - 77 for testing and rating of solar collectors based on thermal performance. Four separate tests are required to be conducted. This paper briefly explains the tests and indicates how they might be adapted for use with concentrating collectors.

1069

A Non Tracking Solar Concentrator with a High Concentration Ratio : Hinterberger H, " Proceedings ISES Conference ", Los Angeles (USA), (1975).

A solar concentrator which achieves the highest concentration possible with a non tracking geometry has been designed. To further increase the concentration ratio, the receiver is connected to the piping leading to power generation equipment in such a way that heat flow to portions of the receiver not being heated by sunlight at that moment is greatly impeded. The design and performance of such a concentrator are discussed in this paper.

1070

Solar Collectors : Hollands K G T, " Proceedings ISES Conference - Sharing the Sun ", Winnipeg (Canada), 2, 1, (1976).

Current research and development activity on solar collectors for heating and cooling of buildings is reviewed. Topics covered include reduction to practice, fundamentals of flat plate collectors, collector testing, selective surfaces ; heat reflecting glass, honeycomb. Collectors, evacuated collectors and concentrating collectors of the V - trough, compound parabolic and linear Fresnel lens types.

1071

Optical Design of Solar Concentrators : Horton T E & McDermit J H, " Journal of Energy ", 4 (1), 4 , (1980).

A generalized technique for the optical design of solar concentrators has been developed. The design technique considers limb darkening effects, collector placement errors and concentrator pointing errors. In addition to giving the designer the ability to explore the sensitivity of a design to these parameters, the technique also allows the designer to prescribe the heating distribution on the collector surface. A detailed derivation of the design technique and illustrative results are presented.

1072

Characteristics of Solar Concentrators as Applied to Space Power Systems : Houck O K and Heath Jr. A R, " Paper at Society of Automotive Engineers of USA and ASME Air Transport and Space Heating ", New York (USA), No. 867, (1964).

Analysis of current solar concentrators integrated into complete space power systems with various electrical conversion methods. Concentrator designs such as inflatable - rigidised, petal, one piece & Fresnel are treated and methods for determining their size, weight and packaging characteristics, when combined with dynamic and static conversion schemes, are described. It is stated that since there are so many design facets involved in integrating concentrators with space power systems, it is unlikely that one concentrator design will be optimum for all space power applications.

1073

Wind Excited Rigid Body Vibrations of the Heliostat Mirror for the Proposed Department of D e f e n c e Solar Furnace : Hoyer S, " Air Force Missile Development Center Report No. AFMDC - TN - 60 - 1 ", A i r Force Base, New Mexico (USA), (1960).

This report presents the rigid - body vibrations of the heliostat mirror. A minimum requirement is that the mirror be designed to withstand the static load produced by the worst possible wind condition. It is pointed out that the mirror can fail even if this requirement is satisfied. Such failure can occur if the mirror is excited at one of its natural frequencies, or if the mirror is dynamically unstable in any possible wind condition. Von Karman vortices can conceivably excite the mirror at one of its natural frequencies. Calculations presented here indicate that this should not prove to be a dangerous source of excitation. It is shown that one must investigate the dynamic stability of the mirror considered as a rigid body.

1074

Use of Calculated Displaced Shapes to Define the Reflected Light Pattern from a Focussed Collector : Hoyle C S, " Proceedings ISES Conference - Sharing the Sun ", Winnipeg (Canada), 2, 208, (1976).

A program has been developed to analyse the reflected light pattern from slightly curved f l a t plate collectors. The deflected shapes of the mirrors can be defined by means of finite element calculations or analytical expressions. The current version allows the comparison of the reflected l i g h t patterns for an externally defined shape , a spherical shape, or either of the two round plate analytical solutions. Besides the deflected shape of the mirror, the input includes direction of the light source and the distance to the focussing point of interest. The resulting reflected light pattern is plotted as it would appear on a plane at the specified position. The program also has the capability to search for the minimum reflected spot size.

1075

Solar Energy Concentrator : Hubbard W M," U S Patent No. 3977773, 31st August (1976) ".

The reflector surface of a concentrator is formed of plastic type material and includes at least one layer of reinforcing material embedded therein. The concentrator has a parabolic concave inner reflective surface and is constructed either as a unitary member or by joining together a plurality of segmented sections. The reflected surface is provided by a plurality of side by side individual rectangular mirror segments at least partially embedded therein. The construction of the reflector is described.

1076

A Solar Furnace Using a Horizontal Heliostat Array : Hughes G," Solar Energy ", 2(3-4), 49, (1958).

A two component solar furnace, condenser – heliostat combination is described in which the condenser faces downwards at 30° towards a heliostat comprised of numerous rows of plane meirrors mounted on a horizontal turntable. It is shown that for a south facing condenser, with the angle of the final flux being limited to 30° below the horizontal, the rows of heliostat mirrors may be mounted so that t h e y overlap, resulting in a reduction of edge losses occurring when heliostat mirrors are all held up in a single plane. The overall side of the heliostat turntable is calcualted for a 6 hr workday throughout the year, & a suggestion is made for using the heliostat control mechanism to provide shutter action. The saving in flux possible by elimination of an independent shutter is estimated at about 8 percent.

1077

A Radiation Flux Meter for Higher Intensity Application : U S Air Force Missile Development Center Report No AFMDC – TR – 59, Solar Furnace Support Studies 2, 145 " Air Force Base New Mexico (USA), (1959).

The high energy concentration and the problem of correct flux measurement for a large solar furnace are discussed. Measuring a known fraction of the total flux to reduce a volume of water is considered. The changes in physical properties of a transparent medium caused by temperature are examined as a possible technique. The sources of error inherent in the instruments are discussed.

1078

Exploratory Study to Determine Favourable Locations in the United States for Construction of a S o l a r Furnace : Hull B B, " Quartermaster Research and Development Center Research Study Report No. QRDC – EA – 4, AD – 699358, (1955) ".

The total receipt of radiant solar energy at the earth's surface is dependent upon the clarity of the atmospheric envelope and the number of sunshine hours. For short wave solar radiation, water vapour and clouds are the two major absorbers and scatterers. The graphs of solar radiation and the average number of sunshine hours shown in this report indicate that the stations in the Southwest United States have the most favourable climatic conditions for the location of a solar furnace that w i l l provide thermal radiation of high intensity throughout the year. On a climatological basis the arid Southwest appears to be the logical site for the solar furnace.

1079

Definitive Design of the Solar Total Energy ; Large Scale Experiment at Shenandoah, Georgia : Hunke R W, Mertz W S and Poche A J, " Proceedings ISES International Congress ", Georgia (USA), (1979).

The U S DOE, with Sandia Laboratories providing technical support and management, is now completing Phase IV Definitive Design of a six phase project for conducting a Solar Total Energy Large Scale Experiment at Shenandoah, Georgia. General Electric Co. is designing the Solar Total Energy S y s t e m s (STES) with capacity to supply 60% of the total electric and thermal requirements of the 42000 sq ft Bluyle of America's knitwear plant, to be served at the Shenandoah site. The system will provide 400 KW of electrical and 3.5 MW of thermal energy. The STES has a classical, cascaded total energy system configuration. It utilizes parabolic dish collectors, trickle oil thermal energy storage and a steam turbine generator. The electrical peak load shaving system is being designed for interconnected operation with the Georgia Power System.

1080

Solar Concentrator : Huskett E R, " U S Patent 4, 011, 858 / March 15th, (1977) ".

An apparatus for collecting solar energy is described. The apparatus includes a parabola shaped reflector around a pipe enclosed in a glass tube, the pipe being located at the focal point of the reflector so that sunlight rays are reflected there and concentrate their heat against it; the pipe extending outwardly at opposite end of the reflector and water passing through the pipe is thus heated for practical uses. The reflector is rotatable and follows the path of the sun in order to o b t a i n maximum efficiency therefrom.

1081

Remarks Concerning Solar Furnaces in Space : Hvez M & Foex M, " Solar Energy ", 13, 417, (1972).

In order to mitigate the difficulty of obtaining high vacuum at elevated temperatures, it has b e e n suggested that a space vacuum be used in combination with a solar furnace. This could be employed on either an experimental satelite solution or the moon. In the present note, some of the problems posed in such an experimentation are examined schematically. Those specific to solar furnaces and high – temperature treatment under space vacuum conditions are considered in particular.

1082

Design of Solar Energy Concentrators for Power Generation in Residential and Non-residential Areas :

142

Ibramsha M, " Proceedings ISES Conference ", New Delhi (India), (1978).

Two novel concentrators are described in this paper. The concentrators can be manufactured   u s i n g
existing technologies that are within the reach of under developed and developing countries. A   flat
concentrator design using 'one way mirrors' is suitable for collection of solar energy incident on the
terraces of buildings. A modular design of a system of concentrators,mirrors and lenses is proposed
for use in the thermal power plant with capacities of more than 100 MW. The solar concentrators   do
not track the sun's movement and implement the concept of distributed focussing.

1083

Solar Heated Air Receivers : Jarvinen P O," Proceedings ISES Conference ", Los Angeles (USA), (1975).

The development of solar heated air / gas turbine electrical power generation systems is considered a
viable alternative to the solar thermal / steam electricity generation approach. One important advantage of such a system is the elimination of the need for water at sites of solar generating plants. In
the solar gas turbine application, a high temperature heated air receiver replaces the normal gas turbine combustion chamber. The feasibility of high operating temperature solar heated air receivers is
investigated in depth because the realization of such receivers is fundamental to the success of the
proposed concept. A number of heated air receiver configurations are considered for absorbing the incident concentrated sunlight and transferring the heat to a working gas. Surface receivers as well as
windowless and windowed cavities are analyzed and compared on the basis of their heat transfer capability. Design considerations for a central receiver / optical transmission solar thermal system are
reviewed and their implications to the design of heated air central receivers are delineated.

1084

Ceramic Solar receivers : Jarvinen P O, " Proceedings IECE Conference ", Boston (USA), p.26, (1979).

The application of ceramic materials to high temperature solar receivers for advanced Brayton   a n d
Stirling thermal electric systems is discussed. Conceptual designs for ceramic cavity receivers employing impingement jet cooled, dome shaped silicon carbide heat exchanger modules are offered. Optically mechanical heat transfer and structural analyses of this novel receiver approach are presented.

1085

Evaluation of Linear Concentrating Collectors : Jeter S M, " Proceedings ISES Silver Jubilee International Congress ", Georgia (USA), p.439, (1979).

A modelling procedure particularly useful for linear concentrating collectors is presented. Thermal,
hydaulic and optical (including off normal degradation) effects can be considered. Exemplar evaluation of performance parameters and collector output projections are presented.

1086

Orientation Studies for Single Axis Tracking Concentrating Collectors : Jeter S M, Craig J I & Grems E G,
" Proceedings ERDA Conference on Concentrating Solar Collectors ", Georgia (USA), p.9 - 49, (1977).

The present paper discusses the orientation problem for single axis concentrators as a means for matching production to load. In the general approach, a criterion function is defined and used to optimize the distribution of concentrator orientation to match a given load. The criterion function employed for this study is an annualized dollar cost of owning the collectors based on the equivalent
cost of the useful heat production, offset by the initial capital investment. The result of the analysis is a concentrator orientation distribution function for a particular load profile t h a t   c a n
then be used to plan the field orientation.

1087

Effects of Haze on Performance of Solar Concentrators : Jeys T and Vant Hull L L, " Proceedings I S E S
Conference ", Los Angeles (USA), (1975).

Attempts have been made to better measure the direct beam intensity and the aureole intensity so as to
know exactly how much more is included in the (conventional) usual pyrheliometric measurements. Three
different tube extensions were added to the Epply Normal Incidence Pyrheliometer resulting i n f u l l
fields of view of 3.82, 2.86 and 2.15 degrees, in addition to the standard viewing aperture (5.77°).
In an attempt to better understand these effects a computer program has been written which anaylses
in detail the light intensity distribution on the reciever for various viewing apertures and steering
accuracies.

1088

A Simulation Model for the Dynamic Performance Analysis of Solar Thermal Electric Power Systems : Johnson G R and Elgabalawi N, " Proceedings ISES Conference ", Los Angeles (USA), (1975).

A simulation model for the dynamic performance analysis of solar thermal electric power plant c o n-
cepts was developed and is in use at Colorado State University. This paper presents a detailed description of the overall program structure, the subsystem modules used to model each of the components,
the control algorithm used to drive the entire electric power plant and presents results for two distinctly different solar thermal electric power plant configurations. The first type of plant a n a-
lysed is of the distributed type while the second is of the central tower receiver type.

1089

Solar Collector for the Generation of Mechanical Energy : Juettner K, " German Patent 2, 617, 6045 / A /
November 3, (1977)".

A solar collector for the generation of mechanical energy is described. First the incoming light i s
focussed by collecting lenses. The spot with the highest light concentration is inside a tube filled
with an easily vaporizable fluid which drives a turbine or a piston engine.

1090

International Energy Agency Solar Power Stations Project : Kalt A C, " Proceedings ISES (U K Section)

Conference (C 16) ", London (U K), p. 57, (1978)

The International Energy Agency (IEA) project, two solar power stations, will be carried out in a re-latively short time period. After design and erection within three years, an operational test phase of two years will follow. The several objectives and project goals are presented in this paper. The basic technical specifications for design of the plants are described. The two plants will be erected on a site already chosen in southern Spain near Almeria.

1091

The International Energy Agency (IEA) Small Solar Power Plant (SSPS) Project : Kalt A C, " Proceedings ISES Silver Jubilee International Congress ", Georgia (USA), p. 1077, (1979).

The SSPS project includes the design and erection of two solar power plants of different technical concepts, on the same site, with the intention to evaluate and compare the relative economic viabili-ty of such plants. One of the plants is a distributed collector system and the other one a c e n t r a l receiver system. There are several objectives and project goals which are presented in this paper, followed by the resulting technical basic specifications. To give an expression for the assigned tech-nical realisation, the main facts of the technical design are given.

1092

Form of Bending of the Surfaces of a Circular Membrane Formed by Electrostatic Tension : Kamildizhanov A Kh, "Applied Solar Energy", 15(3), 74, (1979).

Two axisymmetric uniform films of equal dimensions that are clamped along their contour are consider-ed. It is shown that, as the physical parameters increase, the deflection at the center increases & reflector comes to the long focus type.

1093 ·

Circular Film Concentrator with a Concentric Seam of Constant Curvature : Kamildzhanov A Kh., "Applied Solar Energy ", 15(2), 87, (1979).

It has been found that a circular film solar energy concentrator with a deformable concentric seam of constant curvature yields a reflecting surface whose shape will coincide with that of a paraboloid of revolution, assumed by a circular membrane of radius 'R', attached along its contour, under the action of a drop in pressure 'P'.

1094

Current Status and Future Prospects of Non-tracking Solar Collectors : Kapany N S, " Proceedings SPIE Conference ", 68, 77, (1975).

This paper briefly reviews the state of the art of non-tracking solar collector with various perform-ance criteria in mind. Also described in this paper are developments of some new components capable of producing high efficiency, light weight and low cost non-tracking solar collectors. Windows with 'optical ribs' have been developed to yield high transmission, light weight, ruggedness and efficient radiative and convective trapping. On the other hand, a new component termed 'optical valve' has been developed which is capable of transmitting the incident solar radiation and reflecting back t o t h e absorber much of the emitted infrared radiation. Furthermore an enclosed absorber is shown to achieve almost complete absorption of solar radiation at large angles of incidence.

1095

Solar Collectors Using Total Internal Reflections : Kapany N S, " Proceedings SPIE Conference ", 85, 146, (1977).

A description is given of various configurations of the solar collectors and component employing the phenomenon of total reflection. A double glazed window using optical 'ribs' has been developed to yield high light transmission and ruggedness. Conical Wedges are developed for use as energy concen-trators as well as 'optical valves'. Dielectric compound concentrators with flat, parabolic and el-liptcal reflecting surfaces are described with various applications.

1096

Solar Radiation Energy Concentrator : Kaufman Sr. L I, "U S Patent 4, 148, 300 / April 10, (1974) ".

A solar radiation energy concentrator is disclosed, adapted to convert solar radiation into u s e f u l forms of energy, comprising a base member with a longitudinally extending hollow chamber, said chamber having an open end therein for admission of solar radiation through it; integrally affixed in said open end is a convex glass magnifying lens adapted to receive solar radiation impinging thereon and focus it into a concentrated area on a heat absorbing sensor device located adjacent to that end of the chamber distal from the open end. The concentrator includes a heat exchanging system integrally juxtaposed with refractory lining on the chamber wall, for delivery of the energy input into the de-signated system.

1097

Solar Energy Conversion System; Four Quadrant, Two Dimensional Linear Solar Concentrators : Kelly D A, " Proceedings International Conference on Alternative Energy Sources ", Miami (USA), (1977).

The four quadrant, two dimensional linear solar concentrators have been evolved to meet the need for practical low cost and effective solar energy conversion. The solar concentrators consist of a con-ventional linear parabolic type concentrator unit with the addition of top linear lenses to provide direct solar concentration on the top area of focal zone piping. With this geometrical arrangement, all four quadrants of a focal zone pipe receive a uniform solar heating effect. This new t y p e o f four quadrant linear parabolic concentrator provides concentration ratio of upto 30 : 1 on a continu-ous water - steam focal piping circuit.

1098

Convertible, Tri Mode Solar Conversion System : Kelly D A, " Proceedings 2nd International Conference

144

on Alternative Energy Sources ", Miami (USA), (1979).

The convertible, tri mode solar conversion system has been evolved to provide year round solar conversion with three distinct modes of operation. The basic concept behind the development of this tri
mode system is quite simple. Make all the solar conversion hardware work as hard as possible y e a r
round to improve the cost effectiveness and attractiveness of solar conversion, especially for the
north eastern area of the United States. It utilizes two basic types of solar conversion hardware.The
first being conventional linear parabolic concentrators which are made up in six foot long sections
for convenience in handling and installation. The second basic component is the hot box solar c o l-
lector and storage unit which is designed to both collect and store solar thermal energy during the
winter months. The two basic component types cooperate to both collect and store solar energy during
winter months, while the linear concentrators produce hot water / steam during summer and the hot boxes
store solar heated water.

1099

<u>Sun Tracking Solar Energy Collector</u> : Ketcher J C and Perkins G S, " U S Patent No. 4111184 / Sept. 5th,
(1978) ".

A parabolic reflector is supported so that it can trace the sun. The support for this reflector com-
rises an azimuth frame supported on two wheels and a central pivotal point which are positioned in a
substantially triangular configuration. The two wheels rotate on tracks. On the top of the azimuth
frame there is provided an elevation frame. The elevation frame includes curved rails which define a
portion of an arc and extend vertically. The reflector rides on wheels captured within the curved rails.
The wheels of the azimuth frame are driven by an azimuth actuater. The reflector structure is counter
balanced about its elevation axis by a pendulum cable system which is driven by motor to change elevation. At the focal point of the parabolic reflector, a heat engine or receiver is mounted independently on the reflector. Suitable means are provided for moving the reflector about its two axes in
order to track the sun.

1100

<u>Environmental Influence on CRTF Heliostat Reflectance</u> : King D L anf Myers J E, " Proceedings ISES Sil-
ver Jubilee International Congress ", Georgia (USA), (1979).

A long term dirt build up test program is being conducted at the Central Receiver Test Facility (CRTF).
The Mirror samples are mounted on the end of the elevation axis of the CRTF heliostats, approximately
4 m above the ground level. Three samples are mounted on each of the three orientations; mirror face
up, down and vertical.

1101

<u>Solarthermische Kraftwerke/(Solar Thermal Power Plants)</u> : Kleinrath H, " Elektrotech Maschinenbau
94(1), 19 (1977) ".

The construction of a power plant based on soler energy is considered. Various collecting systems are
discussed and compared, particularly planar collectors, the tower concept. Examples of power p l a n t,
central power plant, capable of generating 250 MW are given. A similar Swiss project rated at 50 MW is
also described.

1102

<u>Development of Solar Collectors for Low Temperature Levels and of Concentrators for Thermal and Photo-
electric Conversion</u> : Kleinwächter H, " Proceedings ISES Conference ", New Delhi(India), 1264, (1978).

Some inexpensive solar collectors and relative devices are described. These include plastic monotube
and multitube collector, and an elastically bent cylindrical concentrator (concentration of 10-20 fold);
a rotating parabolic mirror for operating temperatures of 300 to 800 deg C ; and a concentrator u s e d
for photoelectric conversion when the energy density at the absorber is given. The use of plane mirror
grooves and funnels when operating at the temperature of 200 deg C is considered.

1103

<u>High Temperature Solar Collectors</u> : Kleinwachter H L and Kleinwächter J, "Proceedings Schweizerische
Vereinigung für Sonnenenergie Symposium ", Zürich (Switzerland), p.286, (1975).

A description is presented of investigations which were conducted to develop new economic devices for
the concentration of solar radiation. A magnetohydrodynamic generator with a concave reflector for the
concentration of solar radiation is considered. Attention is also given to cylindrical p a r a b o l i c
mirror with a variable geometry, a pneumatic concave reflector and various mirror types w i t h funnel
design characteristics.

1104

<u>Folding Pneumatic Solar Concentration Mirror</u> : Kleinwächter J," German Patent 2,604,345/A/ Aug.4,(1977).

The invention concerns a mirror for concentrating sunlight. The equipment consists of a torus which
can be blown up and folded with two plastic sheets, one of which has a metallic mirror surface while
the other is clearly transparent. Both sheets are connected as upper and lower cover so that the chamber so formed can have excess or lower pressure applied to it via a valve, and in both cases a c o n-
verging lens of variable focal length is produced by spherical deformation.

1105

<u>System Designs for Low Cost - Low Ratio Solar Concentrators</u> : Knasel T M, " Proceedings SPIE Confer-
ence ", 161, 14, (1978).

This paper explores recent work in low cost - low ratio concentrating solar collectors. Low ratio is
defined here to span the range from 1 to 10 or perhaps more. To keep costs low, stationary systems
are assumed, a concentration of at least two to three is desired. Relatively, even energy distribu -
tion at the reciever, minimum loss of diffused radiation and simple low cost surfaces are important
additional criteria. A brief discussion of stationary imaging and non imaging collectors incorporating
boosters and trapping designs indicate features of potentially cost effective system designs.

1106

Thermal Deformations of Solar Energy Concentrators : Korolev V M, Nazarov A, Sokolov E V Solodovnikova L A, Fokin V G and Machuev Yu. I, " Applied Solar Energy ", 14(2), 13, (1978).

Computational relationships are proposed and a sample calculation given for a concentrator that is not closed at the vertex under symmetric and skew symmetric heating.

1107

A Small Welding Apparatus Using Concentrated Radiant Energy : Korunov Yu.I, Bareskov N A, Sasorov V P, Brander A I, Egorov V D and Genkina R A, " Applied Solar Energy ", 11(3-4), 42, (1975).

A description is given of the optical system (concentrator and light source) of a small welding system using concentrated radiant energy from high pressure xenon lamps. Data on the radiant flux density distribution in the focal spot are given. It is shown that when the power drawn from the mains is 3 KW, the concentrated radiation is sufficient for the welding of steel and titanium a l l o y s of thicknesses upto 2 mm without the use of condenser lenses.

1108

Miniskirt for Dancing Spinner : Kosta S P, Vasagam R M, Patki A V and Ashiya R, " IEEE Trans. Aerospace Electronic Systems ", 15(3), 340, (1979).

An idea of use of a skirt made of optical solar reflectors to concentrate solar energy on the s o l a r cell array of geosynchronous spinning spacecraft is presented. From theoretical considerations it is predicted that the practical realization of this concept will allow significant overall cost & weight reduction of such spacecrafts.

1109

Economic Considerations for Focussing Collectors : Kreith F, " Proceedings Workshop on Solar Collectors for Heating and Cooling of Buildings ", New York (USA), p.179, (1974).

The paper outlines criteria to be used in comparing the cost of useful energy delivered by different types of solar collectors, to summarise available cost projections for SRTA focussing collectors, and to present preliminary results on the performance of a Stationary Reflector / Tracking Absorber (SRTA) solar energy collector.

1110

An Overview of Intermediate Temperature Solar Collector and Energy Storage Technology : Kreith F, Castle J N and Wyman C E, " Proceedings 6th International Heat Transfer Conference ", Ottawa, O n t a r i o, (Canada), (1978).

This paper presents a survey of concentrating solar collector and energy storage technology suitable for the mid temperature range between 370 and 670 deg K. The thermal performance of generic types of concentrating collectors are compared and the various methods for providing energy storage in temperature ranges achievable with these collectors are summarised. The objective of this review is to compile the information necessary for the design of solar energy systems in the mid temperature r a n g e and to indicate areas that require additional research to make solar energy economically viable f o r industrial heat applications.

1111

Heating Characteristics of Concentrators / Receiver System Employing a Selective Absorbing Surface : Kudrin O I and Abdurakhmanov A, " Applied Solar Energy ", 15(3), 26, (1979).

The heating characteristics of a solar concentrator /selective receiver system are obtained as an interrelated complex of the parameters reception efficiency of radiant energy / heating temperature. It is shown that the use of a spectrally selective absorption receiver can be highly efficient in t h e case of an imprecisely manufactured concentrator of light construction. The effect on reception efficiency of profiling of the threshold wavelength over the receiver surface is evaluated in conjunction with the form of the distribution curve of the radiation flux density incident from the concentrator. It is shown that even if the threshold wavelength is taken to be some averaged value that is constant for the entire reception surface, the effect gained from using selective absorption can be considerable.

1112

Selective Radiation Absorption as a Means of Improving Efficiency of a High Temperature S o l a r Power Plant : Kudrin O I, Abdurakhmanov A and Aggeeva I A, " Applied Solar Energy ", 15(4), 36, (1979).

The possibility of improving the efficiency of a high temperature solar power plant with Stirling engine and solar thermal rocket engine by using a receiver with spectrally selective radiation absorbing surface is analysed.

1113

Design and Development of a Paraboloidal Dish Solar Collector for Intermediate Temperature Service : Kugath D A, Drenker G and Koenig A A, " Proceedings ISES Silver Jubilee International Congress ", Georgia (USA), p.449, (1979).

This paper discusses the design and development of a 7 m diameter paraboloidal dish solar collector, which is intended for first application in the Solar Total Energy Large Scale Experiment at Shenandoah, Georgia. Each of the four main subsystems of the collector i.e. (i) reflector, (ii) mount and drives, (iii) receiver and (iv) the controls, is described and trade-off decisions discussed briefly, with major emphasis on the receiver design.

1114

Maximum Solar Flux Concentration Achievable with Axicon Collectors : Kurzweg U H, "Solar Energy", 25, 221, (1980).

The concentration characteristcics of co-axial cone axicon concentrators using the sun as the radiation source are examined. By employing a ray tracing approach the known concentration results for rays entering strictly parallel to the axicon axis, it is shown that the concentration remains finite and that the maximum achievable value is Co = 273 at the optimum reflector cone angle of 90°. All radiation entering the solar tracking collector will strike the central absorber cone as long as the vertex angle of this cone exceeds the angular size of the sun.

1115

Determination of the Temperature Dependence of Material Properties in Image Furnaces : Laszlo T S and Klamkin M S, " Solar Energy ", 4(3), 20, (1960).

A Solar – Fossil Hybrid Power System Using a Combined Cycle : Lam E Y, " Proceedings ISES Silver Jubilee International Congress ", Georgia (USA), (1979).

This paper describes a solar central receiver hybrid power system concept which uses a gas turbine / heat recovery steam generator / steam turbine combined cycle to generate electricity from both solar and fossil energy.

1116

Simulation Exercise of a Cavity Type Solar Receiver Using the Heat Program : Lansing F L , " Proceedings IECE Conference ", Boston (USA), p. 20, (1979).

A computer program has been developed at JPL to support the advanced studies of solar receivers in high concentration solar thermal electric power plants. This work presents briefly the program methodology input data required, expected output results, capabilities and limitations. The program was used to simulate an existing 4 KW(t) experimental receiver of a cavity type. The receiver is located at the focus of a paraboloid dish and is connected to a Stirling engine. Both steady state and transient performance simulation are given. Details about the receiver modelling are also presented to illustrate the procedure followed. Simulated temperature patterns were found in good agreement with test data obtained by high temperature thermocouples. The simulated receiver performance was extrapolated to various operating conditions not attained experimentally. The results of the parameterization study were fitted to a general performance expression to determine the receiver characteristic constants. The latter was used to optimize the receiver operating conditions to obtain the highest overall conversion efficiency.

1117

Measurement and Application of High Heat Fluxes in a Solar Furnace: Laszlo, T S," Solar Energy " 6, 69, (1962).

The suitability of the solar furnace for the generation of high heat fluxes of known, stable intensity has been investigated. Absolute flux measurements have been performed using an absorbing cavity of a new design, which traps a larger portion of the incident flux than the sphere. The results indicate that there is a well defined correlation between normal incidence solar radiation and heat flux in the focal zone which can be used to calculate the heat flux when direct measurements are not possible. On the basis of this correlation, the solar furnace can be used as a radiation standard upto 250 Cal / sqcm / sec intensity. A new coating has been adapted to the Gardon radiometer, which makes it possible to use this instrument to the upper flux limits of the solar furnace. A method has been developed for the calibration of the radiometer in the solar furnace.

1118

Image Furnace Techniques Survey : Laszlo T S, " Solar Energy ", 7, 195, (1963).

The first results of a survey is presented in a tabulated form listing the location of the facility together with a few pertinent data.

1119

Determination of the Temperature Dependence of Material Properties in Image Furnaces : Laszlo T S and Klamkin M S, " Solar Energy ", 4(3), 20, (1960).

A special testing method is described which makes it possible to determine a property of a specimen for narrow temperature intervals for measurements in the wide range temperature distribution of an image furnace. The determination is based on the measurement of the property at several different flux levels of known temperature distribution.

1120

Effects of Regional Insolation Differences Upon Advanced Solar Thermal Electric Power Plant Performance and Energy Costs : Latta A F, Bowyer J M and Fujita T, " ASME Paper No. 79–WA/Sol.–15, (1979).

This paper presents the performance and costs of four 10 MWe advanced solar thermal electric power plants sited in various regions of continental United States. Each region has different insolation characteristics which result in varying collector field areas, plant performance, capital costs and energy costs. The paraboloidal dish, central receiver, cylindrical parabolic trough and compound parabolic concentrator (CPC) comprise the advanced concepts studied. This paper contains a discussion of the regional insolation data base, a description of the solar systems performance costs and a presentation of a range for the forecast cost of conventional electricity by region and nationally over the next several decades.

1121

Focussed Solar Collector Analysis with Axially Varying Input Due to Shadowing From Adjacent Collectors : Lee D O and Schimmel Jr. W P , " Proceedings AIChE Symposium " 73 (164), 187, (1977).

An analytical expression for the shadowing of a collector by adjacent units as a function of collector location, spacing and orientation is presented. Hourly variations of solar input are accounted for, as are the seasonal effects of solar zenith angle. Using the shadowing expression, an axial temperature differential analysis, is used to determine the output of a linear focussed solar collector being shadowed by adjacent collectors. Outputs for typical spacing configuration are presented for a north – south oriented collector field. Calculations are based on a fixed tilt east west tracking system.

1122

Scheme of a Powerful Concentrator with Stationary Flat Mirrors : Linitskhi N V, " Applied Solar Energy", 3(4), 8 (1967).

The paper discusses the creation of a powerful solar installation. The possibility of using stationary flat mirrors on separate holders is considered.

1123

High Temperature Lens Construction : Lorell, K R, "U S Patent 3 - 493, 291 / February 3, (1970) ".

A lens assembly for solar furnace or solar simulator entrances is described. To prevent lens destruction by operational heat, the lens assembly is made up of a multiciplicity of identical quartz lens blocks carefully clustered together in a side by side, continuous relationship within a common plane to form compact lens units. Relief opening between the ends of adjacent segments contain spherical keys to prevent displacements of individual lens blocks.

1124

Effect of Operating Temperature on the Cost of Energy From Solar Thermal Electric Power Plants : Lukens L L, " Sandia Laboratories Report No. SAND - 79 - 0801 ", (USA), (1979).

The operating temperature of a solar thermal electric power plant controls the efficiency of the collector field, the efficiency of the power generation system and the cost of the thermal energy storage system. The effect of these three items, as temperature is varied, on the annualized cost of energy produced by the system is evaluated for both stand - above solar and solar diesel h y b r i d p o w e r plants . The type of solar power plant considered is one using a linear focus distributed collector field and a Rankine cycle power generation system. Systems using different collector performance models, Rankine cycle working fluids and thermal energy storage concepts were included in the evaluation.

1125

The Ramon Areces Concentrating Photovoltaic Array : Luque A, Sala G, Alonso A, Ruiz J M, Araujo G L , Coello - Vera A, Lorenzo I and Sauz J, " Proceedings ISES Silver Jubilee International Conference ", Georgia (USA), (1979).

A prototype of a concentrating photovoltaic array which would supply a power of 1.8 KW peak under AM1 irradiance conditions is being developed and constructed in Madrid (Spain) under the sponsorship of the Ramon Areces Foundation. The prototype array will be connected to a storage unit consisting of acid lead batteries. The array size has been determined taking into account economy and tracking suitability. The mechanical structure has been designed to withstand adverse conditions with combined wind & snow of upto 106 km / hr and 80 kg / sqm respectively. The optimum active area of an azimuth elevation tracking array can be shown to be about 25 sqm when the angle of maximum deformation is fixed to be less than 0.5 degrees. The estimated cost of the iron structure, tracking mechanism, sun sensor a n d installation is about $4400.

1126

Reflective Solar Heat Collector : Lyon F A & Harrison H, "U S Patent 4, 106, 480 / August 5, (1978) ".

A reflecting solar heat collector is disclosed. A stationary cylindrical receiver surface absorbs sunlight. An array of reflecting mirror strips are rotatably mounted in substantially parabolic arrangement having the receiver surface substantially at its focus. Motor means are connected for pivotally rotating the said mirror strips substantially in unison about spaced apart axes p a r e l l e l to said receiver surface and substantially in said parabolic arrangement, to track the sun's apparent motion.

1127

Preliminary Analysis of the Optical Properties of Edge Gripped Reflectors : Lyon R N, " Oak Ridge National Laboratory Report ORNL - TM - 4935 ", (1975).

It is found that in theory an initially flat, uniform, reflective, buckled plate with simply supported edges concentrates reflected light from an infinitely distant point source onto its plane o f symmetry by a factor that increases with increased buckling upto a maximum of about 13 at which point the edge angles are about 50° with respect to the normal to rays of incoming light. That edge angle is obtained by pressing the edges towards each other until the distance between them is about 81 % o f t h e original width of the plate. It was also found further that one can obtain concentration factors o f more than 10,000 on either horizontal or vertical planes by holding the two opposite edges of the plate at fixed angles and adjusting the positions of the edges with respect to each other and incoming light. These findings suggest the possibility of cheap adjustable solar reflectors with concentration factors perhaps approaching 100.

1128

Non - imaging Radiant Energy Collector and Concentrator : Ma H Z, "U S Patent 4, 162, 824 / July 31, (1979).

A collector and concentrator of the non - imaging type for radiant energy and particularly for s o l a r energy, which includes a plurality of longitudinally - extending, generally trough shaped channels having curved facing reflective walls is described. The upper extremity of the channel defines an entrance aperture and the lower extremity defines an exit aperture. The walls are constructed of rectangular, flexible and deformable shuts or vanes attached along one edge to a longitudinal frame member having surfaces which are sloped to reflect a selected collector angle at the exit aperture and which are deformed by being urged against one another along the edge opposite to the attached edge. A r a d i a n t energy utilization means is disposed in the exit aperture.

1129

Experimental and Theoretical Evaluation of a Novel Concentrating Solar Energy C o l l e c t i o n System : Maish A B and Beard J T, " Proceedings ISES Silver Jubilee International Congress ", Georgia (USA), p.511, (1979).

A new concentrating solar energy system was evaluated theoretically and e x p e r i m e n t a l l y at the

University of Virginia Solar Test Facility. The modular, two axis tracking collector combined an aluminized mylar reflector with a flat plate absorber. Radiation collected from the 7 sqm frontal area was concentrated by a factor of 5, making the collector suitable for intermediate temperature applications. The thermal performance of the collector was determined theoretically on a digital computer using nodal approach to model the heat transfer processes. Collector sensitivities to environmental parameters were calculated. An experimental testing program conducted according to the ASHRAE testing standards verified results.

1130

An Infra-red Solar Furnace Pyrometer : Mann K, " Solar Energy ", 9, 136, (1965).

A spectral separation technique of differentiating between emitted and reflected radiation is discussed. In this method the pyrometer is designed to respond to wavelengths absent from the solar spectrum due to atmospheric absorption. This paper discusses the relative advantages of using the ultraviolet and infra-red atmospheric absorption bands and describes a working pyrometer, that is responsive to a narrow band of wavelengths, central at 2.8 microns. This band corresponds to a water vapour and carbon dioxide absorption band in the solar spectrum. Although the pyrometer has certain disadvantages it does overcome the central problem in temperature measurements of solar furnaces.

1131

Program ROE (Reflector Orbital Experiment) Part III ; Preliminary Design ; Inflatable Rigidized Solar Reflector Concept, Volume I ; The Ten Foot Diameter Design : Manning L, " Air Force Aero Propulsion Laboratory Report AFAPL - TR - 64 - 132, Part III ", (USA), (1964).

The conceptual design configuration and a preliminary design analysis are presented for a 10 ft diameter, 60° rim angle inflatable rigidizable concentrator. The design for instrumentation for focal zone, heat flux measurement and reflectance degradation detection is described. An estimated weight summary, mass properties and distribution data with estimated power requirements are presented.

1132

Present State of Efforts in Developing the Solar mirror for Space Stations and Spacecraft : Martin O, " Astronautik ", 4, 2, (1967).

The possibility of employing solar mirrors to provide auxiliary energy for space vehicles is examined. Concentrators of parabolic shape with precise geometrical accuracy and high reflective efficiency are required to harness the energy of the sun properly. Galvanic deposition of nickel from aqueous solutions and of aluminium from certain electrolytes yields suitable geometrical form and high reflectivity. Single piece mirrors are fabricated in large galvanic tanks. Composite reflectors are manufactured from rolled aluminium sheets and then stretch formed and adhesive bonded. Master forming is accomplished by rough milling, shave grinding and spin casting. Vacuum coating on the reflective overlays provides durable surfaces. Methods of evaluating and testing solar power plant components are described.

1133

Complex Static Mirror Solar Concentration Arrays : Mattingly G T, " Proceedings International Conference on Future Energy Concepts ", London (U.K.) p. 244, (1979).

The arrays described in this paper are composed of three elements, a collector and two mirrors. The collector is mounted vertically, one mirror is placed horizontally at its base on the sunward side and the other is inclined. All the calculations presented relate to arrays whose proportions are related by the equation $C = Pl \tan 2\beta$.

1134

Capabilities of High Temperature Materials in a Solar Energy Environment : Matthews L K and Mulholland G P, " Proceedings 86th AIChE Conference ", Texas (USA), (1979).

High temperature materials have been used in foundries and industries for several years. However, these materials have not been used for testing in a solar energy environment. Testing of these materials has been conducted at the White Sands Solar Furnace and the DOE Solar Thermal Test Facility. It is apparent that differences do exist between solar and non solar testing. The results of solar testing are presented.

1135

An Arrangement for Recording the Process of Mass Transport of Materials under the Action of Radiant Fluxes : Mavashev Yu Z, Nechaeva L V, Sokolov S B, Arifov P O, Kulagin A I and Koson A V, " Applied Solar Energy ", 15 (1), 19, (1979).

A thermogravimetric setup that records the change in mass of various materials under the action of large radiant fluxes in a solar furnace is described. The technical characteristics of the setup are given together with the results obtained in recording mass transport of certain materials.

1136

Foldable Solar Concentrator : McCusker T J , " U S Patent 3, 295, 512 / January 3, (1967) ".

A double reflector, consisting of a foldable cone reflector and a collapsible parabolic cylinder reflector located on the axis of the cone, is described. The open cone reflector is made of a flexible sheet having an inner surface highly reflective to solar rays, such as mylar. A heat receiver is at the opening of the cone on the cone's axis, and the parabolic column reflector is connected to the heat receiver so that it extends into the dish substantially to the apex of the cone. The column reflector consists of several parabolic elements with diameters decreasing toward the apex of the cone to permit telescoping for storage. The cone and parabolic reflector are designed so that solar rays incident on the cone are reflected to a parabolic element, back to the cone and then to a focal point on the receiver.

1137

<u>Cone and Column Solar Concentrator Model</u> : McKusker T J, " NASA Report No. NASA - CR - 66012, GER - 12268", (USA), (1965).

Analysis of the cone and column solar concentrator establishes the design feasibility for a 60 ft diameter model. A specific weight of 0.2 lbs/sft and packaging in a 9 ft diameter appear to be attainable objectives. A cone construction was developed consisting of conical film sectors with film stiffeners attached to the outside surface. The construction of a 5 ft diameter prototype optical model concentrator is described.

1138

<u>A Data Analysis Program for Collector Testing</u> : McDowell J H and Uselton R B, " Proceedings ISES Silver Jubilee International Congress ", Georgia (USA), p. 449, (1979).

The program described in this paper was developed to assist in recording and analyzing collector test data accumulated during six months of tests on a Suntec SLATS collector. This concentrating collector is a 400 sft linear focussing moveable SLATS fixed receiver collector which was tested in conjunction with weather recording activities. This data was completed and analysed using this program which includes a very useful subprogram that performs a linear regression on selected parameters.

1139

<u>Solar Total Energy System Evaluation Program</u> : McFarland B L, Nalbandian S J and French E P, "Proceedings 14th IECE Conference ", Boston (USA), p.129, (1979).

A solar total energy system (STES) uses solar energy as the primary source to provide a combination of electrical and thermal power to the user to supplement or replace conventional energy sources.Either photovoltaic or solar thermal collectors can be used. Both can be evaluated by the solar total energy system evaluation program (STESEP). This computer progam is to determine the equipment, sizes and control methodology best sutied to building applications in the commercial sector. This paper describes and discusses application of the STESEP code which can be extended to evaluation of STES in general.

1140

<u>Thermal Performance of a Linear Solar Collector</u> : Meinel A B & Meinel M P, "ASME Paper No.72-WA/(1972)".

A series of graphs present the relationships between the optical system factors and resulting system efficiency. The operating range of temperature is 300 - 700 deg C for a wide range of surface selectivities and optical flux concentrations.

1141

<u>Solar Energy Collector and Concentrator</u> : Meinel A B and Meinel W B, " U S Patent 4, 131, 485 / December 26th, (1978) ".

Modular structures for the collection, concentration and conversion of solar energy to another usable forms such as electrical energy are described. The structures feature three conic section reflective surfaces, two of which focus in front of a receiver element and off the axis of the structure. The third relative surface is utilized to redirect that energy which improves overall efficiency.

1142

<u>Three Dimensional Tracking Solar Energy Concentrator and Method for Making the Same</u> : Miller C G & Pohl J G, " U S Patent 4, 846, 462 / September 6th, (1977) ".

A three dimensional tracking solar energy concentrator, consisting of a stretched aluminized polymeric membrane supported by a hoop, was presented. The system is sturdy enough to withstand expected windage forces and precipitation. It can provide the high temperature output needed by central station power plants for power production in multimegawatt range.

1143 .

<u>Low Cost Solar Energy Collection System</u> : Miller C G and Stephens J B, " NASA Report (N - 77 - 20566) (Contract NAS 7 - 100) ", California (USA), (1977).

A fixed linear ground based primary reflector having an extended curved saw tooth contoured surface covered with a metallized polymeric reflecting material, reflects solar energy to a movable supported collector that is kept at the concentrated line focus of the reflector primary. The primary reflector may be constructed by a process utilizing well known free-way paving machinery

1144   .

<u>Solar Energy Collection System</u> : Miller C G and Stephens J B," U S Patent 4, 149, 521 / April 4, (1979) ".

A fixed linear ground based primary reflector, having an extended curved saw toothed contoured surface covered with a metallized polymer reflecting material, reflects solar energy to a movable supported collector that is kept at the concentrated line focus reflector primary. The primary reflector may be constructed by a process utilizing well known free-way paving machinery. The solar energy absorber is preferably a fluid transporting pipe. Efficient utilization leading to high temperatures from the reflected solar energy is obtained by cylindrical shaped secondary reflectors that direct off-angle energy to the absorber pipe. An arrangement of cylindrical secondary reflector stages and spot forming reflector stages produces a high temperature solar energy collection system of greater efficiency.

1145

<u>Collection Times for Trough Type Concentrators Having Arbitrary Orientation</u> : Miller C W, " S o l a r Energy ", 20, 399, (1978).

A mathematical condition is derived,which determines the time at which acceptance of the sun's beam radiation begins and ceases,for a trough type concentrator having arbitrary orientation. The concentrator considered is a two dimensional ideal concentrator which has a definite range of acceptance angles. Sample calculations are performed which investigate the period of acceptance of a 1.8 X concentrator, for year round optimal collection a south facing 1.8 X concentrator should be oriented with axis horizontal and with slope equal to the latitude. If the concentrator faces slightly away from the south, however, a horizontal axis and a slope equal to latitude will result in severely reduced collection times near a solstice.

1146

High Pressure Steam Generation at 5 MW Solar Thermal Test Facility : Moeller C E and Rush E," Proceedings ISES (American Section) Conference ", Denver, Colorado (USA), 2.1, 811, (1978).

Steam generation and heat rejection system is one of the major operating systems of the DOE's 5 MW Solar Thermal Test Facility, Sandia Laboratories, Albuquerque, New Mexico. The details of this operating system are given with a general overview of the features of the facility. The condensation of steam is acheived by a conventional direct contact condenser and subsequent rejection of heat through a closed cycle subsystem Instrumentation and control are detailed. The schedules for solar testing of 5 MW receivers are presented. The configurations of these solar receivers are given and design steam conditions indicated.

1147

Solar Thermal Electric Power Plant with Distributed Collectors 100 to 1000 MW : Mordchelles-Regnier G, Dahan G and Boy-Marcotte J, " Proceedings ISES Silver Jubilee International Congress",Georgia(USA)(1979).

The development models of solar power plants in the medium range power (from 100 to 1000 KWe) are presented. It has been concluded that optimization of the solar power system, comprising a field of collectors, storage unit and thermal engine, is of capital importance in reducing KWe costs. The optimization depends on peak power and daily energy required by the user; in the 200 to 1000 KWe power range the optimum heat source temperature is between 200 and 250 deg C.

1148

Fundamental Research on Heat Transfer Performances of Solar Focussing and Tracking Collector :Mori Y, Hijikata K, Himeno N and Nakayama W, " Solar Energy ", 19, 595, (1977).

A focussing solar energy cylindrical collector system equipped with a discontinually tracking device with an absorbing pipe of high absorptance surface has been developed. The solar tracking performance and the heat transfer characteristics have been investigated experimentally.

1149

Solar Concentrators for Interplanetary Applications : Naccach A F, " Proceedings ISES (American Section) Conference ", Denver, Colorado (USA), 2.1, 901, (1978).

An analysis is presented of a little studied class of solar concentrators with high potentiality for space application, namely space fabricated gas filled spherical plastic lenses called 'bubbles'. A bubble will behave like a lens, as long as the inside has a higher refractive index than the outside. In the vacuum of space, a gas or vapor will refract incoming light rays and, therefore, can be used as lens material.

1150

Low Cost, Low Technology 3.4 X Concentrating Collector for Industrial / Agricultural Applications : Naccach A F, " Proceedings ISES Silver Jubilee International Congress ", Georgia (USA), (1979).

The ATON - 3.4 X solar concentrating collector is a cylindro-parabolic mirror with a tubular glazed absorber. The collector has an acceptance angle of $12°$(average), needs no diurnal tracking and accepts circumsolar radiation. The operating range is 50 deg C to 130 deg C with efficiencies comparable to that of CPC described in the literature. The minimal mirror curvature makes it easy to manufacture & with a mirror utilization factor of 89%, therefore, the low cost. The ATON - 3.4 X is intended to be used as a feed-water preheater (economizer) of industrial process heat fossil-fuel boiler.

1151

Generalization of the Two Dimensional Optical Analysis of Cylindrical Concentrators : Nicolas R O and Duren J C, " Solar Energy ", 25, 21, (1980).

A two dimensional optical analysis of cylindrical concentrators valid for any incidence angle of the solar ray is described. Unlike previous two dimensional studies, it takes into account (a) the angle 'K' defined by the solar rays and a plane perpendicular to the focal line and (b) the variations of the image width as a function of 'K'. An equation relating 'K' to solar co-ordinates has been obtained. The curves of 'K' as a function of time for several dates and three orientations of the concentrator is presented. The analysis is applied in detail to the cylindrical parabolic concentrator and to the fixed mirror solar concentrator both, with flat receivers. The local concentration factor and its mean value for different values of 'K' are obtained. Using these results and taking into account the useful range of 'K', criteria are given to select the concentrator orientation and the receiver width.

1152

Freezing Points of Lanthanide Oxides Measured with a Solar Furnace : Noguchi T and Mizuno T, "S o l a r Energy ", 11, 90, (1967).

The freezing point of lanthanide oxides, except CeO_2 and Pm_2O_3, has been measured by the specular reflection method with a heliostat type solar furnace.

1153

Study of Umbrella Type Erectable Paraboloidal Solar Concentrators for Generation of Spacecraft Auxiliary Power : Nowlin W D and Benson H E, " NASA Report No. NASA - TN - D - 1368 ", (USA), (1962).

The design procedure for a parabolic erectable rib type solar concentrator employing straight ribs and the effect of design parameters on concentrator weight, contour and vibration are discussed. Experimental results are shown from calorimetric tests of a 10 ft diameter 60 rib concentrator with 3 inch, 6.25 in and 12 in diameter spherical heat receivers, overall and geometrical efficiencies being indicated for various concentration ratio and heat receiver locations. Vibrational data are presented for rib type concentrators of two sizes for atmospheric and vacuum conditions.

1154

Determination of Relative Efficiency of an Equitorial - Mounted Solar Energy Concentrator in Photo - Reduction of a Solution of Lead Tetra-acetate in Acetic Acid : O'Brein K G and Byrne R E," Solar Energy", 12, 285, (1969).

The determination of the relative efficinecy of an equitorial mounted solar energy concentrator as compared with unconcentrated sunlight and an artificial visible ultraviolet light source in the photo-reduction of a solution of lead tetra-acetate in acetic acid is described. The variation of specific reaction rate with time of the day is examined and results obtained are found to be in keeping with other reported methods of measuring the intensity of solar radiation during the day. Various devices which could be deemed to improve the efficiency of the concentrator were tested. The systems used on the solar energy concentrator, in natural sunlight and in conjunction with a 150 W high pressure mercury vapour lamp, are described in detail.

1155

Non Tracking, Concentrating Collectors Utilizing Grazing Incidence Mirrors : O'Meara J E, " Proceed - ings ISES Conference ", Los Angeles (USA), (1975).

A novel approach to concentrating collectors is described. This collector has an array of parabolic surfaces situated such that a large fraction of solar rays strike the mirrors at incident angles greater than 60 degrees. Such large incident angles result in high reflectivity. All reflecting surfaces are segments of parabolas with their foci being the center of the collector tube. The array of mirrors also serves to reduce heat losses from the collector tube. Convective air currents are impeded by the baffle like mirror array. In addition, radiation losses are reduced since a large fraction of the collector tube 'sees' relatively warm mirror surfaces. A working model of the collector has been fabricated and tested. The aperture to collector tube ratio of 8 : 1 and an acceptance angle for incoming solar rays of a pproximately 14 degrees is found.

1156

Analytical & Experimental Study of Total Internal Reflection Prismatic Panels for Solar Energy Con- centrators : O'Neill M J & Gupta Y P," Proceedings ISES (American Section) Conference", Orlando (USA), (1977).

Such panels offer the potential of better performance at low cost for numerous solar concentrator aplications, including heliostats for central receiver (Power Tower) electric power plants and various parabolic concentrators. This paper presents a description of the new reflector concept and results of preliminary theoretical and empirical studies of its performance.

1157

Feasibility of a Concentrating Optical Duct System for Collecting and Transporting Solar Energy to Storage : Oswald R L & Wachtell G P, "Proceedings ISES (American Section) Conference", Denver, Colorado (USA), 2.1, 420, (1978).

Three optical ducting arrangements are briefly analysed for delivering solar energy from roof mounted collecting mirrors to thermal storage at a lower elevation for domestic space heating or cooling. All the three required tracking. In the first, collection is done by an array of two axis tracking mirrors. In the second and third, tracking is done by movement of the target mirror. Collection efficiency should be at least as good as, and possibly substantially better than, an ordinary flat plate collector array.

1158

Effect of Errors in the Shape on the Characteristics of Elipsoidal Concentrators of Radiant Energy : Paderin L Y, " Applied Solar Energy ", 14(6), 9, (1978).

Ellipsoidal mirror concentrators of radiant energy having monotonic and alternating errors in their shape with respect to ideal ellipsoids are investigated. It is shown that the alternating errors of the concentrators considerably reduce the radiant flux density at their foci, compared with monotonic errors.

1159

Comparative Performance of Tracking Type and Non-tracking Type Solar Collectors : Pahoja M H & Nanda S K, " Proceedings ISES Conference ", New Delhi (India), (1978).

This paper reports some results of an experimental investigation on performance of four different types of collectors. These include the flat plate collectors and the parabolic cylindrical, compound parabolic and conical concentrators. A flat plate collector with two glass covers was found to be best suited for use in Delhi provided the required maximum temperature did not exceed 93 deg C. Among the concentrating collectors tested, a conical concentrator with conical absorber, gave significantly higher temperatures for the same concentration ratio. The conical reflector is also easy to fabricate as compared to parabolic reflectors. The requirement of high reflectivity and its continued maintenance for a concentrator, is a major drawback of a concentrating collector.

1160

Multiple Fin Solar Energy Receiver for Line Focussing Collectors : Parker B F, " Proceedings A S A E, Winter Meet ", Chicago (USA), (1977).

A solar energy receiver for line focussing collectors consisting of 24 longitudinal fins made of 5 mil reflective aluminium foil has been designed and constructed. Initial testing at low radiation levels resulted in 44% efficiency at 120 deg C. It is expected to achieve at least 55% efficiency at 150 deg C for heating ambient air when current problems are eliminated.

152

1161

Performance of a Focussing Solar Air Heater : Parker B F, " Proceedings Joint ASAE and Canadian Society of Agricultural Engineers, Summer Meet ", Winnipeg (Canada), (1979).

An air heating focussing solar collector with a cylindrical parabolic reflector and linear receiver was constructed and tested. The results are presented in a modified form of the Hotlel-Whillier-Bliss generalized performance equation by plotting efficiency versus temperature differential divided by direct beam solar insolation. The test unit which has some imperfections in receiver core construction operated at an efficiency of 49 - 57% when heating ambient air, depending on flow rate. The performance is degraded at elevated inlet temperatures.

1162

A Figure of Merit for Solar Collectors with Several Separate Absorber Segments : Patera R P & Robertson H S, " Proceedings International Conference on Alternative Energy Sources ", Miami (USA), (1979).

A figure of merit for solar collectors is presented and examined. The main purpose of obtaining a figure of merit is to aid in comparisons of collectors having several thermally separated absorbing segments. Nevertheless, results are useful for single absorber collectors. The proposed general figure of merit, 'Q', has information about the input distribution, the concentration and the a c c e p t a n c e function built into its definition. In addition, the use of a channel matrix to characterize a collector is proposed. The channel matrix enables the figure of merit to be readily calculated for various input distributions.

1163

Flux Monitor for Natick Solar Furnace : Penniman F G, Petev P H and Davies J M, " Solar Energy ", 10 , 23, (1966).

U S Army Natick Laboratories solar furnace was designed to simulate thermal radiation emitted by an atomic explosion. It was intended for investigating parameters effecting skin burns from high intensity thermal radiation and to evaluate protection afforded by experimental fabric systems. A heliostat concentrator system is used to produce maximum irradiance of 125 Cal / sqcm sec. at the center of a 4 in diameter image. The attenuator can be set to control irradiance and also serve as a safety d e v i c e since it falls shut rapidly in case of malfunction.

1164

Pulse Shaper for Natick Laboratories' Solar Furnace : Penniman F G, Goff R J and Davies J M, " S o l a r Energy ", 12, 85, (1968).

A pulse shaper which produces pulses corresponding to a range of weapons from 25 kilo ton to the megaton range is described. It consists of an array of 16 radial vanes, each about 1.5 ft long and placed in a converging beam about 40 in in front of the focus. These vanes which are driven at variable speeds can control irradiance at target.

1165

Programme THEK de Centrales Solaire, a Collectors Distribues : Peri G, " Entropie ", 15(85), 43, (1979).

The general studies done by the CNRS on solar energy plants have raised the problem of building energy production units in the medium electrical power range, around 100 KW. Among the possible solutions, the principle of using distributed heliothermal converters has been chosen as the most advantageous solution, in the present state of things. At the moment, the THEK program consists of three stages. The first stage, the THEK 1 project, involves the design construction and testing of two prototype modules for converting solar energy into heat, with a view to drawing up the specifications for pilot series and then, mass production. The second stage, the THEK 2 project consists of designing, building and testing a field of 100 solar panels that will be connected to an existing thermal loop. The t h i r d stage covers the various applications that can be imagined on the basis of the heliothermal conversion modules developed for THEK 1.

1166

Le Compromis " Selectivite / Concentration " dans la Captation de L' energie Solaire : Peri G, Papini F and Pasquetti R, " Proceedings International Conference on Solar Electricity ", Toulouse (France), p. 429, (1976).

In the design of solar thermal collectors and convertors, three principal parameters are to be considered (i) working fluid temperature, (ii) incoming energy or radiation concentration ratio and (iii) absorber selectivity. Taking into account that usable energy has to be as large as possible, these parameters are not independent, namely, for a given working temperature, we can determine t h e value for the absorbing surface " selectivity " parameters as a function of the concentration ratio ; higher than these values. The usable energy does not increase in a significant way. This paper presents the results given by such a study, that shows the compromise which lies between " selectivity " and concentration when the concentration ratio is relative. These results are shown after the radiative b a l - ance of the absorber including all the parameters mentioned before.

1167

Solar Thermal Electric Power in the Range of Medium Size Units / The French Program THEK : Peri G, Desautel J, Imbart B, Audibert M, Pasquetti R and Battistelli J, " Proceedings ISES Silver Jubilee International Congress ", Georgia (USA), p.1121, (1979).

This project is concerned with the design of heliopower units, in the small (or medium) range, that is 300 to 3000 KW or 50 to 500 electrical kilowatts. It is a distributed collector system using a parabolic dish as a basic unit. The first step is the study, design, realization and experimentation of one module protoype (subsystem) for the solar to thermal energy conversion. In the second step a number of such modules connected to a thermal loop with a storage tank a r e studied. Finally, the design of a complete heliothermal system adapted to a particular utilization (steam production , desalination, mechanical and electricity generation) is undertaken.

1168

A Method of Shaping Mirror Elements by Elastic Deformation : Perrot M & Peytavin A, " Solar Energy",10,3,(1966).

This paper presents a method for testing the accuracy during the manufacture of mirrors, having t h e form of elements of a paraboloid of revolution. These mirror elements can be put at some d i s t a n c e from the vertex of the paraboloid. The mirror is obtained by elastic deformation of a plane mirror, for example, a plate of pure hardened, polished and aluminized aluminium.

1169

Comparative Technical Evaluation of Solar Collectors : Peters P J, " Proceedings SPIE Conference ", San Diego, (USA), 68, 128, (1975).

A comparative technical evaluation of the four primary solar collecting configurations, viz the central receiver, parabolic trough, paraboloidal dish and Winston collector is made. Subsystem & overall efficiencies are presented and technical descriptions and performance characteristics are discussed. Comparisons are made on a solar thermal conversion basis. Applicability of the ' Winston ' collector for photovoltaic conversion is summarised.

1170

Experimental Researches Concerning the Use of Solar Energy in the Generation of Electric   E n e r g y : Petrescu S, Damescu Al., Baran N and Dimitriu S, " Proceedings 2nd International Solar Forum ", Hamburg (Germany), p. 233, (1978).

The paper presents theoretical and experimental aspects, concerning the cylindrical parabolic type solar radiation receivers ; the paper comprises the concentrated radiation receivers by whose aid superheated steam has been obtained. The steam jet is actuating the pilot turbine, coupled to a small electric generator.   ·

1171

Practical Method for Including Material Scattering Effects in Determining the Amount o f Intercepted Sunlight in Solar Concentrators : Pettit R B, Vittitoe C N and Biggs F, " Proceedings ISES Silver Jubilee International Congress ", Georgia (USA), (1979).

In imaging solar concentrators, the amount of solar radiation incident on a receiver surface depends upon both the overall concentrator slope and the angular Distribution of light rays (sunshape)   that reach the receiver. Sunshape broadening effects, which include the specular reflectance or transmittance properties of mirrors of glazings, image degradation caused by surface slope errors, & tracking errors are combined into an effective error cone. Broadened sunshapes for a variety of effective error cone is approximately 2 to 3 times the RMS width of the incident sunshape, the broadened sunshape can be adequately described by a circular normal distribution.

1172

Solar Image Characteristics of Concentrators : Phillips G and Bayazitoglu Y, " Proceedings 2nd International Conference on Alternative Sources of Energy ", Miami (USA), (1979).

A mathematical model is developed to find the characteristics of the solar image for a reflecting type concentrating collector. The model encounters the effect of the size of the solar disc and determines the intensity distribution of the concentrated beam on the receiver. The computer implementation o f the model is such that the geometry of the reflector and the receiver do not have to be specified b y explicit mathematical relations. The reflector and the receiver do not need to· be axisymmetric.Therefore, surface imperfections can be considered. Finite planer elements to represent the system for the reflector and the receiver are represented by nine nodes. The piecewise ray tracing method is adapted to study the incident and reflected energy. This includes the shading effects by the absorber & the reflector, and also takes into account the multiple reflector.

The intensity distribution for a conical reflector with its base as a receiver is studied. The effect of misorientation is compared with previous studies to check the accuracy of the model. The intensity distributions for conical reflector with cylindrical absorber and spherical reflector with conical absorber are calculated for various sizes. The effect of misorientation and surface irregularities in the reflector due to manufacturing errors in the intensity distribution are presented.

1173

Efficient Low Cost Concentrating Solar Collectors : Pierce N T, " Solar Energy", 19, 395, (1977).

Some novel concentrator receiver designs are proposed to reduce material costs, simplify construction and improve efficiency. Operating efficiencies were better than reported flat plate performance and material costs were less. Advantages and disadvantages are discussed as well as performance and cost information. A multiple reflector "venetian blind" array which will be useful in south facing windows of various types and sizes was also tested and found to perform better than the reported   flat plate collectors.

1174

Solar Energy Reflecting System : Polley J A, " U S Patent 4, 106, 485 / August 15, (1978) ".

A solar energy reflecting system is disclosed which can modify its reflector configuration in accordance with the movement of the sun across the sky. The system includes a circular dish like housing with upper and lower circular hoops in spaced relationship within the housing which can be displaced relative to each other. A series of reflector panels are sequentially spaced around the   peripheral wall of the dish like housing and controlled by the displacement of the upper and lower hoops to lie in an outwardly flamed direction and pivot as the hoops are displaced relative to each other. A sensing device detects the position of the sun in the sky and accordingly displaces the hoops to thereby pivot the reflector panels appropriately to give the proper reflector configuration which will provide the best focussing of the sunlight and concentrate it onto a heating coil where the heat can be absorbed and transferred to a storage tank.

1175

Elastically Deformed Linear Focusing Reflector : Powell R A, " Proceedings IECE Conference ", Boston (USA), p. 7, (1979).

A low cost, high quality linear focussing reflector can be produced by applying twisting moment loads to the longitudinal edges of an initially flat sheet of reflector material. This causes it to elastically deflect to a cylindrical cross section. Supplemental loads force the cross section to that of a parbola. Experiments using a test bed that simulates a two foot (o.61 m) long section of a six foot (1.83 m) wide reflector made by this method aupports the theory. A stable and structurally stiff reflector is produced having a geometric concentration ratio of 21 in cylindrical configuration and 88 in the parabolic configuration. A production design that is simple, does not require large capital outlays, and has material costs in the range of $7 / sft ($75 / sqm),is discussed.

1176

Development of a Flexible Optical Solar Reflector Stable under Space Environment : Preuss L, "Messerschmidt – Boelkow – Blohm Gmb H. Report No. MBB – UR – 194 – 73 ", (1973).

For the development of flexible optical solar reflectors, stable under space. requirement a comprehensive materials selection was performed concerning suitable polymer foils, polymer lacquers,reflecting metals, corrosion protection layers, and interference filters. Outgassing tests were performed on polymer foils and lacquers. Twenty six different flexible optical solar reflectors were manufactured and tested under simulated ground and space environmental conditions. Test results provided the basis for the final selection of two systems for a qualification phase.

1177

High Temperature Solar Energy Conversion System : Price K M, " Solar Energy ", 25, 187, (1980).

This paper identifies certain key parameters 'P' and 'Q' and establishes a relation between them that can be used to exhibit the interplay between the conversion efficiency ηcon the converter temperature 'T', parameters of the optical subsystem and the parameters of the converter entrance in a high concentration, solar energy system. The results are displayed on one universal diagram. A locus of maximum efficiency is then deduced which is of general applicability to the whole range of high concentration systems. The choice of converter temperature 'T' proves to be critical and a design formula for optimum 'T' is derived.

1178

Radiation Transfer Through Specular Passages : Rabl A, " International Journal Heat Mass Transfer ", 20, 323, (1977).

A new technique is developed for calculating exchange factors for specular radiation passages of high reflectivity ρ. It is shown that for a large class of configurations, even with curved reflector surfaces, the average number of reflections< n >can be calcualted by a simple analytic formula and without any ray tracing. The exchange factor can then be approximated by $\rho^{<n>}$. The method is illustrated by many examples, including V-troughs and compound parabolic concentrators which are relevant for solar energy collection. The $\rho \Rightarrow 1$ limit of radiation cavities can be treated in a similar manner.

1179

Yearly Average Performance of the Principal Solar Collector Types : Rabl A, " Solar Energy Research Institute Report No. SERI / TR – 631 – 716 ", (USA), (1981).

The results of hour by hour simulations for 26 meteorological stations are used to derive universal correlations for the yearly total energy that can be delivered by the principal solar collector types; flat plates, evacuated tubes, CPC,collectors that track about two axes and central receiver. The corelations are polynomials of first and second order in yearly average insolation, latitude and thres – hold (= ratio of heat loss and optical efficiency). With these corelations the yearly collectible energy can be found by reading a single graph and multiplying coordinates by the collector parameters. This simple method reproduces the results of hour by hour computer calculations with an accuracy (rms error) of 2% for flat plates and 2% to 4% for concentrators.

1180

Reflector System for Yield of Solar Energy : Radons U, "German Patent 2, 511, 740 /A/Sept. 30, (1976).

The reflector system consists of a number of reflecting strips which can be tilted on their longitudinal axis. The reflected solar radiation is collected on a collector surface strip. To protect against the weather, the reflector strips are provided with a transparent cover. The settings of the reflectors takes place by a time dependent control arrangement and an overriding control arrangement controlled by the solar radiation.

1181

Luminescent Solar Concentrators : Rapp C F and Boling N L, " Proceedings 13th IEEE Photovoltaic Spec. Conference ", Washington (USA), p. 690, (1978).

A photovoltaic concentrator of novel designs has recently been suggested. In its simplest form, this concentrator is a flat plate of some luminescent material. A different concentrator configuration that offers several distinct advantages utilizes a thin luminescent film ($\approx$1 nul) deposited on an undoped substarate ($\approx$1/8 in thick). Several 6 inch square plates have been constructed and collector conversion efficiencies of several per cent have been achieved.

1182

Design Options for Solar Total Energy Systems : Rapp D, " Proceedings ISES (American Section) Conference ", Denver, Colorado (USA), p. 967, (1978).

A total energy system is an on site local electrical power plant in which the rejected heat from the turbine exhaust is used as fully as the load permits for low level heat applications such as space heating, domestic hot waters and absorption chillers. Such systems have a special appeal for solar energy applications because the collected solar energy is more fully utilized than in a central power plant at the same operating temperatures,and because the long piping runs required for distributed fields of solar collectors do not lead to economies of scale,for very large solar electric power plants. '

1183

Utilization of Heavy Fill Gases in Annual Solar Receiver Geometries for Heat Loss Reduction : Ratzel A C, " ASME Paper No. 79 – WA / Sol. – 18 ", (1979).

Analytical and experimental work has investigated reducing thermal conduction and natural convection heat losses in annular solar receiver geometries using high molecular weight fill gases. Gases analyzed in a Sandia Laboratories prototype receiver design include nitrogen, argon, krypton, xenon,sulphur hexaflouride and Freon C – 318^r. Experimental results indicate that high molecular weight monoatomic gases can reduce receiver heat loss by nearly 50%, comparable to annulus gas evacuation to the 5.0 to 5.0×10^{-1} Pa range. Computer simulation studies show that heavy gas utilization in the annular space can improve overall collector performance by 4.5 to 13.5%, depending upon the gas and the annulus pressure.

1184

Techniques for Reducing Thermal Conduction and Natural Convection Heat Losses in Annular Receiver Geometries : Ratzel A C, Hickox C E and Gartling D K, " Journal of Heat Transfer (Trans. ASME), 68, 101 , (1979).

Analytical and experimental work has been undertaken to analyze thermal conduction and natural convection heat losses in annular receiver geometries. Techniques studied for reducing conduction heat losses include evacuation of the annulus gas, oversizing of the annular space while maintaining slight vacuum levels and use of gases other than air in the annular spaces. The geometry considered, total heat loss reductions of 10% to 50% may be obtained depending upon the means by which the conduction heat loss is limited. In addition, natural convection studies considering the effects of non–uniform temperature distributions and eccentric cylinders are discussed. The numerical analysis performed indicates that highly non–uniform temperature distributions are required to appreciably affect the natural convection process between concentric cylinders, and that rather large eccentricities cause only a slight increase in natural convection heat transfer.

1185

Apparatus for Making Curved Reflectors : Reader A F, Russell W E and Werner E A, " U S Patent 3, 541, 825 / November 24, (1970)".

Fabricating and forming each curved segment of a light weight mirror by polishing one surface of a flat magnesium plate and then machining stiffening ribs in the other side is described. Each plate is accurately formed on a single mold with the application of heat and pressure after the machining.

1186

Applications of Solar Concentrators in Domestic Cooking : Reddy M S, Krishan H and S a b h e r w a l S P, " Proceedings National Solar Energy Convention ", Annamalai Nagar (India), (1980).

This paper describes a manually trackable concentrator comprising mirror strips of 18.75 cm X 3.75 cm approximately arranged as a Fresnel concentrator of variable focal length. This whole arrangement can be adjusted to a predetermined position giving maximum concentration for different positions of the sun. The fixed receiver is essentially a multipurpose hot box designed and positioned in such a way as to catch most of the rays reflected from the mirror strips, thus utilizing most of the solar energy falling on the mirror. The design details for such an arrangement are briefly described. The performance results and the experience in ' jhuggi–jhonpries' (poor peoples' settlements) in Delhi (India) are reported. This may be one of the unique cases for multicraft cooperation of low income group people living in a semi rural or slum environments in suburban New Delhi (India).

1187

New Low Concentrating Nonfocussing Mirror : Regalado E, Alvarez I and Mayer E, " Proceedings I S E S (American Section) Conference ", Denver, Colorado (USA), 2.1, 1023, (1978).

There has been an increased interest for low concentrating mirrors which do not require the use of heliotrope devices, and allow fluid outlet temperatures above 100 deg C. Such concentrators generally have a definite axial orientation like in the cylindrical parabolic type; however, they cannot be installed in a fixed position, due to critical angular acceptance to direct radiation, so the maximum shifting angle of the sun during the year must be considered. For optimum collecting purposes, the plane profile of the curved mirror discussed in this paper consists of two symmetrical elements of a logarithmic spiral, and was chosen in order to develop a concentrator which would not require orientation during the year.

1188

A Solar Thermal – Electric Experimental Plant : Rubeck M and Berely W D, " Proceedings ISES Silver Jubilee International Congress ", Georgia (USA), (1979).

One approach to generating electricity from solar energy is to utilize an open Brayton – Cycle in a central receiver type solar thermal electric – power plant. An important next step in the development of this technology is the design construction and operation of an experimental plant. Boeing Engineering and Construction Company, under contract to Electric Power Research Institute, conducted a nine month study to develop a conceptual design and a definition of an open Brayton cycle experimental plant of above 1.5 MW electric output. This paper describes the assumptions, analyses and results of the conceptual design effort.

1189

Weight Minimization of Sandwich Type Solar Collector Panels : Reuter Jr. R C, " Proceedings IECE Conference ", Boston (USA), p. 1, (1979).

Solar collector modules of various types (e.g. line focussing,heliostats) require stiff, light weight structural panels to support and protect their reflective surfaces. Symmetric, sandwich types   panel construction fulfills these requirements. Analytical methods are utilized in the paper to study   the existance, utilization and practicability of minimum weight & adequately stiff design for sandwich panel construction in the solar collector application.

1190

Calorimetric Evaluation of Two Cone – Column Solar Energy Concentrators : Rhodes M D and Willis C M, " NASA Report No. NASA – TN – D – 5109 ", (USA), (1969).

Results of the investigations on two cone column solar energy concentrators are presented in this paper.

1191

Heat Transfer Analysis of Receivers for a Solar Concentrating Collector : Rice M P , Modest M F, Barton D N and Rogers W E, " ASME Paper No. 79 – WA / Sol. 20, ", (1979).

The heat transfer characteristics and performance of conically wound, single layer, monotube receivers for a concentrating solar collector are investigated. A discussion of methods and assumptions used to formulate the basic energy balances is presented. Results obtained from this analysis indicate the degree of sensitivity of performance to variations in tube radius, inlet temperature, ambient temperature,absorptivity of the tube coating , and cone half angle within receiver size and mounting limitations.

1192

Nontracking Moderately Focussing Heat Pipe Solar Collector : Roberts Jr. C C and Naperville I, " P r o-ceedings 3rd International Heat Pipe Conference ", Pala Alto (USA), p. 114, (1978).

A design for a moderately focussing heat pipe solar collector is presented that incorporates advantages of both flat plate and concentrating collectors. A prototype collector was constructed & tested using four gravity assisted heat pipes. The concentration ratio was 3.1 to 1. Test data demonstrated collector efficiencies near 60% at low loss factor. Absorber temperatures ranged from 40 degC to 55 deg C.

1193

Evaluation of a Faceted Mirror Solar Concentrator : Roger W E and Borton D, " Proceedings ERDA Conference on Soncentrating Solar Collectors ", Georgia (USA), p. 4 – 37, (1977).

The design and performance aspects of a 33 sqm (360 sft) solar concentrator, constructed at the NYSERDA research station in upstate New York are presented in this paper. The concentrator was designed as a research tool to evaluate the performance of this heliostat design and of receivers at its focus a s well as to ascertain characteristics of insolation at the site.

1194

Focussing Solar Collector Development ; Final Report : Rogers W E, Borton D N and Burr A A, " Report PB-275042 ", (1977).

This report discusses design modifications and engineering tests on a dual tracking focussing solar collector concept. The 366 sqft concentrator was constructed to evaluate performance of a heliostat design and of receivers at its focus, as well as to ascertain characteristics of insolation at site.

1195

Economic Analysis of a Point Focussing Concentrating Collector System : Rogers W E, Borton D N R i c e M P and Rogers R J, " Proceedings ISES Silver Jubilee International Congress ", Georgia (USA), p. 454, (1979).

A brief description of a 70 KW thermal parabolic Fresnel reflector point focussing concentrating solar collector is given. This paper explores factors such as risk, use of land, capital costs, working capital requirements, solar system performance and solar availability. An example is shown to illustrate how cost, allowances and other financial information can be presented in discussion of solar apparatus as well as conventional equipment.

1196

10 MW Solar Thermal Electric Power Plant Design for Solar Day Operation : Romero A F, " Proceedings International Conference on Solar Electricity ", Toulouse (France), p. 821, (1976).

A complete design is presented to supply power to the city of Hermosillo, Sonara in the northwest part of Mexico. Even though the design covers only solar day operation (8 to 12 hours daily), its main objective is to increase fossil fuel savings. Further improvements can be made to include night and day operation provided that a good design on energy storage is made. However, continuous operation is disregarded in this paper, due to the extremely high increase in initial investment. An analysis of available solar insolation data for the zone is included and also some experimental measurements, as well as some other geophysical characteristics which led to the decision of selecting an open Brayton cycle (gas turbine), using air as the working fluid.

1197

Accurex Concentrators on Solar Energy : Rossiter Ed., " Proceedings ERDA Conference on Concentrating Solar Collectors ", Georgia (USA), p. 8 – 1, (1977).

Several thermal applications for solar systems are described, including irrigation pumping,   process hot water and process steam. The design and construction of the 25 hp solar irrigation system at Willard, New Mexico, is discussed in some detail. Some aspects on the economics and costs for systems are presented.

1198

Concentration of Sun's Rays Using Catenary Curves : Rottigni G A, " Applied Optics ",17(6), 969, (1978).

An examination has been carried out of the difficulties a horizontally placed parabola would encounter

in concentrating the sun's rays when the angle of incidence between the axis of the parabola and parallel incident rays grows from 0 to 50°. The results obtained with two catenary curves formed by hypothetical reflecting adjustable sheets were compared with the parabola, showing that the concentration can be held at interesting levels as regards technical application.

1199

Performance and Analysis of a Concentrating Collector for Heating Air : Rumsey T, Coffelt J & Dobie J,
" Proceedings Joint ASAE & Canadian Society of Agriculture, Energy Summer Meeting," Winnipeg (Canada),
(1979).

A concentrating solar collector to be used to heat air for crop drying has been constructed. T h i s
paper presents results of initial performance tests of the collector along with a mathematical model
to predict collector performance. The concentrating collector's performance has been demonstrated in
two sets of test runs. In its full configuration, with transparent cover and aluminized mylar reflector, the collector has operated at 45% efficiency with a temperature rise of 34 deg C. T h e   primary
source of losses are the reflection loss and the heat losses from the receiver tube.The p r o p o s e d
model predicts temperature rise and efficiency quite well ; within 3% at operating flow rates.

1200

A General Design Method for Closed Loop Systems Having Flat Plate or Concentrating Collector : Ryan W A
and Klein S A, " Proceedings ISES Silver Jubilee International Congress", Georgia (USA), (1979).

The purpose of this paper is to demonstrate a method of estimating the monthly average performance of
closed loop solar energy systems having flat plate or concentrating solar collectors. The method requires relatively few calculations (possible on a hand held calculator), yet yields results of sufficient accuracy for design and economic evaluation. The method is a marriage of the Collares-Pereira
and Rabl's utilizability method for concentrating solar collectors and Klein and Beckman's g e n e r a l
design method for closed loop systems.

1201

Concentration of Solar Rays by a Conical Mirror : Saksena B M L, Gupta V P and Pande C S, " Bulletin of
Seismological Society of America ", 13(4), 371, (1964).

A conical mirror, for getting a high concentration of uniformly distributed solar intensity o v e r   a
comparatively large area, is examined theoretically. Expressions are obtained for the concentration,
the amount of energy and temperature available as a function of the parameter of the cone i.e. t h e
semi vertical angle, the vertical height ' h ' and the reflection coefficient 'r'. A cone of semi vertical angle of 15 degrees and having a vertical height of 5.1 times the collector diameter, can under
ideal conditions, give a concentration of about 14 times and a temperature of more than 400 deg C.

1202

Solar Concentrators : Salariya K S & Singh M, " Proceedings ISES Conference ", New Delhi (India), p.
1322, (1978).

Solar energy can be concentrated most efficiently by using lenses. Three types of inexpensive lenses
have been developed (73 & 90 cm diameters) in double convex, plano convex and concavo convex f o r m s .
These lenses are made from thin round acrylic sheets which are bolted together along t h e i r circum-
frences between two mild steel rings. Thin small cylindrical spacers of acrylic are placed between
the sheets at their center to give them the shape of the lenses. These hollow structures act like lenses when filled with clear water or any other transparent liquid. Their performances relating to solar energy are presented.

1203

Inexpensive Way of Concentrating Solar Energy : Sandhu B S and Chauhan R S, " C h e m i c a l Engineering
World ", (India), 11(5), 35, (1976).

An inexpensive way of concentrating solar energy is described which involves the use of a water lens
i.e. water supported over a polyethylene sheet stretched over a rim such that the water assumes t h e
shape of a plane convex spherical or cylindrical lens. The efficiency of solar energy collection depends on the quality of the plastic sheet and the average water depth over it. Reasonable collection
efficiency can be obtained with an average water depth of less than 10 cm. In experiments w i t h   a
spherical water lens of 61 cm (2 ft) diameter, a collection efficiency of 40 to 50% at 100 deg C   w a s
obtained. With a cylindrical configuration (46 cm wide), at stagnation temperatures, steam can v e r y
easily and efficiently be generated with either configuration. The possibilities of solar energy collection with spherical and cylindrical water lenses on small and large scales are discussed.

1204

Cost Effective Total Energy System Using a Faceted Mirror Sunlight Concentrator and High   Intensity
Solar Cells : Sater B L, Goradia C, Rogers W and Borton D, " Proceedings ISES Conference ", New Delhi
(India), p.1271, (1978).

The design of an integrated photovoltaic / Thermal High Intensity Solar Energy System (HISES)which can
supply a major portion of the electrical and thermal needs of a residential / commercial site is presented. A novel suntracking faceted mirror sunlight concentrator is described in detail. A preliminary cost analysis is presented and capital investment and conservation aspects are also discussed.
Energy costs of roughly 1 ¢ / KW(th) and 2¢/KWe may be possible with the large scale use of HISES.

1205

A Simple Technique for Measuring Slope or Surface Error of a Concentrator : Sayler W and S a n c h e z
J, " Proceedings ISES Silver Jubilee International Congress ", Georgia (USA), (1979).

A technique developed to measure both sag and slope error of a solar concentrator is described. T h e
test method uses a Helium neon laser as a light source and two sheets of transparent acrylic w h i c h
when suspended vertically and weighted, define parallel planes. The laser is aligned so that its beam
is perpendicular to the two planes. The concentrator is then placed so that a hypothetical normal to
the bottom of the collector is parallel to the direction of the incident laser light. In other words,

if the laser is simulating light from the sun, the concentrator is aligned to face the l a s e r as it would normally be aligned to face incident solar radiation. With the positioning of test pieces completed, one can design measurements and analysis.

1206

<u>Solar Collection Limitations for Dynamic Convertor ; Simulation of Solar Thermal E n e r g y Conversion System</u> : Schrenk G L, " Paper at AGARD Conference ", Cannes (France), (1964).

A mathematical model for analysis of actual solar collectors has been developed that allows the calculation, without making numerical approximations, of energy flux on any arbitrarily shaped focal surface from any arbitrarily shaped collector surface. Typical results from this model are presented to show the effects of surface and orientation errors.

1207

<u>The Role of Simulation in the Development of Solar Thermal Energy Conversion Systems</u> : Schrenk G L, " Proceedings ERDA Conference on Concentrating Solar Collectors ", Georgia (USA), p. 9 - 39, (1977).

A comprehensive approach to the design, analysis, testing and operational use of solar thermal systems has been developed and is presented. It involves the use of simulation procedures wherein a delicate balance is achieved between mathematical simulation and actual laboratory simulation. These procedures are more than just another series of isolated mathematical models of idealized physical systems ; they are a simulation philosophy that covers all aspects of design, construction, testing and operational use of the system. The accompanying computer codes, originally written as large FORTRAN programmes, have recently been rewritten in APL on the APL - PLUS Scientific Time Sharing Corporation System. As a result, an extremely powerful and versatile interactive mathematical solar simulator has been created. APL - PLUS service is readily available on a nationwide basis : thus it is quite easy for anyone, even with limited programing experience to utilize these existing APL routines.

1208

<u>Generation " Dispatch " Modeling for Central Solar Thermal Generation Design</u> : Schweppe F, " Proceedings DOE Workshop on System Study for Central Solar Thermal Electricity ", Houston (USA), (1978).

The design of an optimum central station solar plant depends on the explicit characteristics of the rest of the power system (generators, load, storage). Ideally, the design optimization i n v o l v e s searching over both, system expansion, 4 system operation (generation dispatch) variable for the whole system, to minimize the sum of capital and operating costs. This paper discusses one part of the overall problem; the " simulation " of the generation dispatch to predict operating cost and reliability. This modeling methodology is called production costing. The choice of production costing methodology depends upon computer time, accuracy and type of output trade - off.

1209

<u>Stability of Plasma Sprayed Coatings Tested at the White Sands Solar Facility</u> : Schreyer J M, Schmitt C R, Hays R A and Farewell D, " Proceedings ISES Silver Jubilee International Congress ", Georgia (USA), (1979).

Plasma sprayed coatings on steel plates were tested at the White Sands Solar Facility at temperatures from 200 deg C to 1000 deg C. Analysis of the specimens before and after testing showed erbium dedecaboride (ErB_{12}), yttrium hexaboride (YB_6), titanium diboride (TiB_2) and chromium oxide (Cr_2O_3) to be stable above 600 deg C. A heat balance on the water cooled specimens of these coatings showed 73% to 97% heat recovery efficiency.

1210

<u>Test and Analysis of a Northrup Collector Controller</u> : Scott D R, Kissel R R and Reid H, " NASA Report No. DOE / NASA - CR - 78153 ", (USA), (1978).

The collector controller is examined as a functioning control system that drives the Northrup collector from east to west to follow the sun then back to east at sundown in readiness for the next   sunrise. The major components are examined separately with particualr emphasis placed on an analysis of the electronic drive circuit. The collector is a concentrating collector using a Fresnel lens. Results are plotted from hardware testing and analysis with recommended changes to improve the system.

1211

<u>Preliminary Technical and Economic Data on Solar Thermal Power Conversion Systems</u> : Selcuk M K, Caputo R S and Truscello V C, " Proceedings ISES Conference ", Los Angeles (USA), (1975).

Preliminary information is presented on solar thermal power conversion systems, as developed   during a NASA sponsored program to assess space based power systems and their merit as compared to   e a r t h based systems. Discussion in this paper is limited to terrestrial solar thermal conversion systems. A preliminary characterization is presented for a number of potential solar thermal power conversion systems using various types of collectors. Attempts are made to establish baseline designs & define performance of cost effective systems. Total system and component costs are validated and the m o s t accurate, uptodate predictions are used. Existing concepts are reviewed and screened from practical operation, lifetime thermal efficiency and cost effective performance standpoints. Some original concepts are also included. Power plants with power ratings in the 100 - 1000 Mw range and no t h e r m a l storage are evaluated during the first phase of the study.

1212

<u>Evaluation of Hardware and Software Requirements for the Advanced Components Test Facility Real Time Data Collection System Experiences from Two Solar Experiments</u> : Seminario C and Henderson D, " P r o - ceedings ISES Silver Jubilee International Congress ", Georgia (USA), p.1401, (1979).

The Advanced Components Test Facility is a DOE owned solar thermal test facility operated by Georgia Institute of Technology. The facility is primarily concerned with the R & D effort in the high temperature solar technology field.

1213

Heat Transfer Characteristics of a Linear Solar Collector : Seraphin B O," Applied Optics ",12, 349, (1973).

The heat transfer characteristics of a linear solar energy collector are calculated as functions of dimensions, spectral quality of the selective absorber surface, optical flux concentration of the optical configuration, thermal parameters and flow rate of the heat transfer medium. Carnot efficiency exit temperature and an upper limit to the amount of heat extracted are determined for the systems in which liquid sodium serves as the heat transfer medium. The performance is evaluated for selective absorber surfaces representing the state of the art as well as for surfaces requiring a more mature thin film technology.

1214

Solar Thermal Electric Power Systems : Comparison of Line Focus Collectors : Shaner W W and Duff W S , " Solar Energy ", 22, 49, (1979).

Three types of line focussing collectors , parabolic trough, fixed slats with movable absorber and movable slats with fixed absorber, are evaluated to find those systems that are capable of producing the lowest costs of electrical energy. Minimum costs per Kw / hr are found using sequential optimization techniques that consider variation in rim angle, reflectance aperture width, length, orientation, tracking, contour error, slat width, slat curvature, tangent slat angle, slope installation methods, materials, fabrication methods, absorptance, emittance, cover transmittance, field shape, layout, pipe sizes, insulation thickness, turbine generator cooling tower efficiencies and designs. This approach provides a uniform treatment of both cost and performance for a solar thermal electric power system. This uniform treatment of solar thermal electric power system for all collector types ensures that valid comparisons are made.

1215

Instrumentation for the 108 Foot Diameter Cloudcroft Solar Furnace : Shank M E, " U S Air Force Missile Development Center, Solar Furnace Support Studies Report No. AFMDC - TR - 59 - 15 ", Air Force Base, New Mexico (USA), 2, 261, (1959).

Instrumentation problems for the Cloudcroft solar furnace are considered for the areas of temperature measurement, atmosphere and focal spot aperture control and flux measurement. Basic considerations of design and development are discussed and specific recommendations set forth. Little of the necessary instrumentation is commercially available, but possible sources are cited for that which can be obtained. For the remainder, sources for design and development are indicated.

1216

Band Facet Collectors : Sharafi A Sh. and Kleyn G A, " Applied Solar Energy ", 1(1), 667, (1965).

A brief discussion is given of metallic faceted solar collectors used for increasing the concentration of solar radiation by a factor of 4 to 6. These are made of electropolished aluminium or metallized film and are shown to be good reflectors of the ultraviolet section of the radiation spectrum. Consideration is given to the calculation of the facet width and the amount of facet increase from center to the edges.

1217

Distribution of Solar Energy Flux Density at the Boundary Between Light and Shade : Sharafi A Sh., " Applied Solar Energy ", 2(5), 36, (1966).

Formulas are derived and used to determine the distribution of solar energy flux density in the boundary zone between light and shade (in the penumbra of a screen and on the edge of a spot produced by a plane mirror) caused by the sun's rays. The derived formulas may be used to design various types of faceted collectors of solar energy and various types of suntracking devices ; and may also be used to calculate energy distribution on the spot formed in the reflection of solar rays from unit areas of a solar collector including concentrators whose surfaces deviate from the ideal (taking mirror surface inaccuracies into account).

1218

Caustic Surfaces and Irradiance for Reflection from an Ellipsoid, Elliptic Paraboloid and Elliptic Cone : Shealy D L and Burkhard D G, " Applied Optics ", 12, 2955, (1973).

General formulas are derived for the caustic surface and irradiance over an arbitray receiver surface for point source radiation on collimated rays that are reflected or refracted by a curved surface. Specific formulas are obtained for light from a point source that is deflected by an ellipsoid, an elliptic paraboloid and an elliptic cone. As a numerical example caustic surfaces are calculated for a concave spherical surface and a concave paraboloid.

1219

Flux Density for Ray Propagation in Discrete Index Media Expressed in Terms of the Intrinsic Geometry of Deflecting Surface : Shealy D L and Burkhard D G, " Optica Acta ", 20, 287, (1973).

An exact, analytical formula for the flux density (energy per unit area per unit time) over an arbitrary receiver surface for rays which have been reflected from or refracted through an arbitrary curved surface is derived. The change in flux density is associated with the geometrical concentration or spreading of the beam produced by the curvature of the deflecting surface. The formula is expressed in terms of the intrinsic geometry of the deflecting surface (Gaussian curvature, mean curvature and normal curvature). An equation for the caustic surface is also derived. The general formulas are applied to calculate the flux density on an image plane when light from a point source is reflected from and refracted into a spherical surface. Equations are also given for the associated caustic surfaces. These yield, for paraxial rays, the standard mirror and lens formulas.

1220

Optical Investigations in Solar Energy Engineering : Sheklein A V and Rekant N B, " Semiconductor Solar Energy Convertors (Ed. V A Baum) ", p.193, (1969).

Theoretical and practical problems, associated with optical measurements of the reflection coefficients of mirrors and of transmission coefficients of solar radiation are considered. The optical properties measured by various devices are discussed and the methods are considered for finding parameters representing the total characteristics over the spectrum and overall directions. The possibility of the solution of some problems specific to solar energy engineering by means of standard optical equipment is considered, particularly the measurement of diffusely reflected solar radiation and the measurement of the specular reflector coefficient from the first face of a plane parallel sample.

1221

<u>Thermostatically Controlled Solar Energy Concentration Equipment Without Solar Tracking</u> : Shimada K, " German Patent 25, 52092 / A / May 26th (1976) ".

The Solar Energy Concentrator consists of a cylindric lens fixed in east west direction. In the concentration area of the solar radiation there are several long flow channels (ducts) also in east west direction. Each duct is controlled by a thermostat which opens the duct only when the temperature of this absorber rises above a predetermined value. To improve the efficiency of the system the ducts are covered with a heat absorbing material. The whole system can also consist of several lenses especially Fresnel lenses arranged in a rectangular field.

1222

<u>The Proof of Certain Laws of Optics Used in Solar Furnace Calculations</u> : Simon A W, "U S Air Force Missile Development Center Report No. AFMDC - TR - 59 - 15, Solar Furnace Support Studies 2, 187 ", Air Force Base No. 4 Hollman (USA), (1959).

Several laws of optics dealing with calculation on the solar furnace are proved. They are (1) that the fractional part of the light reflected at each point of refraction of a plane parallel glass plate is the same, and (2) that the sum of the fractional part of the light incident on any surface of a plane parallel glass plate reflected therefrom and the fractional part refracted therefrom is equal to unity. The laws of reflection of a metallic surface with a glass superstrate are deduced.

1223

<u>Application of Quasi Fixed Arrays of Adjustable Focus Reflectors in Solar Collectors</u> : Singh R & Sahgal P N, " Proceedings ISES Conference ", New Delhi (India), (1978).

In the present work, a linear array of 8, 50 X 125 cm reflectors has been constructed. Each panel consists of an aluminium channel frame in which strips of 5 X 50 cm thick are flexibly mounted. Along the center of each panel is located a pressure bar which rests on the center of each of the mirror strips. On loading the pressure bar by adjusting screws in the frame, the mirror strips undergo elastic deformation to give accurate single curvature parabolas. The advantages of employing such constructions are (a) cheap plane surface mirrors are used, (b) the focus of each plane can be adjusted on-site by simply adjusting two screws and (c) an array can be assembled consisting of panels of different focal lengths using the same construction modules. Concentration ratio of about 3 can be obtained.

1224

<u>Wide Angle Lenses and Image Collapsing Subreflectors by Nontracking Solar Collector</u> : Sletten C J, Holt F S and Herskovitz S B, " Applied Optics ", 19, 1439, (1980).

This paper presents new optical methods for the design of nontracking solar energy concentrators with acceptance angle of 60° in the elevation (altitude) plane and ±50° in azimuth sectors. A two point corrected cylindrical stepped prism lens (SPL) with 30.48 cm aperture height and $F/D \cong 1$, which focusses well over the acceptance interval has been designed. Image collapsing subreflector (ICS) surfaces are synthesized that reflect the incident illumination refracted by the lens onto a small fixed absorbing area or shelf ≈ 7.6 cm wide resulting in near maximum theoretical concentration ratios for these broad acceptance angles. Nearly 100% of the incident optical rays intercept the absorber shelf. The wide angle and image collapsing optical properties were confirmed by laser and solar experiments. Roof top thermal tests on a 30.5 X 30.5 cm collector section using selectively absorbing tubes with water as circulate were conducted that indicate aperture efficiency of $\approx 60\%$ could be expected on large area collectors based on this design.

1225

<u>A Method of Solar Thermal Generation of Electricity</u> : Smith R H, " Proceedings ISES Conference ", Los Angeles (USA), (1975).

This paper provides technical background information related to a solar thermal power generating system. The system considered would use an optimized secondary reflecting concentrator at the focal line of a cylindrical array of one axis heliostat mirror.

1226

<u>SOLERGY Collector Concept</u> : Smith R H, " Proceedings ISES Conference ", Los Angeles (USA), (1975).

The SOLERGY collector concept is essentially the use of a non-moving reflective surface optimized for the purpose of directing sunlight on to an absorber pipe. The reflector is composed of two shapes which flow together forming a single continuous curve. A spiral portion of the curve serves to efficiently direct light onto the absorber pipe. A parabolic section of the reflector gives the collector a degree of flux concentration. In the basic SOLERGY design the flux concentration is increased to the maximum level consistent with acceptance of sunlight during at least 6 day light hours every day of the year. The SOLERGY collector offers the performance advantages associated with flux concentration, combined with the manufacturing simplicities of a flat plate collector. There are no requirements for moving parts, highly critical manufacturing tolerances, or scarce materials in the SOLERGY collector.

1227

<u>Properties of Binary Oxide Systems Produced by Melting in a Solar Furnace</u> : Smokovdina G S, Shermazanyn Ya. T, Efimovskaya T V & Shakhparonyan V V, " Applied Solar Energy ", 11, (3 - 4), 20, (1975).

The high temperature region of the diagram of the state has been investigated in a solar furnace in the case of binary systems based on high melting oxides. It is shown that solar systems can be successfully employed in studies of the phase composition and the synthesis of compounds throughout the temperature range upto the melting point. Data obtained in this way can be used to synthesize new high temperature materials.

1228

A Note on Performance of Solar Concentrators : Sodha M S, Pal B P and Thyagrajan K," Physics Education", 4(4), (1977).

In this paper the variation of concentration efficiency ' X ' with angle θ which the solar radiation makes with the direction corresponding to the maximum value of the concenetration efficiency is examined. Four concentrators analysed are (i) Winston Collector, (ii) Cylindrical Parabolic Trough,(iii) a Compound Wedge and (iv) a Conical Concentrator. Results are represented graphically and discussed.

1229

Performance of Solar Concentrators ; a Theoretical Study : Sodha M S, Umesh G, Thyagrajan K, Pal B P & Ghatak A K, " Proceedings ISES Conference ", New Delhi (India) , p. 1309, (1978).

In this paper the authors report a theoretical study on the performance of three types of solar energy concentrators, viz the compound parabolic, the cylindrical parabolic and the compound wedge concentrators. For fixed exit and entrance aperture widths, they have calculated the relative energy distribution at the exit aperture of the concentrators, at different times of the day for three typical days of the year. The sun is assumed to be a moving point source and the collectors to be fixed on the ground at the equator. The energy distribution is calculated numerically by tracing a large number of rays.

1230

Determination of the Surface Shape of Film Solar Energy Concentrators : Solodynikov Yu. A, Vorotni - kov V I and Gafurov A M, " Applied Solar Energy ", 11(2), 93, (1975).

The linear problem of determining the deformed shape of a circular mirror film membrane with a diametric seam ; the membrane is acted upon by a uniform transverse load and arbitrary linear load that is distributed over the diameter is considered. The problem of determining the bending of the seam when acted upon by an arbitrary linear load is also solved.

1231

Application Analysis for the Solar Industrial Process Heat Market : Stadjuhar S A, " Proceedings 10th Model Simulation Annual Conference ", Pittsburgh (USA), p.935, (1979).

The importance of the industrial process heat market,in terms of energy consumption and amenability of this market to solar thermal technology, are examined. An analytical method for evaluating solar industrial process heat systems has been developed and implemented in a flexible, fast calculating, computer code - PROSYS / ECONMAT. The long term average performance model PROSYS predicts annual energy output for several collector types, including flat plate, non tracking concentrator, one a x i s tracking concentrator. The comparison computer program ECONMAT calculates the solar equipment cost and generates a life cycle cost analysis. Analytical results demonstrate the software flexibility for use in feasibility and parametric sensitivity studies.

1232

Optimized Concentrating / Passive Tracking Solar Collector; Final Report : Sterne K E, Johnson A Z and Grotheer R H, " Report No. TID - 29377 ", (1979).

A concentrating solar collector having half the material cost of other collectors with similar performance characteristics was developed and tested. The selected design is a compound parabolic concentrator (CPC). The output is a fluid heated to 100 deg C with good efficiency. The optical design of the reflector surface was optimized, yielding a 2.0 : 1 concentration ratio with a 60° acceptance angle and a low profile. Double glazing was chosen consisting of a polyester film outer glazing and an inner glazing of glass tubes around the absorber. The selective coated s t e e l absorber tubes are connected in series with flexible plastic tubing. A laminate of metallized plastic film over plaster was chosen for the reflective surface. The reflector is made rigid by attaching filled epoxy h e a d e r plates at each end. Aluminium side nails and an insulating back completes the design. T h e finished design resulted in a material cost of $21.40 per square meter in production quantities. Performance testing of a prototype produced a 50% initial efficiency rating.

1233

A Low Concentration Collector for Hot Water Production : Straatman J T H, Van Wieringen J S & Venis J,"Solar Energy", 25, 571, (1980).

A method of reducing collector price by using a weak concentrating cylindrical mirror of cheap metallized material is proposed. Absorber plate size is also reduced by using it on both sides. Cylinder axis orientation is kept east - west.

1234

Irradiance for Skew Rays Incident Upon a Trough Like Solar Collector of Arbitrary Shape : Strobel G L and Burkhard D G, " Solar Energy ", 20, 25, (1978).

A formula is derived for the irradiance at the base of a trough like solar concentrator for rays incident at any arbitrary angle upon the curved sides of the concentrator. Two angles are r e q u i r e d to specify the direction of the incoming rays. It is shown that one of the angles, ϕ, enters the expression for the irradiance, only through the multiplicative factor $\cos\phi$ and therefore plays the r o l e of obliquity factor associated with cosine of the angle of incidence for a flat plate collector.

1235

A Status Report on the Sandia Laboratories Solar Total Energy Program : Stromberg R P, " Solar Energy ", 17, 359, (1977).

A feasibility study of a combination of solar energy collection and total or cascaded energy systems has been underway at Sandia Laboratories, Albuquerque, New Mexico (USA), since the summer of 1972. This paper summarizes results of the program through June 1975 by providing system analysis, design and test data .

1236

Concentrators for Solar Energy Utilization : Swarup G and Balasubramanian V, " Proceedings Indian National Solar Energy Convention ", Bombay (India), p. I - 1 (1979).

A review of concentrating collectors for the photothermal utilization of solar energy is presented. Various types of tracking and nontracking concentrators such as vee troughs, Fresnel lenses, Winston collectors, paraboloids, parabolic troughs, segmented mirror troughs, control tower heliostat arrays, and spherical mirrors are described and their advantages and disadvantages enumerated. A short description is given of the recently proposed thermophotovoltaic system using paraboloided reflectors. The problems of heat transfer at the absorber, tracking requirements and energy storage are briefly discussed. Major experimental units under construction at different places are listed. Overall efficiency and cost effectiveness of solar thermal systems are discussed and the probable methods of improving the cost effectiveness explored. The need to develop cheap concentrators is addressed.

1237

Fiber Optical Solar Collector : Swet C J, "U S Patent Application 247, 744, (1972) ".

The invention is an improved solar collector comprising a fiber optical solar receiver which passively concentrates incident solar energy for delivery as an intensified flux to an absorptive target. In one embodiment of the invention, the present solar collector comprises a boule of fibers shaped into an arcuate collecting surface at one end, the fibers tapering to a flat output plane at the opposite end of the boule. Solar radiation entering the collector at the collecting surface is concentrated within the tapered portion of the boule and delivered as an intensified flux to an absorptive target disposed in operative relation to the output plane, the absorptive target being either a cookpot, a thermal storage mass or the hot junction of a thermoelectric generator.

1238

Heliotropic Thermal Generators : Swet C J, " Proceedings 8th IECE Conference ", Philadelphia (USA), p. 348, (1973).

A design concept is presented for relatively inexpensive line focussing collection systems characterised in particular, by reflector rotation about a polar axis to maximize energy interception, and by passive sun tracking, powered and controlled by directed solar radiation on a bimetal helix. Some applications are described, problem areas are examined, and indications of technical feasibility and economic attractiveness, are presented.

1239

Turntable Solar Arrays : Swet C J, " (Unknown) ".

A method of assembling and orienting extremely compact arrays of linear concentrators is described in which the entire array is rotated as a unit in turntable fashion to follow the sun. The performance of such arrays is examined parametrically in terms of direct energy interception per unit area as a function of geometry, geographic latitude, time of day and season of year, and is also shown to exceed that of arrays of individually oriented linear collectors that rotate about the north south polar axis. An adaptation based on the use of tilted Fresnel reflectors is described, which enables the collection of diffused sunlight while concentrating the direct component. The feasibility & utlity of this adaptation are preliminarily assessed. Applications of turntable arrays to irrigation pumping, electric power generation and total energy systems are depicted.

1240

Solar Strip Concentrator : Szulmayer W, " Solar Energy ", 14, 327, (1973).

A solar concentrator, capable of focussing solar heat on a rod or a water pipe, consists of a transparent extruded plastic strip 6 in wide and 0.40 in thick. Four patterns of prismatic parallel grooves and ridges produce the required beam refraction. Temperatures between 140 deg to 290 deg F were produced in exploratory heating experiments which indicate the possibility of using the strips for water heating, steam production, regeneration of desiccants (silica gel), as well as thermoelectric power generation. Some aspects of design, production and operation of the strips are discussed.

1241

Stationary Mirror Systems for Solar Collectors : Tabor H, " Solar Energy ", 2(3 - 4), 27, (1958).

An angle in solar geometry termed as ' EWV altitude' is defined, and its variation with time and season is shown. This variation indicates the necessary acceptance angle of a stationary mirror system for solar collectors. It is shown that a completely stationary mirror cannot give any useful concentration, but if the tilt is varied with the seasons, an east west cylindrical parabolic mirror without diurnal movement can yield a concentration of approximately 3. This may be increased to about 4 with the aid of a small auxiliary (fixed) mirror to provide a second stage of optical concentration.

1242

A Description and Assessment of Large Solar Power Systems Technology : Tallerico L N, " Sandia Laboratories Report SAND - 79 - 8015 ", (USA), (1979).

This document summarises the systems being developed by the Department of Energy Large Solar Thermal Central Power System Program. Included are the technical concepts upon which the systems are based & to the extent possible, estimated cost, performance and assessment of typical systems. The intent is to provide potential users with an overview of present technologies that will be available within the next few years. Sandia Laboratories assessments of the strengths and weaknesses of each technology have been included in the hope that developers of technology will be able to improve component and system designs to the point where they become fully competitive with alternative energy sources.

1243

Cylindrical " Elastical " Concentrating Solar Collectors : Tan J F, Langridge D, Letham C D & McCormick P G, " Proceedings ISES Silver Jubilee International Congress ", Georgia (USA), p. 522, (1979).

The paper includes a theoretical analysis of the elastical shape and gives examples of computed r a y traces plotted for various rim angles. Also included is a theoretical plot of incidence factor versus concentration ratio for three different angles. Experimental performance figures for an elastical collector are presented for two different concentration ratios.

1244

Solar Thermal Electric Power Systems in Japan : Tanaka T, " Solar Energy ", 25, 97, (1980).

The major subjects of this paper are to report the outline of recent basic research and technical development for solar thermal electric power systems in Japan. Solar thermal electric power systems are presently being developed as one of the most important systems in ' Sunshine ' projects which were initiated in 1974 to develop the utilization systems of new energy resources. Conceptional designs of solar thermal power systems were already done on the basis of the results of supporting researches & two pilot plants of solar thermal electric power systems of a capacity of 1000 KWe are under construction on the basis of the conceptional and detailed designs, and would be constructed by 1981. The present condition of these pilot plants and the major researches which are thought to be most important subject in the basic research and technical developments for solar thermal power systems are described in this paper.

1245

Economic Analysis of STEPS in Japan : Tani T, Sawata S, Tanaka T, Kamimoto M, Sakuta K and Ezawa N , " Proceedings ISES Silver Jubilee International Congress ", Georgia (USA), (1979).

With the objective of evaluating the STEPS (Solar Thermal Electric Power System) from the viewpoint of technological and economical possibilities, various kinds of research and developments are being conducted within the 'Sunshine' project. In order to clarify problems which must be solved for practical application of a large scale commercial plant, two types of pilot plants, central receiver system and distributed system of STEPS with electric capacity of 1 MWe are being designed in detail. In this paper, analysis of primary system designed is made of configurations to concept feasibility of STEPS in Japan using the experimental data of model systems and relative distribution of direct normal solar flux density which has been measured by the Japan Weather Association. The simulation program was developed for the estimation of electrical energy cost of STEPS using hourly normal solar flux data, various collector area storage capacities, various turbine inlet temperatures and a cost model.

1246

Solar Heating Device : Tarcici A, " U S Patent 3, 797, 496 ; Filed March 19th, (1973) ".

A solar heating device is described comprising an upright tubular post, a rigid support protruding perpendicularly from the post and a generally paraboloidal reflector, having a rigid arm extending along the arms of the reflector from the center towards the focal point. A pivot perpendicular to a free end of the arm is selectively pivot mounted and secured in one of a plurality of notches in the support to provide a compensation for the lateral displacement of the focal point as the reflector is tilted. A height adjustable grid, mounted on a tube slidably received within the post enables the support of an object to be heated. A rod indicating the focal point of the reflector can be disposed along the axis.

1247

Metal Mirrors of Large Size : Taylor H D, " Optical Engineering ", 14, 559, (1975).

The wide state of the art in manufacturing large one piece metal mirrors has been reviewed.

1248

Application of the Similarity Theory to Investigation of Heliotechnological Systems with Reflectors : Teplyakov D I, " Applied Solar Energy ", 3(6), (1967).

The application of physical similarity for solving typical heliotechnological problems (transfer and distribution of radiant energy in the concentration field, reception of energy in cavity receivers , etc.) is well based theoretically and experimentally. Criteria of heliotechnological similarity are proposed, which consist of the concentrator and target temperatures. Experimental and theoretical examples are given to explain the nature and the efficiency of the application of the similarity theory to investigation of the heliotechnological concentrating systems.

1249

Pyrometers for High Temperature Optical Furnaces : Teplyakov D I, Aparisi R R and Kolos Y G, " Applied Solar Energy ", 8(4), 101, (1972).

A pyrometer which has been designed for measuring the temperatures of samples heated within the focal point of high temperature optical furnaces (solar furnaces in this case) of the laboratory type is described. This pyrometer measures the temperature of the irradiated side of the sample. Its design is based on the two colour principle of temperature measurement in which the sensor reacted to the so called ' red blue ratio ' within the spectrum of sample's own radiation.

1250

Comparative Performance Characteristics of Cylindrical Parabolic and Flat Plate Solar Energy Collector : Tester J W, Mayer R M and Frass A P, " ASME Paper No. 74 - WA / Ener - 3 ", (1974).

The performance of a flat plate and a cylindrical parabolic focussing solar energy collector was measured concurrently and compared for collector temperatures ranging from 150 to 185 deg F under Oak Ridge, Tennessee weather conditions with a view to their use in a residential total energy system. The flat plate collection system was a conventional design while the focussing collector was a low concentration factor fixed orientation device that employed a finned tube receiver. The importance of seasonal and diurnal variation in the sun's alignment was critic1 to performance of the focussing collector and was considered in detail. Performance models were developed to predict monthly operation under Oak Ridge weather conditions. The economies of utilizing either system for domestic space heating and air conditioning were investigated briefly with the resulting data, but further data for collector temperatures upto at least 250 deg F are needed for a definite comparison.

164

1251

Wind Loads on Para – Cylindrical Solar Concentrators : Thomas A and Soundarnayagam S," Proceedings Indian National Solar Energy Convention ", Bombay (India), p.110, (1979).

A knowledge of the magnitude and distribution of wind pressures on a concentrator surface is essential for ensuring structural integrity of solar collector units. Tests have been conducted on a scale model having a rim angle of 52°, for various angles of attack of wind. Results of the above tests are reported in this paper and their implications are discussed.

1252

Thermal Efficiency Test Procedures and Results for a Cylindrical Concentrating Solar Collector : Thomas W C, " Proceedings ERDA Conference on Concentrating Solar Collectors ", Georgia (USA), p.6 – 9, (1977).

The applicability and limitations of established test standards for evaluating cylindrical concentrating (CC) collectors are examined. A CC solar collector manufactured by Martin Processing Inc. is described. Differences between test procedures for CC and other collector designs are considered. Existing standards make no distinction between the ' hour ' angle and ' acceptance ' angle as related to the incident angle. Since test results should provide information for determining day long response, a new test procedure for the angular response tests is suggested. The efficiency of a CC collector is generally much more sensitive to the directional properties and magnitude of the scattered component of solar radiation. Results for thermal efficiency and angular response under various environmental conditions are shown and compared with theory.

1253

Comparative Ranking of 0.1 to 10 MWe Solar Thermal Power Systems : Thornton J P, Brown K C, Edgecombe A L, Finegold J G and Herlvich F A, " Proceedings 14th IECE Conference ", Boston (USA), p.190, (1979).

Recent studies have suggested a significant potential for small solar thermal electric power generating plants to be used by small communities, rural areas and remote load centers. The Solar Energy Research Institute (SERI) has been contracted by the Department of Energy to perform a comparative analysis and ranking of 8 generic solar thermal systems in the 0.1 to 10 MWe capacity range.   The paper describes the methodology adopted for this comparative analysis, including a discussion of the study ground rules, system options evaluated, performance and cost simulations and the formal ranking decision analysis utilized.

1254

Annual Performance Comparisons of Parabolic Trough and Flat Plate Collectors Based on Measured Insolation : Treadwell G W and Grandjean N, " Proceedings ERDA Conference on Concentrating Solar Collectors ", Georgia (USA), p.9 – 25, (1977).

Parabolic trough and flat plate nontracking collectors in various orientations have been analytically compared for annual performance capabilities at two locations and at various delivery temperatures. These comparisons have been made using hourly measured direct and total horizontal insolation and weather data. The comparisons indicate that the parabolic trough will outperform the flat plate collector at delivery temperatures as low as 80 deg F. This results as the thermal losses are lower and the tracking of the sun more than compensates for the trough's inability to use the diffuse components of radiation.

1255

Self Regulating Automatic Heliostat Reflecting Mirror Device : Trombe F, "U S Patent 2, 712, 772 / Jan.12,(1955).

The device is for maintaining a beam of rays, and in particular, sun rays in a fixed direction,giving a continuous and progressive adjustment without oscillation. A flat mirror (heliostat) is used to reflect the rays in a fixed direction ; the power elements which give the heliostat the desired position in relation to the sun are operated by liquid under pressure. The feed of this liquid is controlled by a valve which is itself operated in response to the currents supplied by photo electric cells under the effect of the sun's rays. The flow rate of the liquid under pressure therefore depends on the intensity of the current supplied by the light sensitive cells.

1256

Precision Machining (Diamond Turning) of Optical Reflectors : Tunnell H A, " Proceedings ERDA Conference on Concentrating Solar Collectors ", Georgia (USA), p.5 – 113, (1977).

Diamond turning is a rapidly developing technology being utilized to produce both, large and unique optical reflectors in a cost effective manner. The process is especially suited for the high reflectivity metals such as copper, gold, silver and aluminium. The capabilities of the advanced machine systems developed at the Oak Ridge Y – 12 plant include a broad range of shapes and sizes, flat or curved reflectors upto 24in in diameter, have been machined with better than 40 microinch figure accuracy and 2 microinch surface finish, while reflectors upto 78 in in diameter have been machined 75 microinches figure accuracy and 3 microinch surface finish. The efficiency and capability of this process to produce large numbers of complex reflectors or precision masters for replication may make it useful in manufacture of reflectors for future solar energy concentrator systems.

1257

Epoxy Matrix Originals Produced by Centrifugal Casting : Tveryanovich E V, "Applied Solar Energy",8(1), 19,(1972).

A study has been undertaken to investigate deformations in the matrix originals produced by centrifugal casting method. These deformations occur during manufcaturing and also while galvano copies are being made.

1258

Calibration of Photometric Curves Obtained During Tests on Solar Concentrators : Tveryanovich E V, Madaev V V, Shermazanyan Ya. T, Vartanyan A V, " Applied Solar Energy ", 11 (3 – 4), 10, (1975).

A method is described for calibrating fiber-optic microphotometers using photometric and calorimetric curves and also estimates of total radiation incident on the concentrator with known parameters. Data are reported on the distribution of 'h' for a concentrator 1.4 m in diameter, obtained by both methods. Good agreement between the results is reported.

1259

On Solar Energy Concentration : Umarov G Ya., "Applied Solar Energy", 3(5), (1967).

A review on uptodate problems of solar energy concentration is given. Advantages and defects of some large concentrators are analyzed. Works are also considered which deal with specific irradiation distribution in the mirror plane and in the focal volume. Treated separately, are works on concentrating areas manufactured from films and different slow solidifying resins. Classification is made for installation using concentrated solar energy.

1260

Determination of Radiation Field of an Ellipsoidal Radiation Furnace with a Xenon Lamp Source : Umarov G Ya., Zakhidov R A and Sokolova Yu. B, " Applied Solar Energy ", 10(3), 27, (1974).

Formulas are derived for the radiation flux incident on the receiver in the case of an ellipsoidal radiation furnace. The point radiation source is placed in the primary focus and the receiver in the secondary focus. The 'Uran' furnace is considered as an example.

1261

Multifacet Solar Field Kitchen : Umarov G Ya., Alimov A K, Alavutdinov Dzh. N and Khodzhaev A Sh., " Applied Solar Energy ", 12(3), 81, (1976).

Results are reported for the development, manufacture and testing of a multifacet solar field kitchen. Utilization of an equitorial rotation system simplifies servicing of the unit. Structure is dismountable to permit transportation. The technique developed for assembling the concentrator eliminated the need to adjust the facets. The multifacet solar kitchen supplies the needs of 12 to 15 people for hot water and may be recommended for use by teams, field camps etc.

1262

Method of Increasing Cotton Yield by Using Concentrated Pulsed Light : Umarov G Ya., Ibragimov Sh. I, Alimukhamedov A A, Alimov A K and Alavutdinov D N, " Applied Solar Energy ", 15(3), 63, (1979).

As a light source a DKST - 20000 - 2 Xenon lamp was used, while the concentrator was a special two mirror system consisting of ellipse - cylindrical and cylindrical reflectors. Field germination was from 12.1 to 22.6% higher than for controls, experimental variants matured 4 to 7 days earlier than controls while the yield of raw cotton was greater by 6.8 to 7.8 centners per hectare. The stimulating effect of CPL on cotton yield is a stable one.

1263

A Study of Constant Power Regulation of Solar Furnace Heat Flux : Vartanyan A V and Shermazanyan Ya. T, " Applied Solar Energy ", 1(4), (1965).

Shading geometries for collector surfaces of solar furnaces to regulate their radiation flux & achieve a stable heat field are studied. Various structured shading patterns are evaluated for their heat regulating power by utilizing the basic laws of regulation to derive a family of hyperbolas for the boundary conditions of regulating systems. Relationships are derived for lower type of regulators, cylindrical regular screening off exterior portions of the reflected flux and a cylindrical regulator screening off interior portions of reflected flux.

1264

Comparative Characteristics of Solar Furnace Power Regulators : Vartanyan A V, " Applied Solar Energy ", 2(6), 8, (1966).

Comparative characteristics of power regulators of solar furnaces and the techniques of qualitative analysis of separate systems are investigated. The dependence data allow the choice of the regulator type according to concrete investigations on solar furnaces.

1265

Tracking Angle and Rates for Single Degree of Freedom Solar Collectors : Veneruso A F, " Sandia Laboratories Report no. SAND - 76 - 0027 ", (USA), (1976).

This report describes the motion required by tracking system for single degree of freedom solar collectors such as those used in the Sandia Laboratories' Total Energy System. Specific consideration is given to east west and north south oriented collectors consisting of parabolic trough reflectors. The graphs and equations of motion given in this report provide inputs for identifying tracker system requirements independent of the specific type of tracker hardware used.

1266

Use of Spacecraft Parabolic Antenna Reflector for Obtaining Electrical Power : Venkataraman S T, Kosta S P, Lakshmeesha V K and Mathur R S, " IEEE Trans. Aerospace Electronic Systems ", AES - 14(1), 209, (1978).

It is proposed that by covering the back surfaces (partially or fully) or the earth pointing parabolic antenna of a geosynchronous satellite with flexible solar cell panels, an important premium of useful raw electrical power can be achieved. Merits and demerits of the proposals are discussed with various diagrams and figures.

1267

Apparatus for Focussing and Using Solar Energy : Vigourex F E, "US Patent 4084, 581 / 18th April (1978) ".

For focussing and using solar energy, an apparatus comprises outer reflectors and inner back reflectors distributed symmetrically with respect to an axis, which is maintained towards the sun during day light hours. A heat collector is positioned along this axis beyond the reflectors and back reflectors. In section, by any plane passing through the axis, the reflectors form a continuous broken line and back reflectors, a discontinuous broken line. The heat received in the collector can be used for heating water.

1268

Survey of Generic Types of Focussing Collectors for Large Scale Applications : Wachtler W J, Price J H, Henry R L and Hall B L, " Proceedings ERDA Conference on Concentrating Solar Collectors ", Georgia , (USA), p. 8 - 35, (1977).

This paper describes the pertinent features and characteristics of focussing collectors and, from a generic standpoint, compares and evaluates their large scale applicability. Five generic type of focussing collectors were surveyed; the parabolic trough ; parabolic dish ; linear segmented array ;Fresnel lens and fixed mirror. Generic types are discussed in terms of their optical and structural characteristics. Components considered include the concentrator , receiver drive mechanism, pylons and tracking device. Design considerations are given to the auxiliary components used, such as flexible hoses, couplings, insulation and construction materials. The collectable energy is compared for both single axis and two axis tracking modes. Further, the collector field layout and orientation are discussed regarding annual performance and heat losses. Collector selection criteria are identified as well as methods for making predictions of both performance and cost. A summary of the significant findings is presented, which consists of the availability of each type of collector in terms of size, cost, production and development.

1269

Method of Making Solar Reflectors : White G, " U S Patent 4, 127, 926 / 5th December (1978) ".

A method is discussed of making a solar reflector from a flat rectangular sheet of reflecting material into a substantially parabolic cross sectional shape to form a part of the solar heat collecting system. The reflector is supported to concentrate the rays of the sun on a pipe through which liquid is circulated. The sheet material is bent permanently along its logitudinal center line to give the two half portions thereof a mirror angle with respect to a horizontal plane, bending the longitudinal edges back upon themselves to form slotted hook portions along the longitudinal edges of the sheet material and bending the two half portions towards each other below the breaking point of their elasticity to receive the longitudinal edges of a transluscent plate in the slots of the hoop portions where by the parts are held in assembly.

1270

Effects of Angular Misorientation on the Performance of Conical, Spherical and Parabolic Solar Concentrators : Wijeysundera N E, " Solar Energy ", 19, 583, (1977).

The present note describes a ray tracing technique, which is applied to study the behaviour of conical, spherical and parabolic reflectors when angular deviation between the direction of the solar beam and the axis of the reflector target system on the concentrator shape, its size and other target dimensions, is studied.

1271

Geometric Factors for Plane Specular Reflectors : Wijeysundera N E, " Solar Energy ", 20, 81, (1978).

A general formula for calculating the geometric factor between a specular reflector and a collector surface, for a given orientation and a given position of the sun is obtained. The formalism is based on the matrix transformations and is therefore especially suited for use in digital computer programs.

1272

Efficiency and Stability of Experimental Fluorescent Planar Concentrators (FPC) : Wittwer V, Heidler K, Zastrov A and Goetzberger A, " Proceedings 14th IEEE Photovoltaic Spec. Conference ", San Diego (USA), p.760, (1980).

After a summary of the theoretical and practical limits of the fluorescent planar concentrators (FPC) experimental results of stability and efficiency measurements are given.

1273

Ground Based Solar Energy Technology Advances : Woodcock G R and Gregory D L, " Proceedings 10th IECE Conference ", New York (USA), p.607, (1975).

Competitive busbar output costs for intermediate of base load solar power plants can be achieved only when the costs of the concentrating heliostats is below \$4.00 per square foot of reflecting area. In addition the heliostats must resist various weather factors and have a long, relatively maintenance free, life. A concept is derived for a ' weather decoupled ' heliostat, fabricated primarily from tensioned plastic films. The plastic film enclosure (dome) over the reflector reduces the heliostat efficiency ; however, this is easily compensated by increasing the total heliostat area.

1274

Design, Performance and Architectural Integration of Solar Heating System Using Reflective Pyramid Optical Condenser : Wormser E M, " Proceedings ISES Conference ", Los Angeles (USA), (1975).

An optical system has been devised which condenses both direct and diffused solar radiation from a relatively large entrance aperture area onto flat plate collectors, one third to one fifth the area.The optical system is of simple design, made up entirely of plane surfaces and employs reflective surfaces having low tolerance requirements and made of inexpensive materials. It consists of a truncated pyramidal reflector with a moving reflecting flap. A standard 2 ft X 8 ft flat plate collector is mounted near the apex of the pyramid. In the prototype test unit, the reflective pyramid concentrates the solar energy incident on an entrance aperture of 55 ft^2 by a factor of 3.5X onto the flat plate having an area of 16 ft^2. The reflective flap moves to close the cavity when there is no sun and opens when there is sun, adjusting its angle to compensate for seasonal and geographical variation of sun elevation.

1275

Pyramidal Optics Solar Concentrating System : Wormser E M, " Proceedings ERDA Conference on Concentrating Solar Collectors ", Georgia (USA), p.3 - 133, (1977).

A low gain concentrating system, the pyramidal optics solar concentrating system is described. The system has been incorporated in a model home of 1400 sqft heated area near Rehoboth, Delaware. Performance of the system through the winter of

1976 - 1979 is described briefly.

1276

Design and Testing of a Solar Receiver for a Stirling Engine : Wu Y C, Moynihan P I, Day F D & Selcuk M K, " Proceedings ERDA Conference on Concentrating Solar Collectors ", Georgia (USA), p. 5-63 , (1977).

A highly efficient solar receiver is a key component of a solar electric power generation system that uses a suntracking paraboloidal collector, a solar receiver, and a stirling engine linear alternator combination. As one part of the efforts in the development of such a system, a solar stirling receiver experiment program was undertaken and the results are reported in this paper. The paper presents the design, the testing and the test results of an experimental receiver, suggestion of advanced receiver design concepts and considerations of the integration of the receiver with the stirling engine.

1277

Solar Radiation Calculation for Semi Cylindrical Receivers : Yakubov Yu. N, Baibutaev K B & Khozhiev A Kh., " Applied Solar Energy ", 10 (6), 91, (1974).

Analytic relations are derived for determining the radiation incident on and passing through the transparent surface of semi cylindrical receivers.

1278

Solar Energy Collector : Yoke J H, " U S Patent 4, 129, 119 / December 12th, (1978)".

A solar energy collector is disclosed herein which comprises a pipe interconnecting lower and upper headers, the upper header structurally supporting the pipe. The pipe extends within an opening in the lower header and includes an exterior shoulder which rests upon the lower header. The other end of the pipe includes an exterior shoulder, the pipe extending into an opening in the upper header. A spring rests on the shoulder on the upper end of the pipe and holds a combination and fluid seal against the upper header. A parabolic reflector is supported by the pipe, and a fluid is movable through the pipe from one header to the other for receiving the solar energy focussed upon the pipe by the reflector. The pipe is observable to rotate so as to direct the reflector towards the sun.

1279

Health & Safety Hazards with Solar Concentration Systems : Young L L, " Solar Energy ", 22, 329, (1979).

This paper discusses the primary health and safety hazards associated with solar concentration systems . The limiting hazard is corneoretinal damage. The article also discusses the methodologies derived to compute the unique safety and health hazards associated with solar energy collector and storage systems. It is concluded that research in this area is continuing, especially for eye hazards, with more extensive work planned.

1280

Design, Construction & Operation of a Low Pressure Solar Steam Facility : Youngblood S B, " Proceedings ISES (American Section) Conference ", Delaware (USA), p. 686, (1980).

Accurex Corporation has designed, constructed and is now operating a solar facility that provides 345 deg F process steam to the Johnson & Johnson manufacturing plant in Sherman, Texas. The facility consists of 11,250 sft of parabolic trough concentrating collectors, 95,000 gals flash boiler, and 20-up circulating pump. Experience gained through start up of the facility is summarized, and projected economics are discussed.

1281

Conceptual Design of Solar Thermal Power Plants : Zahn R, " Proceedings ISES (U K Section) Conference (C16)", London (U K), p. 24, (1978).

Thermal electric conversion methods are discussed with respect to a broad power range. The following topics are covered. Components for solar thermal power plants, performance of systems & cost of electricity. It was concluded that photovoltaic generators using present technology are superior in the low power level. Concepts of power stations using distributed collector systems are recommended for the medium power range (about 50-500 Kw). The MW range is governed by the central receiver concept under all conditions of the study.

1282

Analysis & Classification of Concentrating System Design Methods : Zakhidov R A, "Applied Solar Energy", 13(4), 1, (1977).

The historical development of computational methods for the radiant field & structural parameters of single mirror, multimirror and facet type concentrating systems is considered. The fundamental concepts of a general method developed for the design of multielement systems are presented and the most important computational results are shown.

1283

General Principles of Multielement Concentrating System Design : Zakhidov R A, "Applied Solar Energy", 14(1), 17, (1978).

The basic principles of the general theory of multielement system design are presented and the elements of the analytical model and the computer algorithm structure are analyzed. The fundamental suitability of the method for the design of systems of any complexity and mission [multimirror, multifacet, and multiheliostat solar electric power station (SEPS) systems and so on] is demonstrated.

1284

Evaluating the Performance of Reflecting Concentrating Systems : Zakhidov R A, " Applied Solar Energy", 15(5), 31, (1979).

Several devices and methods for evaluating the performance of mirror elements of concentrators are described, they have been developed and employed at the Central Planning, Design and Technological Bureau for Scientific Instrument Construction, Academy of Sciences of the Uzbek SSR.

1285

Evaluation of Deformation Techniques for Forming Solar Concentrator Reflecting Surface : Zakhidov RA, Dudko Yu. A, Petrosov OP, Zmener G S & Dubrovskii L A, "Applied Solar Energy", 13(5), 33, (1977).

The results of an evaluation of the concentrating capability of facets by the deformation technique are presented. The basic characteristics of the use of this facet forming technique are noted. A comparison is made of the concentrating capability of deformed planar & paraboloidal facets which are identical in plane form.

1286

Performance of Low Cost Solar Reflectors for Transferring Sunlight to a Distant Collector : Zentner R C, " Solar Energy ", 19, 15, (1977).

An experimental method has been utilized to determine beam spreading due to microscopic surface irregularities on a heliostat used in a central collector solar electric plant. The test method provides a nearly independent measure of the effect of surface imperfections. Data are presented for 5 candidate materials and, as reference, an optical quality first surface mirror.

Addendum

Cylindrical Parabolic Trough Concentrator

1287

The Parabolic Trough Plants Using Black Body Receivers ; Experimental and Theoretical Analysis: Barra O A and Franceschi L, " Solar Energy " 28,163 (1982).

A 50 m² experimental prototype of a solar parabolic trough plant has been realized and is at present functioning to produce industrial process heat in an Italian brewery. The prototype, which is of the long focus with a black body receiver, is used to carry out both static and dynamic experimental analysis of the cavity receiver itself and to determine a series of experimental parameters to be used in setting up mathematical models. These models simulate the plant performance under any working conditions with particular attention to the fluid dynamic, hydraulic and heat transfer aspects. In the present paper the experimental results are presented and the theoretical results forecast by the computer model are outlined.

1288

LINSOL : A Model for Predicting the Optical Performance of Parabolic Trough Solar Thermal Systems : Dellin T A, " Proceedings 16th IECE Conference ", Atlanta, Georgia (USA), 2, 1746, (1981).

A detailed model has been developed to predict the optical performance of parabolic trough solar energy systems. The model is one - two orders of magnitude faster than previous, less complete calculations and makes tractable investigations of a wide range of design & application alternatives for trough systems.

1289

Parabolic Trough Concentrators : Giuffrida M, Tornielli G P, Pidatella S, Rapetto A, Bellafrontes E and Zani P E, "Proceedings 3rd Photovoltaic Solar Energy Conference ", Cannes (France), (1980).

A photovoltaic unit utilizing parabolic trough concentrator has been realized and tested by Ansaldo. Rated power capacity of the single unit is about 4.5 Kw at 1 Kw/sqm (AM1). The reflector is back silvered glass, 0.92 reflectivity, continuous surface hot sagged, or divided into elementary stripes. The supporting frame is made of a sandwich of steel and PU foam. Solar cells, manufactured under an original design in Ansaldo are 33 X 33 mm single crystal silicon cells. The unit global efficiency in real operating conditions is about 9% configuration is alt-azimuthal and the movement is controlled by sun trackers.

1290

Calculation of Parabolocylindrical Concentrator Concentration Factor : Kolomoets N V, Markman M A and Shmatok Yu.I, "Applied Solar Energy", 16 (5), 22 (1980).

The concentration factors and optimal aperture angles of parabolocylindrical concentrators for heat receivers of six different shapes under the condition of complete capture by the heat receiver of the reflected solar radiation, and also the concentration factors and efficiency of the concentrator as a function of the relative size of its entrance aperture with incomplete capture of reflected solar radiation are determined.

1291

Theoretical and Experimental Study on Heat Transfer and Thermal Performance of Concentration Solar Collector of Horizontal Coaxial Cylinders : Kunitomo T, Aizawa K, Marumoto K and Tamehiro H,"Bulletin of the JSME", 25 (199), 68 (1982).

Heat transfer and thermal performance of the concentrating solar collector of horizontal coaxial cylinders for a solar thermal power source are studied theoretically and experimentally. The design parameters considered in the theory are the concentration ratio, the diameters of a semi-transparant glass tube and a selective absorber tube, the effective length and the optical property of an absorber tube, the storage temperature, the pressure in the annulus and the inlet temperature and flow rates of a heat transfer medium. The thermal performance of the absorber system is evaluated by the outlet temperature of the fluid, the collector efficiency, the energy efficiency and the final conversion efficiency. The experimental results confirm the calculation method of heat transfer and thermal performance.

1292

Experimental Investigation of Parabolic-cylinder Solar Concentrator with Tubular Heat Receiver : Markman M A, Kolomoets N V, Simanovskii L M, Shmatok Yu.I and Zakharov O P,"Applied Solar Energy",16(6),61(1980).

The focussing characteristics and efficiency are investigated for a parabolic cylinder concentrator with tubular heat receiver fabricated by elastic deformation of a sheet of electropolished anodized aluminium.

1293

Geometrical Optical Performance Studies of a Composite Parabolic Trough With a Fin Receiver : Mathur S S, Kandpal T C, Singh R N and Singhal A K, "Applied Energy", 9, 223 (1981).

The geometrical optical performance characteristics of a composite parabolic trough (CPT) with a fin receiver has been studied. The variation of geometrical concentration ratio with mirror element size and the rim angle of the parent parabola has been studied and the results are presented graphically. The distribution of local concentration ratios over the receiver plane has also been determined for some typical concentrator designs.

1294

Mean Wind Loading on Parabolic Trough Solar Collectors : Randall D E, McBride D D and Tate R E, "Journal of Solar Energy Engineering ", 103, 313 (1981).

Two wind tunnel force and moment tests have been conducted on parabolic trough solar collector configurations. The two tests were conducted in different flow field environments, one a uniform flow infinite air stream, the second a simulated atmospheric boundary layer flow with the models simulating a ground mounted installation. The force and moment characteristics of both isolated single module troughs and of trough modules within array configurations have been defined over both operational and stow attitudes. The influence of various geometric design parameters for collector modules and arrays has been established. Data indicate that forces and in general the pitching moment increase with mounting height and with trough aspect ratio. Collector modules interior to large arrays experience wind force reductions as high as 50-65%, while appropriate fences or berms surrounding the arrays can provide exterior modules with protection of this order.

1295

Potential Economic Benefit Using Parabolic Trough Collectors to Supplement Power Cycle Boilers: Schimmel Jr.W P & Lukens L L , "ASME Paper No: 81-WA/Sol-17",(1981).

An analysis has been made of the potential for using a parabolic solar collector system for preheat and boiling in a power generation cycle. The advantage of such a system is that solar energy is used to heat the water in a steam Rankine cycle device upto the superheat regime, thus displacing the fossil fuel usually required.

1296

Generalized Optical Performance Characteristics of a Parabolic Trough Using a Fin Receiver: Sharma M S, Kandpal T C, Mathur S S,Singhal A K and Singh R N, "Proceedings National Solar Energy Convention ",Bangalore (India),(1981).

This paper presents the generalized optical performance characteristics of a focussing parabolic trough using a fin receiver. Results of some numerical calculations are represented graphically and discussed.

1297

Fabrication Techniques for Sandwich Construction of Paracylindrical Concentrators: Thomas A, Rao M R and Ramani R V, " Proceedings National Solar Energy Convention", Bangalore (India), (1981).

A paracylindrical concentrator shell has been designed and fabricated using scrap aluminium honeycomb core, thin stainless steel face skin,(the free surface of which acting as the reflecting surface) and aluminium backup skin. A precision scanning jig has been fabricated to scan the paracylindrical shell in order to assess the accuracy of its geometry.

1298

Aerodynamic Behaviour of a Linear Focussing Parabolic Solar Concentrator: Thomas A, Rao M R, Rao K P and Reddy K R, "Journal Institution of Engineers India ", Part ND 61, 51 (1981).

Design of linear focussing parabolic concentrators has to adhere to stringent specifications on their basic structures in terms of stability, stiffness and overall length. The design demands very small slope errors of the parabolic surface under various loading conditions and also that the strength of the structure must be sufficient to withstand gusty winds without catastrophic failure. The present study is on the behaviour of a collector under different wind loadings. This paper describes some experiments conducted on a linear focussing semi-monocoque parabolic collector of 2.5 X 1 m size in 4.3 m X 2.7m (14 ft X 9 ft) open circuit wind tunnel at the Department of Aeronautical Engineering, Indian Institute of Science (Bangalore,India). Three configurations are tested,(i) j(AR=2.5) with reflecting side of the concentrator facing the wind stream, (ii) (AR=2.5) with reflecting side of the concentrator facing the wind stream,(iii) (AR=2.5) with stiffened side of the concentrator facing the wind stream and,(iv) (AR=0.4) with reflecting side of the concentrator facing the wind stream. Emphasis is placed mainly on the study of aerodynamic moments that are developed due to incidence as well as wind speed.

1299

Systematic Rotation and Receiver Location Error Effects on Parabolic Trough Annual Performance: Treadwell G W and Grandjean N R, " ASME Paper No: 81-WA/Sol-12 ", (1981).

This paper deals with the effects of certain systematic errors on performance, and therefore, their influence on the design of troughs. Systematic rotation error is the angle between the reflector vertex-focus axis and the vertex-sun axis; systematic receiver location error is the vectorial deviation of a receiver from focus. The technique for calculating the influence of systematic errors on performance is outlined and the methods of identifying and minimizing these errors are suggested.

1300

An Analysis of the Influence of Geography and Weather on Parabolic Trough Solar Collector Design: Treadwell G W, Grandjean N R and Biggs F , "Journal of Solar Energy Engineering (Trans.ASME)",103,98 (1981).

The potential performance of single axis tracking parabolic trough solar collectors is a function of optical energy distribution and receiver size has been calculated for 11 sites using typical meteorological year input data. A simulation based on SOLTES code was developed which includes the three dimensional features of a parabolic trough and calculates the thermo optical tradeoffs. The capability of the thermo optical model has been confirmed by the comparison of calculated results with the experimental results from an all day test of a parabolic trough. The results from these 11 site analyses indicate a potential performance superiority of a north-south horizontal axis trough and in addition, a high quality (optical error, σ system ≤ 0.007 radian) collector should be of the same geometric design for all of the sites investigated and probably for all regions of the country.

Fresnel Lens and Reflector Systems

1301

Efficiency of Fresnel Lenses : Gupta P K, " Applied Energy ", 9,173 (1981).

This paper discusses the efficiency of Fresnel Lenses with respect to optical (reflection & transmission) losses. The efficiencies of lenses of different step width i.e.2,3,4 and 5 mm are the same (91.9%) in the case of reflection losses and the maximum efficiency of a lens of 5 mm step width is 95.8% with respect to transmission losses.

1302

Theoretical Analysis of the Fresnel Lens as a Function of Design Parameters: Gupta PK, "Applied Energy", 9,301 (1981)

A study is made of the theoretical performance of the Fresnel lens, a cheap and lightweight system for solar concentrators, which provides the design parameters i.e. the radius to the center of the steps, the focal distance from the back of the lens to the plane of the image with the object(i.e. the sun)of infinity, the thickness of the lens plate, the step width and the refractive index of the material (with respect to air) used in its fabrication. Numerical calculations have also been carried out for a Fresnel lens of Perspex sheet.

1303

Direct Conversion of Solar Radiation to Mechanical Power Using Bimetal and Linear Fresnel Lenses:Hellickson M L and Pritchard V R, " Trans. ASAE " p.1026 (1981).

Concentration of direct solar radiation on bimetallic helical coils caused a simple sunpowered motor to produce rotational shaft power on an intermittent basis. Solar radiation was concentrated on to the bimetallic coils with linear Fresnel lenses. Data accumulated during actual and simulated operations allowed performance of a detailed heat balance which indicated that continuous rotational motion of the motor might be attained in one of two ways: reduction of bimetal mass; and/or use of a bimetal with greater deflection at lower operating temperatures.

1304

Alignment and Performance of a Single Axis Solar Tracking System : Jain P K, Gupta R and Sathe A P,"Proceedings National Solar Energy Convention", Bangalore (India), (1981).

A Fresnel lens which consists of a special sawtooth structure on one face, when moulded from transparent plastics using appropriate dies does not get good surface finish resulting in poor optical quality. In this paper the authors have presented a novel profile of a Fresnel lens which consists of a single continuous on one side and simple rectangular steps on the other. The dies with this profile would be easier to generate, maintain and polish which would result in the improved surface finish and better optical quality of the moulded lenses.

1305

A Fixed Fresnel Lens With Tracking Collector : Kritchman E M, Friesem A A & Yekutieli G,"Solar Energy",27,13 (1981).

The application of a wide angle concentrating Fresnel lens to a linear solar energy system in which the optical concentration is stationary while the absorber follows the locus of best foci, is investigated. The two substantial direction possibilities of the linear axis, east - west and polar are compared to each other. It is shown that such a concentrator may operate about six hours a day throughout the year with an average effective concentration exceeding 10. Specifically a polar installation, including a fixed lens and a fixed assembly of separate absorbers behind it, may enable sufficient concentration for residential heating and air conditioning without any moving parts.

1306

Convex Fresnel Lens with Large Grooves : Kritchman E M, Friesem A A & Yekutieli G, "Solar Energy", 27,129 (1981).

The concentration capability of linear Fresnel lenses is limited mainly by coma and chromatic aberrations and by the lenses. The authors have considered new designs for these lenses in which the concentration is almost independent of the groove's size, and which are free of coma at the edge of the acceptance field specifically, a concentration factor as high as 34 is achieved for monochromatic radiation coming from an acceptance angle of $\pm 1°$.

1307

Chromatic Aberration Effect on Solar Energy Systems Using Fresnel Lenses : Lorenzo E, " Applied Optics", 20, 3729 (1981)

In concentration systems using Fresnel lenses the effect of the chromatic aberration can become important. A method is proposed to take this effect into account for designing purposes. Also defined is a parameter that allows one to estimate the degradation of the thermodynamic quality of the concentrator due to this effect. This parameter follows a hyperbolic law, with the acceptance angle showing that it is important to consider chromatic aberration when modeling concentrators with a high concentration factor. However, this complexity is unnecessary for moderate or low concentration factors.

1308

Fresnel Lens Analysis for Solar Energy Applications : Lorenzo E & Luque A, "Applied Optics", 20, 2941 (1981).

An analysis is presented of an arbitrarily shaped linear Fresnel lens acting either as a sole concentration stage or as the first stage of a two stage concentration system in which the second stage concentration considers the first as a Lambertian source. The gain and the position of the lens is determined for all possible configurations, and it is demonstrated that a curved lens with a refractive index approaching infinity and with a given profile turns out to be an ideal concentrator.

1309

Analytical / Experimental Studies of the Optical Performance of a Transmittance Optimized Linear Fresnel Lens Solar Concentrator: O'Neill M J and Walker R A, "Proceedings ISES (American Section) Conference Phoenix, Arizona (USA), 3.1, 510 (1980).

A transmittance optimized linear Fresnel lens solar concentrator has been developed. The optical performance of the lens has been analytically predicted, using the method of cone optics, to define the radiant flux profile in the focal plane. This paper presents a brief description of the lens, its predicted performance and its measured performance.

1310

A Modified Linear Fresnel Lens : Singhal A K, Singh R N, Kandpal T C & Mathur S S, "International Journal of Energy Research", 6, 191 (1981).

A modified linear Fresnel lens design has been proposed which produces a more uniform illumination over the receiver than the conventional linear Fresnel lenses.

Mirror Boosters / V—Trough / Polygonal Troughs

1311

Bi-yearly Adjusted V-Trough Concentrators : Chiam H F, "Solar Energy", 28, 407 (1982).

Previous analyses of V-trough concentrators have resulted in a range of trough geometries and different concentrator adjustment routines being recommended. While these configurations produce moderate yearly averaged concentration of solar energy, the concentrator performance during the year may vary greatly. This paper presents a simple approach for optimizing the trough configuration, in particular to produce uniform monthly flux enhancement with minimum adjustment of the concentrator. The method which is based on reflector characteristics identified from an analysis of single reflector - receiver system, successfully realizes configurations that provide satisfactory performance while requiring either collector tilt or the concentrator to be adjusted only at the equinox. The comparative performance of symmetrical and asymmetrical trough configuration is given and, in addition, the inter-relationship between the trough opening angle, the reflector/cover width ratio and the frequency of adjustment.

1312

Performance Evaluation of Three Step Compound Wedge Stationary Concentrator : Mannan K D and Cheema L S, "Proceedings National Solar Energy Convention", Bangalore (India), (1981).

The optical characteristics of cylindrical compound wedge collectors have been analysed. The design of a three step collector having an aperture area of $4.3\,m^2$ and using glass mirrors as reflectors has been described. The thermal performance for collecting solar energy at 100 to 175 deg C has been experimentally evaluated.

1313

Convective Heat Transfer in V - Trough Linear Concentrators : Meyer B A, Mitchell J W and El-Wakil M M, " Solar Energy ", 28, 33 (1982).

Natural convection heat transfer coefficients have been experimentally determined for trough type collectors. The effects of Rayleigh number, tilt angle and ideal concentration ratio on the Nusselt number have been experimentally determined over ranges representative of collector operation. The Rayleigh number range tested was upto 10^7, the tilt angle was varied from 30-90°and ideal concentration ratios of 2,3,4 and 5 were tested. The experimental results are supported by finite element solutions to the governing equations. A method of extending the results to truncated trough collectors and C P C collectors is suggested. The convective losses are compared to those for conventional flat plate collectors . The critical Rayleigh number increases with concentration ratio. Nusselt number increases more rapidly at high concentration ratio than at low values. Correlation equations are developed over the range of parameters tested.

1314

The Effect of Length on Absorption for a Trapezoidal Groove Collector : Puri V M, "Solar Energy", 27, 463 (1981).

The effect of collector length on the concentration ratio and collection efficiency of beam radiation insolation components have been studied for an optimal V-groove collector. In this paper, collector base to reflector depth (b/d) is considered a parameter. A complete set of nomographs are presented for an unshaded collector showing collector length effect (l/b) on collection efficiency as a function of relative sun-collector position for several b/d ratios. Collection efficiency for each of the b/d ratios for an infinite V-groove collector have also been indicated on these nomographs. For a partially shaded collector, the extent of shading of reflector(s) is an additional variable, therefore nomographs are presented only for a representative case. The collection efficiency values for a finite V-groove collector on a typical winter day are compared with an infinite collector.

1315

Development and Study of a Flat Mirror Multivalent Concentrator : Schwartz L M, Louat R and Menguy G, "Proceedings 2nd International Alternative Energy Sources Conference ", Miami (USA), (1979).

The concentrator described in this paper is made up of square flat mirrors each measuring 25 cm X 25 cm. The total reflecting area is $4\,m^2$. The advantage of plane mirrors is their low cost. A 'blind' tracking type is chosen. The captor is oriented by data directly calculated from the equations of the Earth's movement around the sun. A dc motor is used, mounted on a big reductor with a pulley system whose speed output is about 1/30 rpm. This sort of concentrator could be useful in a lot of studies involving the use of direct solar radiation, particularly in not very well insolated areas.

Compound Parabolic Concentrator

1316

Design and Testing of a Uniformly Illuminating Nontracking Concentrator : Gupta A, Kumar S, Murlidhar and Tewary V K, " Solar Energy ", 27, 387 (1981).

The transverse parabolic shape of the reflecting surface in the Winston non tracking solar concentrator has been modified so that the illumination at the receiver is exactly uniform for certain angles of incidence and reasonably uniform for other angles. An exact mathematical expression has been derived for the transverse shape in the limit of extreme non-uniformity. The performance of the proposed design has been analysed theoretically as well as tested experimentally. Experimental studies show that the illumination at the receiver remains uniform to within 10 - 12% with no lateral shadows throughout a typical solar day. The proposed design should therefore, be quite suitable for photovoltaic applications.

1317

Investigation on the Prediction of Thermal Performance of Compound Parabolic Concentrators :Hariprasad C R, Natarajan R and Gupta M C, "Proceedings 2nd International Alternative Energy Sources Conference ", Miami (USA), (1979).

This paper focusses attention on the prediction of thermal performance of a compound parabolic concentrator for different values of insolation and mass flow rate of collector medium (water) under steady state conditions. The analysis involves an iterative scheme and a method is proposed by which the absorber temperature,outlet temperature and glazing temperature can be predicted for given insolation and mass flow rate.

1318

Thermal Analysis of CPC Collectors : Hsieh C K, "Solar Energy ", 27, 19, (1981).

Mathematical formulations were developed to study thermal processes in a compound parabolic concentrator (CPC) collector. The system under investigation consists of a CPC cusp fitted with a concentric, evacuated double pipe to serve as a heat absorber. Heat is transmitted to the circulating fluid flowing inside a 'U' tube via the heat getter slipped inside the inner pipe. The collector has a cover for dust protection. Four nonlinear, simultaneous equations were derived to predict heat exchange among various components in the system. Collector efficiency equations were also developed following the Hottel - Whillier Bliss formulisms. These equations were subsequently used in a computer programme to test the collector performance under varied operating conditions. Test results indicate that, because of the high thermal resistance between the receiver jacket and the envelope, the collector performance is quite stable and is nearly independent of many parameters tested. The efficiency of the collectors is high and shows only a very slight drop at high operating temperatures. . The predicted results were also compared with experiments.

1319

Design and Analysis of a System Using CPC Collectors to Collect Heat and to Produce Industrial Process Steam : Hsieh C K, Reed K A and Schertz W W, "Proceedings ISES (American Section) Conference ", Phoenix (USA), 3.1, 654 (1980).

A system has been designed to use CPC (Winston) collectors to collect solar energy and to provide industrial process steam. The system is composed of two loops of collectors. In the collector loop the energy collected is used to operate a preheater. Hot pipe receivers are used in collectors in the boiler loop to minimize scaling problems. An attempt is made to use computer modelling to study the collector performance.

1320

Experimental Test of a Faceted Non-imaging Solar Concentrating Collector : Lopez A M, Ortabasi U, Atienza J A and Ozakcay L M, "Proceedings ISES (American Section) Conference ", Phoenix (USA), 3.1, 610 (1980).

Preliminary tests of an experimental collector designed to produce process heat upto 300 deg C are reported. The collector is a modular non-imaging trough concentrator (5.25 X) whose mirror surface consists of planer facets sealed inside low cost glass tubes. A transient behaviour test was used to measure the collector overall heat loss coefficient throughout a wide range of temperatures. Preliminary measurements of the collector optical efficiency were carried out and are compared to theoretical predictions.

1321

Convective Heat Transfer in Trough and CPC Collectors : Meyer B A, Mitchell J W and El Wakil M M," Proceedings ISES (American Section) Conference ", Phoenix (USA), 3.1, 437 (1980).

Natural convection heat transfer coefficients have been experimentally determined for trough type collectors. The effects of Rayleigh number, tilt angle and ideal concentration ratio on the Nusselt number have been experimentally determined over ranges representative of collector operation. The convective losses are compared to those from conventional flat plate collectors.

1322

Seasonally Adjusted Concentrator with Modifications of Absorber Shape : Mullick S C and Nanda S K," Applied Energy ", 9, 257 (1981).

The heat losses from concentrating solar collectors can be reduced by thermally isolating the absorber with an optical air gap between the absorber and the glass cover. However, in reflectors attempting to attain maximal optical concentration, some of the reflected rays will escape through this gap, resulting in increased losses. This paper describes a design procedure for an absorber which avoids the optical losses through the air gap. The results for the modified absorber are compared with those for a plane tubular absorber developed previously. The intercept factor is improved from 0.8 to 0.92. The surface area of the modified absorber is about 9% larger than that of a plane tubular one, but the shape is such that the overall heat loss factor is much lower. As a result the total heat loss for the modified absorber at a temperature 100 deg C above ambient is only 0.2% higher than that of the plane tubular one. The improved performance curves are shown.

1323

Least-cost Cusp Concentrator Design : Nanda S K and Mullick S C," Proceedings National Solar Energy Convention ", Bangalore (India), (1981).

Cusp concentrators require larger concentration areas, but can be designed for larger acceptance angles allowing larger mirror tolerances. Design procedures are outlined to compute the optimum combination of acceptance angle and maximum mirror slope for any required concentration ratio, taking into account the material as well as the fabrication costs. The cusps are compared with the parabola.

1324

Optical Losses and Heat Transfer in an Evacuated Tubular Absorber under 5 X Nonimaging Concentration : O'Gallagher J J, Hilgart M, Snail K & Winston R," Proceedings ISES (American Section) Conference ", Phoenix Arizona (USA), 3.1, 605 (1980).

In this paper the authors summarize the results of extensive operating experience with a 5.25 X parabolic concentrator. This collector used General Electric Evacuated tubes and was built with Sheldahl silver foil mirrors. A detailed analysis of the individual contributions to the optical losses shows that the observed efficiency is in agreement with material properties and ray trace analysis.

1325

<u>A Stationary Evacuated Tube with Integrated Concentrator for 100 deg C to 300 deg C</u> : O'Gallagher J J, Snail K, Winston R, Peek C and Garrison J D, " Proceedings ISES (American Section) Conference ", Philadelphia (USA) (1981).

A substantial improvement in optical efficiency over contemporary external reflector evacuated tube collectors can be achieved by integrating the reflector surface into the outer glass envelope. The design, fabrication and preliminary test results for a prototype collector based on this concept are described in this paper. Efficiencies about 40% upto nearly 300 deg C may be achieved.

1326

<u>Optical Collection Efficiencies of Arrays of Tubular Collectors with Diffuse Reflectors</u> : Window B and Zybert J, " Solar Energy ", 26, 325 (1981).

The fractions of solar radiation collected by arrays of cylindrical absorbers with diffuse reflectors of various shapes have been found using a Monte - Carlo ray tracing technique. The fractions of solar radiation collected for arrays of modules with outer glass envelopes and selective absorbers, situated above a plane diffuse reflector, have been calculated for variable module spacings in the range 1.5 - 3 absorber tube diameters. The calculations include the effects of polarisation and angle of incidence on the reflectance of the outer envelope and the selective absorber, the effect of absorption in the glass envelope and by the reflector. Multiple reflections or second chance effects play an important role in increasing the fraction absorbed over what is predicted with simple theories. The curves derived for the model with realistic properties of the components are useful for designing collectors. Tube spacings less than two diameters produce only small increases in collector performance. Collectors made with the module aligned in E-W rather than N - S will be less efficient, but the difference will be small ($\cong$ 5%).

1327

<u>Optical Collection Efficiencies of Tubular Solar Collector with Specular Reflectors</u> : Window B & Bassett I M , " Solar Energy ", 26, 341 (1981).

The optical collection efficiencies of arrays of evacuated tubular collectors with specular involute reflectors have been studied using computer ray tracing techniques incorporating all the major physical features. The optical collection efficiencies under various representative assumptions concerning the angular distribution of incident radiation are presented as functions of the tube to tube spacing. The hemispherical collection efficiency obtained with collector aperture equal to absorber surface is in accordance with the expected loss resulting from the truncation of the reflector, and the losses associated with the mirror, with the envelope and with the absorber. At smaller apertures, the collection efficiency for hemispherically incident radiation agrees with a recent theory of O'Gallagher et al.

1328

<u>Solar Receiver Enclosure Enhancement by Controlled Directional Scattering</u> : Winston R, " Proceedings ISES (American Section) Conference ", Phoenix,Arizona (USA), 3.1, 596 (1980).

Efficient solar collector design was applied to an investigation of two-dimensional (trough like) cavity structures. This paper describes a novel principle for designing cavity enclosures. The method maintains high transmission at the expense of some concentration in the presence of gaps as large as the radius between reflector and receiver.

1329

<u>Non-Imaging Concentrators for Photovoltaic Arrays in Space</u> : Winston R, Greenman P and Rogkeij D, " Proceedings 16th IECE Conference ", Atlanta, Georgia (USA), 1, 390 (1981).

A study of two stage concentrators was made for the purpose of designing an optimum concentrator for photovoltaic arrays in space. The study was directed at designs with two dimensional geometry as they are more easily deployed in space than three dimensional (point focussing or cone light) designs, because they are better suited to the moderate concentration.

1330

<u>Non-Imaging Concentrators for Solar Thermal Energy</u> : Winston R & O'Gallagher J J," ASSET ", 3(4),17 (1981).

Compound Parabolic Concentrators (CPC's) permit the use of low to moderate levels of concentration for solar thermal collectors without the requirement of diurnal tracking. When used in conjunction with evacuated absorbers with selective surfaces a fully stationary CPC (designs with concentrations 1.1 X 1-4 X are now commercially available) has a typical efficiency of about 40% at 150 deg C above ambient with available conventional materials. An experimental higher concentration 5 X CPC requiring approximately 12 tilt adjustments annually, when used with a similar available evacuated absorber, has been built having a measured efficiency of 60% at 200 deg C above ambient and is capable of efficiencies near 50% at 300 deg C.
With such thermally efficient absorbers, higher concentrations are neither necessary nor desirable.The temperature capabilities for CPC's with non-evacuated absorbers are somewhat more limited.However, with proper design, taking care to reduce parasitic thermal losses through the reflectors,non-evacuated CPC's will outperform the best available flat plate collectors at temperatures above about 50 - 70 deg C.

Paraboloid Of Revolution / Solar Furnaces

1331

<u>Les Fours Solaires de Recherche du Laboratoire d'Energetique Solaire d' Odeillo /(Research Solar Furnaces of the Solar Energy Laboratory of Odeillo)</u> : Arnaud G, Flamant G, Olalde G & Robert J F, " Entropie ", 17 (97), 139 (1981).

The solar concentrators and the program on solar energy research at the ' Laboratoire d' Energetique Solaire ' at Odeillo are described. Concerning the applications of concentrated solar energy the following subjects are described ; performance of solar concentrators, results of computation and experimental ones, high temperature solar receiver concepts, etc.

1332

Optical Analysis of Point Focus Parabolic Radiation Concentrators : Bendt P & Rabl A, "Applied Optics ", 20, 674 (1981).

A simple formulism is developed for analysing the optical performance of point focus parabolic radiation concentrators. To account for off-axis aberrations of the parabola, an angular acceptance function is defined. The radiation intercepted by the receiver of a real concentrator is obtained as a convolution of angular acceptance function, of optical error distribution and of angular brightness distribution of the radiation source. For numerical calculations this method is more accurate and less time consuming than the ray tracing method.

1333

Parabolic Reflectors Formed by Inflation : Bracewell R N & Price K M, "Solar Energy ", 27, 535 (1981).

Paraboloids of revolution over 1 m in diameter have formed from flat sheets by inflation without the use of a mold or template. Quality is adequate for microwave use and for most high concentration applications for focussing solar energy. The process is suitable for making reflectors in ones or twos to special focal lengths and diameters, but could also be automated for production runs.

1334

Progress in Point Focussing Solar Concentrator Development at JPL: Carley W J, "Proceedings ISES (American Section) Conference ", Phonix, Arizona (USA), 3.1, 548 (1980).

Point focussing receiver technology is being developed by the Jet Propulsion Laboratory as part of the Solar Thermal Power System Project for the US Department of Energy. Two first generation concentrator designs, the test bed concentrators (TBC) and the low cost concentrator (LCC) are described in this

1335

Small Community Solar Thermal Power Experiment : Marriott A T and Kiceniuk T, "Proceedings ISES (American Section) Conference ", Phoenix, Arizona (USA), 3.1, 524 (1980).

The first solar thermal power plant specifically designed as an alternative source of electric energy for small communities in this country (USA) is described in its second phase of development. As a result of Phase I concept definition studies, a decision was made to pursue a parabolic dish system using distributed generation.

1336

Shadowing Effects in Fields of Paraboloidal Collectors : Nehru K V K, "Proceedings National Solar Energy Convention ", Bangalore (India).

It is recognized that development of solar collectors in fields of given area unavoidably involves shadow casting. The pattern of shadowing of a collector by the surrounding ones in a field of paraboloidal dish collectors is analysed. It is shown that the design of the optimum spacing of the collectors is to be based not only on the maximum total energy yield but also on the maximum average 'capture factor'. Rectangular and staggered arrays of collectors are studied, the latter show marginal advantage over the former. Field dimensions are found to have only a weak influence on the optimum spacings. Optimum spacings for various latitudes are recommended.

1337

Dispersed Solar Thermal Generation Employing Parabolic Dish Electric Transport with Field Modulated Generator Systems : Ramakumar R and Bahrami K, "Solar Energy ", 27, 7, (1981).

This paper discusses the application of field modulated generator systems (FMGS) to dispersed solar thermal electric generation from a parabolic dish field with electric transport. Each solar generation unit is rated at 15 KWe and the power generated by an array of such units is electrically collected for insertion into an existing utility grid. Such an approach appears to be most suitable when the h e at engine rotational speeds are high (> 6000 r/min) and, in particular, if they are operated in the variable speed mode and if utility grade a.c. is required for direct insertion into the grid without an intermediate electric energy storage and reconversion system. Predictions of overall efficiencies based on conservative efficiency figure for the FMGS are in the range of 25 percent and should be encouraging to those involved in the development of cost effective dispersed solar thermal power systems.

1338

Defocussing Caused by Obliquely Incident Rays on Parabolic Reflectors : Ren-Xue Xu, "Solar Energy ", 28, 235 (1982).

In this paper a method of trajectory of intersecting points of neighbouring reflected rays from a reflecting surface is presented. Based on this method, the defocussing effects caused by obliquely incident rays on a parabolic reflector are studied. The analytical form of the defocussing curve is obtained and its relation to the incident angle is shown.

1339

Les Applications du Four Solaire de 1000 Kw du CNRS a Odeillo (Applications of the CNRS 1000 Kw Solar Furnace of Odeillo) : Royere C, "Entropie ", 17 (97), 147 (1981).

The French activity on large solar thermal test facilities is checked against the main milestones in this field in the world, this provides arguments for the prominent role played by France. The characteristics of point focussed solar energy are summarized either as a source of process heat or as a source of heat at medium and high temperature with some specific feature, then a short review is presented on the applications of the 50 KWth Montluis solar furnace from 1952 to 1967. A few general aspects of the Odeillo 1000 Kwth CNRS solar furnace are looked at shortly to bring a better understanding of the operation and use of this facility. The various fields of applications and utilizations since 1969, of the 1000 KWth solar furnace are described and explained by precise and actual examples; refractory oxides processing, ores processing metallurgy, thermal shocks, physical measurements and testing of s o l a r power plant components.

1340

Comparison of Electrochemical and Thermal Storage for Hybrid Parabolic Dish Solar Power Plants : Steele H L and Wen L, "ASME Paper No. 81 – WA/Sol – 27 ", (1981).

The cost of storage systems which can compete with the use of fuel in hybrid parabolic dish solar power plants is identified for one set of specific assumptions. The hybrid plants burn fuel to increase the hours of usage each day. The cost and performance characteristics of concentrators, receivers and power conversion units are based on estimates by the contractors developing this hardware under the direction of the Department of Energy and the Jet Propulsion Laboratory. Thermal storage systems are not yet designed and only the cost goal which would make them competitive is known.

1341

Heat Transfer Analysis on Metal Melting in a Foundry Solar Furnace : Suresh D and Rohatgi P K," Solar Energy", 26, 87 (1981).

In this paper the heat transfer analysis with a model calculation for zinc melting in a foundry solar furnace is presented. The use of a programable calculator in these calculations is elucidated. The theoretical melting curve is compared with the experimental curve.

1342

Use of Solar Furnaces - I ; Materials Research : Suresh D, Rohatgi P K and Coutures J P, " Solar Energy", 26, 377 (1981).

In this paper a detailed review of the use of solar furnaces in materials research is presented.

1343

Working Experience with a Foundry Solar Furnace : Suresh D Kishore & Rohatgi P K, "Solar Energy", 27, 457 (1981).

In the present work, the performance of a solar furnace capable of melting small quantities of foundry grade metals and alloys has been studied under various conditions. Crucibles of different materials & shapes were tried and the effects of having different heat shield materials was also studied.Al-bronze crucible with cavity and well polished stainless steel heat shield were found to be most effective in enhancing the efficiency of the furnace. Many important industrial applications of the present solar furnace such as the recovery of metallic zinc from slags, had also been realized.

1344

Use of Solar Furnaces - II ; Thermophysical Properties : Suresh D, Charters W W S and Rohatgi P K, " Solar Energy", 28, 273 (1982).

In this paper the authors have given a review of work done on the measurement of high temperature thermophysical properties of materials using solar furnaces. This includes the measurements of thermal expansion, thermal conductivity and diffusivity, thermal energy absorption, heat content, a b l a t i o n characteristics, mechanical properties, optical characteristics and electrical properties of materials.

1345

Parabolic Dish Collectors; a Solar Option : Truscello V C, " Sunworld ", 5 (3), 80 (1981).

A parabolic dish solar collector as a device designed for capturing solar radiation and converting it to thermal and electric power is described. Advanced technologies in parabolic dish systems are discussed.

1345A

Designing a Solar Energy Concentrator on the Basis of Paraboloid of Revolution : Umarov G Ya., Sharafi A Sh.and Abduazizov A, "Applied Solar Energy", 4 (6), 105, (1968).

The dependence of focal point diameter of a concentrator made of wedge mirror strips, strengthened on a reinforced concrete substrate of rotating paraboloid shape (a) from the strips width; and (b) from temperature changes of the reflector at rigid strengthened strips are determined.

Central Tower Receiver Systems

1346

Simplified Method for Calculating Radiation Density on the Surface of Tower Type Solar Power Station Receivers : Aparisi R R, Teplyakov D I & Khantsis B G, "Applied Solar Energy", 16 (5), 27 (1980).

The calculation of the radiation densities on the surfaces of flat and cylindrical receivers of tower type solar power stations (SPS) are examined. The data used for the calculation are compared with previously obtained experimental results, which are subjected to additional analysis.

1347

Central Tower Solar Test Facility (RM/CTSTF): Bivilacqua C and Gislon R, " Proceedings 16th IECE Conference ", Atlanta, Georgia (USA), 2, 1742 (1981).

An experimental program is discussed which is concerned with the following topics : facility characterization, advanced heliostatics evaluation, materials component tests, high concentration photovoltaic experiments and advanced solar thermal cycles.

1348

A Consideration of Possible Receiver Designs for Solar Tower Plants : Boese F K, Merkel A, Stahl D and Stehle H, "Solar Energy", 26, 1 (1981).

A 500 Kw solar tower plant in Almeria, (Spain), will be provided with a sodium primary circuit. In the present paper detailed information is given on the basic investigations and on the design of the solar receiver of this circuit. Considerations concerning the optimum geometry as well as thermodynamic behaviour are described. Different design possibilities and the respective behaviour of the r e c e i v e r with respect to short term transients due to cloud passage, heliostat field or mass flow failure,as well as with respect to optimum operational conditions are discussed. Special stress is placed upon a comparison between the receivers with or without a large thermal mass.

1349

Measurement Challanges in Solar Central Receiver Systems : Brumleve T D and Gibson J C, " Proceedings 7th Energy Technology Conference ", Washington D.C.(USA), p.1550 (1980).

In the development of solar central receiver systems, a number of unusual diagnostic needs is discussed.

The most challenging of these are in the measurement of certain heliostat and receiver characteristics. For heliostats, techniques are being developed for measuring (1) the specular and spectral reflectivity of candidate mirror surfaces, (2) the contour accuracy of mirror modules, (3) the alignment or canting accuracy of mirror modules on a heliostat, (4) the tracking accuracy of heliostats and (5) the angular deflection or backlash of heliostat drives and structure under gravity or wind loadings.

1350

An Analysis of Convective Losses from Cavity Solar Central Receivers : Clausing A M, " Solar Energy ", 27, 295 (1981).

An analytical model is presented which enables the estimation of convective losses from cavity recei - vers. Evidence from solar experiments is used to test the hypothesized mechanisms. The analytical results and experimental evidence indicate that the convection loss from cavity receivers is appreci- able. The model indicates that the influences of wind on the convective loss at normal operating con- dition are minimal. It also shows that the internal thermal resistances, i.e. the ability to heat the air inside the cavity, are of greatest importance. Buoyancy induced flows are, on the other hand, very effective in transforming energy across the aperture.

1351

Solar Central Receiver in Perspective : Dellin T A, " Proceedings ISES (American Section) Conference ", Phoenix, Arizona (USA), 3.1, 573 (1980).

The variation in the energy costs of solar central receiver systems as a function of power level, recei- ver geometry, heliostat size and heliostat canting option are presented. The results were obtained us- ing a new version of the DELSOL computer model.

1352

Preliminary Evaluation of the Volumetric Air Heating Receiver : Drost M K and Eyler L L, " ASME Paper No. 81 - WA/Sol - 26 ", (1981).

This paper reports the results of a preliminary evaluation of the Volumetric Air Heating R e c e i v e r (VAHR). The VAHR is an innovative air heating receiver designed for application with point focus cent- ral receiver power generation system or industrial process heat systems. The VAHR concept consists of an array of fin shaped pins arranged in concentric cylindrical rows around an inlet manifold.

1353

Performance Analysis of a Windowed High Temperature Gas Receiver Using a Suspension of Ultrafine Car- bon Particles as the Absorber : Fisk W J, Wroblewski Jr. D E and Hunt A J, " Proceedings I S E S (American Section) Conference ", Phoenix, Arizona (USA), 3.1, 553 (1980).

The paper summarizes the results of an analytical study of the efficiency of single and double window- ed high temperature solar receiver suitable for use with small central tower or point focus dish col- lectors. A detailed window energy balance is used to predict the window temperature.

1354

La Centrale Solaire Themis 2500 Kw, Le Champ d' Heliostats Le Recepteur Solaire - THEMIS/(THEMIS - The 2500 KW Solar Power Plant ; The Heliostats ; Its Solar Collector) : Hillairet J, Collon A, Lecrec A, Gravrand J M and Pouget - Abadie X, " Entropie ", 16 (961), 78 (1981).

Electricite de France (French Electricity Board) is building the 2500 KW Themis solar power station at Targassonne in the Eastern Pyrenees. A 201 heliostats field collects the solar radiation to direct it into a collector located on a 100 meter high tower. A molten salt circuit carries the thermal energy to a stockage at 450 deg C corresponding to a sunlit day. A turbo-alternator is producing 2500 KW tor a 11.70 KW collected power. This expensive kind of power station should however concern the export trade when the implantation and operation conditions are favourable to it. The optical focusing of solar ra- diation and the concept of the solar collector are described.

1355

Heliostat Characterization at the Central Receiver Test Facility : King D L and Arvizu D E, " Journal of Solar Energy Engineering (Transactions ASME)", 103, 82 (1981).

This paper summarizes the work that led to the current state of heliostat evaluation capability at the Central Receiver Test Facility (CRTF) operated for the Department of Energy by Sandia Laboratories in Albuquerque N M (USA). It includes, a description of the CRTF heliostat, measurements of environmental beam measurements with an instrument sweeping bar, beam quality and tracking accuracy data obtained with the newly developed Beam Characterization System (BCS) and comparisons of measured beam data with the heliostat computer model HELIOS.

1356

Short Term Effect of Cloud Transits on the Barstow Central Receiver Pilot Plant : Laurence C L, " Pro- ceedings ISES (American Section) Conference ", Phoenix, Arizona (USA), 3.1, 586 (1980).

A cloud transit study has been performed to determine the impact of clouds on receiver designs. Because of lack of significant statistical data on cloud events, the cloud formations are chosen from actual data being taken at the Barstow site. Selection is made with the purpose of representing both typical and extreme situations. The results from the simulation for five cloud events are presented. The power levels and rates of change of power levels are determined.

1357

High Flux Solar Experimental Facilities : Marshall B W, Mulholland G B, Brown C T and Hays R, "High Tem- perature Science ", 13(1 - 4), 5 (1978).

The solar facilities at White Sands are operated for the United States Army and consist of two comple- mentary facilities, an operational 30 KW solar furnace (WSSF) and a 250 KWth High Intensity Solar Ther- mal Facility (HISTF) whose installation is scheduled for late 1981. The WSSF is a focussing type solar facility consisting of a single heliostat, an alternator, a concentrator and a control chamber. A des- cription is given of the Central Receiver Test Facility.

1358

Test Experience at the DOE/Sandia Mid Temperature Solar Systems Test Facility : Mc Culloch W H, " Proceed- ings ISES (American Section) Conference ", Phoenix, Arizona (USA), 3.1 488 (1980).

Sandia Laboratories Albuquerque N.M, (USA), has been involved in the study, design, construction, testing and evaluation of solar energy systems. This report presents some of the insights and experience acquired during this period.

1359

Solar Thermal Test Facility for 60 – 300 deg C Located in Tropics : Ozakcay L M, Lopez A M and Ortabasi U M, " Proceedings ISES (American Section) Conference ", Phoenix, Arizona (USA), 3.1, 485 (1980).

A test facility was designed to measure the parameters determining and affecting the thermal efficiency of concentrating collectors operating in the 60 to 300 degC range under tropical conditions. Data acquisition is done by a microprocessor based data logger & the data analysis is done by a desk top computer.

1360

High Flux Sodium Cooled Solar Absorber Concept for Central Receiver Power Plant : Pomeroy B D, Roberts J M and Narayanan T V, " Proceedings ISES (American Section) Conference ", Phoenix, Arizona (USA), 3.1, 558 (1980).

An absorber concept is described which uses liquid sodium coolant and a three header configuration to capture solar power efficiently. The mechanical design of this absorber is discussed and thermal performance estimates are presented showing the solar capture efficiency over a range of solar intensities. A thermal stress analysis is presented which shows that a limiting factor on the flux capability may be tube-wall creep/fatigue failure and not the heat transfer capability of sodium.

1361

Solar One : A 10 Megawatt Solar Thermal Central Receiver Pilot Plant : Reeves J N and Schweinberg R N , " Proceedings ISES (American section) Conference ", Phoenix, Arizona (USA), 3.1, 578 (1980).

A 10 Megawatt solar thermal power plant is described using central receiver technology. The pilot plant will be tested for five years & information derived will be used for future commercial solar generating stations.

1362

Design and Optimization of 100 MW Closed Cycle Gas Turbine Solar Power Plant for Arid Land Area : Sabbagh J A, " IEEE Conference Publication No. 192 ", p. 173 (1981).

The possibility of supplying electricity for 24 hours from solar energy is studied. The extra cost is in the thermal storage, the central receiver tower, cavity and the heliostat field. For the rare cloudy days the plant has to be shut off. In isolated areas an open cycle gas turbine may be used to give required power.

1363

Central Receiver Solar Energy System for an Oil Refinery : Sommerlad R E, Raghavan R and Pichnarcik R A, " Proceedings ISES (American Section) Conference ", Phoenix, Arizona (USA), 3.1, 568 (1980).

The paper discusses the conceptual design of a central receiver solar energy system that will provide practical and effective use of solar energy for an oil refinery currently being designed by F o s t e r Wheeler Energy Corporation for the Provident Energy Company. The design is for a baseline system consisting of a tower mounted natural circulation water/steam receiver with an 18.4 Kg/s (146,000 lb/h) capacity.

1364

Definition of the Heliostat Array for the Barstow 10 MWe Pilot Plant : Vant- Hull L L, " Proceedings ISES (American Section) Conference ", Phoenix, Arizona (USA), 3.1, 581 (1980).

The Barstow Pilot Plant has been designed to simulate a 100 MWe commercial power plant. A radial -stagger array of 1818 heliostats has been defined for the Pilot Plant, subject to the site specific c o n- straint, which retains all the characteristics of a commercial scale surround field.

1365

A Simplified Method for Calculation of Concentration Characteristics of a Solar Tower System : Wei L Y, " Solar Energy ", 26, 559 (1981).

A simplified method for calculation of the concentration characteristics of a solar tower system i s developed for a heliostat system of quantized mirrors and arbitrary shape. It is shown that the proposed method yields results almost indistinguishable from those calculated by the exact methods.

Two – Stage Solar Concentrators

1366

Design of Non Imaging Concentrators as Second Stages in Tandem with Image Forming First Stage Concentrators : Greenman P, O'Gallagher J J, Winston R and Welford W T, " Proceedings ISES (American Section) Conference ", Phoenix, Arizona (USA), 3.1, 544 (1980).

The flux concentration of paraboloidal mirrors of short focal ratio may be enhanced to near the thermodynamic limit by the addition of low concentration (2 - 4 X) nonimaging concentrators such as compound elliptical concentrators or hyperbolic trumpets. An outline of the design procedure and the results of ray trace analysis are presented.

1367

Secondary Concentrators for Parabolic Dish Solar Thermal Power Systems : Jaffe L D & Poon P T, " Proceedings 16th IECE Conference ", Atlanta, Georgia (USA), 2, 1752 (1981).

One approach to production of electricity or high temperature process heat from solar energy is to use point focussing, two axis tracking concentrators in a distributed receiver solar thermal system. This paper discusses some of the possibilities and problems in using compound concentrators in parabolic dish systems.

1368

Optimized Second Stage Concentrator : Kritchman E, " Applied Optics ", 20, 2929 (1981).

The incorporation of a second stage trumpet like reflective element at the focal plane of a paraboloidal dish reflector is analyzed to determine the improvement in solar concentration capability. The optimally adapted trumpet compensates for the optical aberrations of the dish and for large f/nos. The consequent concentration capability is comparable to the ideal.

1369

Second Stage CEC Concentrator : Kritchman E M, " Applied Optics ", 21, 751 (1982).

The incorporation of a second stage CEC reflective element at the focal plane of a paraboloidal d i s h reflector is analyzed, and the improvement in solar concentration capability is computed. By compensating for insufficiently large relative aperture of the primary, the second stage was found to increase the concentration to a close proximity of the ideal limit.

1370

High Temperature Solar Concentrator Design : Wientzen R, Davis W J & Forkey R E, " Proceedings SPIE Conference ", 237, 281(1980).

An optical sub-system design for a solar concentrator having applications to high temperatures thermal photovoltaic concepts has been developed. The baseline system consists of a spherically segmented primary mirror optically coupled to a compound parabolic concentrator and a hemispherical receiver. S e c ondary concentrators having conical, parabolic and hyperbolic cross sections are considered parametrically. Performance tradeoffs relative to concentration ratio, system configuration and manufacturing sensitivities have been evaluated using geometric ray tracing techniques. The advantages of using a two stage concentrator over a single collecting aperture are discussed. Performance of typical systems are reviewed covering transmission, flux distribution and tolerance sensitivities.

Photovoltaic Concentrator Systems

1371

Advanced Concepts for Photovoltaic Concentration : Bloss W H, Griesinger M and Reinhardt E R, " Proceedings 3rd Photovoltaic Solar Energy Conference ", Cannes, (France), (1980).

A new type of a concentrating system is presented which allows high concentration and simultaneous splitting of the spectral regions. This dispersed concentrating (DISCO) system is based on phase holograms which exhibit only minimum absorption. In dischromated gelatin films on glass substrates, local hardening effects by spatial variation of high light intensity (laser) can be achieved which leads to optical density profiles and results in a selective Bragg reflection at direct illumination. The characteristics of this type of Transmission Volume (TV) hologram (diffraction efficiency, spectral sensitivity) are described and some applications as dispersive concentrators are discussed. By the use of . t h e s e plane elements, concentrators with different focal planes or with special image patterns which are selective to different spectral regions have been realized.

1372

Performances of Photovoltaic Concentration Module Consisting of Silicon Solar Cells with a Facet Cylindrical Mirror ' COSS ' : Bonal R & Ragot Ph., "Proceedings 3rd Photovoltaic Solar Energy Conference ",Cannes (France),(1980).

A concentration device using silicon solar cells which work under a sunlight concentration upto 50 suns is described. The solar cells are located on the optical focus line of a facet cylindrical concentrator mirror ' COSS ' of the same type as the device used for the 100 KWe French solar power plant in Corsica ('COSS' type). This concentrator consisting of reflecting plane facets lies out along a cylindric surface generating line. The pencils of light rays reflected by every elementary mirror converge to a plane focal surface centered around a generating line of the cylinder. The mirror device is fixed and east west oriented. In photovoltaic conversion, this conception with concentration gives some advantages relative to a classical optical design.

1373

Operational, Reliability and Maintenance Experience with Photovoltaic Concentrator Arrays : Burgess E L and Shafer B D," Proceedings 3rd Photovoltaic Solar Energy Conference ", Cannes (France), (1980).

The operational, reliability and maintenance experience with photovoltaic concentrator arrays in t h e United States is discussed. Steady progress in performance has resulted over the past 5 years. Concentrator arrays with efficiencies as high as 14 - 15% have been tested. No unsolvable r e l i a b i l i t y or maintenance problems are expected. The gap between concentrator and flat plate technology is rapidly closing .

1374

Gallium Arsenide Photovoltaic Dense Array for Concentrator Applications : Cape J A, Harris Jr. J J a n d Wiczer J , " Proceedings ISES (American Section) Conference ", Phoenix, Arizona (USA), 32, 1072 (1980).

This paper reports on the design, fabrication and testing of a " dense array " consisting of 16 overlapped linear modules each containing 16 contiguous 1 cm X 1.25 cm GaAs concentrator cells in a row. The results of tests carried out at Sandia's Central Receiver Test Facility at concentrations upto 1500 suns a n d temperatures to 185 deg C are presented.

1375

Le Programmes SOPHOCLE /(The SOPHOCLE Program) : Clavierie M, Dupas A and Esteve D, " Proceedings 3rd Photovoltaic Solar Energy Conference ", Cannes (France), (1980).

The SOPHOCLE program has for its purpose the development and the experimentation of a family of photovoltaic generators using concentration of solar rays. The SOPHOCLE generators make use of an altazimuth mounting with two axis tracking. The conception of the system is modular, with silicon solar cells, & concentrating Fresnel lens integrated 6 X 6 in sealed modules providing also for passive cooling.   The concentration ratio is about 45. The peak power output is in the range 100 W_e - 5000 W_e.

1376

Intermediate Concentration Systems for Terrestrial Photovoltaic Panels : Cluff C B and Call R L, " Proceedings ISES (American Section) Conference ", Phoenix, Arizona (USA), 3.2, 1086 (1980).

Experiments using a patented process of intermediate concentration (2 to 10 X sun) on off-the-shelf terrestrial photovoltaic panels are described. This intermediate concentration is obtained with segmented parabola collectors.

1377

Photovoltaic Performance with Concentration : Theory and Experiment : Cobble M H & Wabrek R M, " Proceedings 7th Annual UMR DNR Conference on Energy ", Rolla, Missouri (USA), p.129 (1980).

A model of a photovoltaic cell having a resistance in series with the load, for use with solar concentration, has been analyzed and tested over a range of concentration.

1378

Autonomous Photovoltaic Converter with Linear Focussing Concentrator : David J P, Floret F, Guerin J, Paiva J C and Aiche L, " Solar Cells ", 4(1), 61 (1981).

An autonomous photovoltaic convertor with a linear focussing concentrator allowing an effective concentration gain of 20 was designed, realized and tested. The photo cells are special silicon cells; their cooling system is passive. The concentrator is a Fresnel type segmented mirror. Part of the electrical power supplied by the device is used in its own suntracking system, the other part charges at 12V, 30 Amp lead battery and the energy is stored during the day is used at night to light a room by neon tube. The realization and performance of this prototype converter are described. Although cost estimation does not seem meaningful, further developments of such modules are discussed.

1379

Passive Cooling Measurements for a Concentrating Photovoltaic Array : Deaver F K and Edenburn M W, " ASME Paper No. 81 - WA / Sol - 10 ", (1981).

Solar cells in a concentrating photovoltaic array absorb heat that must be removed. One method is to use a finned aluminium heat exchanger whose modelling requires estimating the coefficient of heat transfer from fins to surroundings ; transfer depends on such parameters as wind speed and direction and exchanger design. A transient analytical model has been constructed and applied to experimental data.

1380

Multibandgap Cells in a Thermophotovoltaic System : Demichelis F and Minetti - Mezzetti E, " Proceedings 3rd Photovoltaic Solar Energy Conference ", Cannes (France), (1980).

In this paper the design of a thermophotovoltaic (TPV) converter which operates with conventional multilayer nonabsorbing filters as selective mirrors and conventional multibandgap cells is reported. The TPV consists of two facing radiators placed near the focus of two parabolic mirrors. Sunlight is focussed on the two radiators and heats them. The spectral region of the radiation emitted by the radiators, corresponding to the maximum spectral response of the cells reflected at selective mirrors, is focussed into multibandgap cells. The IR radiation is exchanged between the two radiators to keep them at steady temperature. The calculations of the time dependent energy balance of the converter predict efficiencies of approximately 45 % .

1381

Aplanatic Double Reflection System for Thermophotovoltaic Applications ; Design : Demichelis F, Ferrari G and Minetti - Mezzetti E , " Applied Optics ", 20 , 4190 (1981).

The design of a solar concentrator is presented in this paper. It consists of a spherical mirror and a field of Fresnel mirror facets deployed on a spherical surface so that the sine condition is satisfied , eliminating both spherical aberration and coma. This particularly easy to construct optical system yields high concentration ratios and has the distinct advantage of having a narrow beam aperture near the receiver. These design features make the concentrator especially suitable for thermophotovoltaic applications.

1382

Application Potentials for Flat Plate & Concentrator Photovoltaic Collectors : Ferber R R, Costigue E N and Thornhill J W, " Proceedings ISES (American Section) Conference ", Phoenix, Arizona (USA), 3.2, 1136 (1980).

The two major PV system concepts described are (a) flat plate solar collectors, which have a power capacity of 80 - 120 watts/sqm and (b) concentrator enhanced solar collectors, with possible capacities of 150 KW/sqm. This paper examines the performance characteristics of these two photovoltaic technologies discusses the merits and limitations of each, and attempts to define areas where they will be competitive for ultimate commercial use.

1383

Study and Tests of a Hybrid Photovoltaic/Thermal Concentrating Collector : Gibart C, " Proceedings 3rd Photovoltaic Solar Energy Conference ", Cannes, (France), (1980).

A hybrid photovoltaic/thermal collector prototype using concentrated sunlight has been studied, realized and tested. The prototype design is described and comments are given on the chosen technological solutions. Experimental peak electrical and thermal efficiencies are presented as a function of temperature and concentration ratio, and are compared to the efficiencies of specific collectors. Technical feasibility is demonstrated but economical viability cannot be presently established.

1384

Study of and Tests on a Hybrid Photovoltaic Thermal Collector Using Concentrated Sunlight : Gibart C , " Solar Cells ", 4 (1), 71 (1981).

A hybrid prototype photovoltaic thermal collector using concentrated sunlight was studied theoretically, realized and tested. Reasons are given for choosing this concept. This prototype design is then described and explanations are given for the different technological solutions adopted. The test set-up, instrumentation and measurements are described further. The experimental peak electrical and thermal efficiencies are then presented as functions of temperature and concentration ratio and are compared with the efficiencies of specific collectors. Technical feasibility is demonstrated but economic viability cannot be established at present. Trends for future developments are then indicated.

1385

Manufacturing Cost Projections for Photovoltaic Concentrators : Hodge R C, " Proceedings ISES (American Section) Conference ", Phoenix, Arizona (USA), 3.2, 1141 (1980).

Photovoltaic (PV) concentrator array designs have been evaluated in terms of their performance, low production and life cycle cost potential. This paper describes the designs to include their projected direct manufacturing and total installation costs.

180

1386

<u>PV Receiver Assembly for PRDA Concentrator Systems</u> : Kauffman W R and Higbie J H, "Proceedings I S E S (American Section) Conference", Phoenix, Arizona (USA), 3.2 1148 (1980).

This paper describes the design of a photovoltaic concentrator applications experiment. This system denoted as CAPVC (Commercial Application of a Photovoltaic Concentrator) is a nominal 50 KW$_p$ system designed for the roof of a building in Albuquerque, New Mexico.

1387

<u>System Simulation of a Fresnel/Photovoltaic Concentrator Application Experiment for the Dallas – Fort Worth Airport</u> : McDanal A J, "Proceedings ISES (American Section) Conference", Phoenix, Arizona (USA), 3.2, 1081 (1980).

A summary of the system simulation model for the DFW Fresnel/Photovoltaic power system is presented. Major components are described and annual system electrical and thermal performance are predicted.

1388

<u>Photovoltaic Concentrator Power Station for a Village in Saudi Arabia</u> : Nazer M O, "Proceedings Third Photovoltaic Solar Energy Conference ", Cannes (France), (1980).

The Saudi Arabian National Center for Science and Technology has initiated a project under t h e S a u d i Arabia-United States Program for Cooperation in the Field of Solar Energy (SOLERAS) to build and test an experimental photovoltaic power station capable of delivering 350 KW peak d.c. power in an agricultural village in Saudi Arabia. The proposed system is based on Martin Marietta Fresnel lens concentrator silicon cell arrays, coupled with 100 KW/h lead acid cells storage and the necessary power conditioning and controls.

1389

<u>Automated Concentrator Cell Module Assembly</u> : Olah S and Sampson W, "Proceedings 3rd Photovoltaic Solar Energy Conference ", Cannes (France), (1980).

The objectives of the present work were to reduce the cost of photovoltaic concetrator cell modules by reducing the cost of cell interconnection, encapsulation and module assembly, and to design and fabricate equipment suitable for semiautomation. An automated soldering machine has been designed, fabricated and proved out. Test modules were fabricated to perform the necessary accelerated   environmental testing such as humidity testing, thermal testing and dielectric insulation. Cost analysis was made to show the cost reduction of the manufacturing of linear concentrated modules using this machine.

1390

<u>Low Cost High Concentration Ratio Solar Cell Array for Space Applications</u> : Patterson R E, Rauschenbach H S, Cannady M D, Whang U S and Crabtres W L," Proceedings 16th IECE Conference ", Atlanta, Georgia (USA), 1, 383 (1981).

The analyses and rationale that led to the development of a miniaturized Cassegrainian type concentrator solar array concept for space applications are described. In orbit cell, operating temperatures near 80 deg C are achieved with purely passive cell cooling and a net concentration ratio of 100 .

1391

<u>Construction and Testing of Experimental Plants Using Solar Cells and Concentrators</u> :Pfeiffer H,"P r oceedings 3rd Photovoltaic Solar Energy Conference ", Cannes· (France), (1980).

Results of the development work for photovoltaic conversion with concentrators are presented. A prototype module was constructed and tested consisting of 4 solar guided parabolic troughs with an aperture area of 32 sqm and a power output of approximately 2.5 KWe. One photovoltaic generator delivers electrical energy with an overall efficiency of 9.3% . In the present hybrid system both heat and electricity are produced. At a temperature of 90 deg C the thermal efficiency is 45% and the electrical efficiency 6%. The temperature of the solar cells mounted on the specially developed cooling channel is approximately 10 deg C above the coolant temperature. An indoor test facility was constructed and examined to simulate sunlight with respect to wavelength distribution, intensity and parallelism. This light was concentrated to 8000 W/sqm on a focal line 1.1 m in length. Non-uniform illumination of s o l a r cells, unavoidable in concentrator systems, was simulated and a small efficiency decrease of less than 3% measured.

1392

<u>Optical Characteristics of the Dispersion Concentrator and Its Influence on Photovoltaic Systems</u>: Sassi G, "Proceedings 3rd Photovoltaic Solar Energy Conference ", Cannes (France), (1980).

The characteristics of the chromatic dispersion concentrator are analyzed. The actual radial distribution in the focal zone of the various monochromatic components of the solar spectrum and the extent of their partial overlapping and moreover the radial distribution of the energy density is taken into special consideration. It is shown how it is possible to determine the most convenient disposition of the various photovoltaic elements in the focal zone and the most useful electrical connection among t h e same. Finally, an indication of the linear dispersion concentrators is given.

1393

<u>Photovolatic Concentrator Development</u> : Shafer B D and Boes E C, "Proceedings ISES (American Section) Conference ", Phoenix, Arizona (USA), 3.2 , 1067 (1980).

A number of PV concentrator array designs have been fabricated to date. The point and linear focus Fresnel lens designs and linear focus parabolic trough are described. A review of the status of present generation arrays is given as well as some insight into what are described as promising component a n d module developments.

1394

<u>Bivalent Thermal/Electrical Use of Solar Energy with Photovoltaic, High Concentration Heat Absorber & Spectrum Splitting Filters</u> : Simon M, "Proceedings ISES (American Section) Conference ",Phoenix,Arizona (USA), 3.2, 1067 (1980).

The author presents the development work for a 3 KWe concentrating silicon cell module with a 4 heliostat mounted cylinder parabolic troughs and a concentration factor of 20 to 40 with bivalent electrical a n d thermal energy use. Future preparation work with spectrum splitting filters for high c o n c e n t r a t i o n ratios on a parabolic dish with photovoltaics and a secondary thermal absorber for high   temperatures are presented.

Tracking of Solar Concentrators

1395

Dynamic Response Analysis of a Solar Powered Heliotropic Fluid Mechanical Drive System : Cope N A, Ingley H A and Morrison C A, " Proceedings 2nd International Alternative Energy Sources Conference ", Miami (USA), (1979).

This paper provides a summary of work performed during the design, construction and subsequent analysis of a solar powered tracking mechanism. This device is a fluid-mechanical system which utilizes solar radiation as its power source. The mechanism utilizes basic mechanical and thermodynamic principles in its construction and operation. This paper discusses these principles and describes the construction of the tracking device.

1396

Line Focus Sun Trackers : Gee R, "Proceedings ISES (American Section) Conference", Phoenix, Arizona (USA), 3.1, 501 (1980).

A variety of improved sun trackers have been developed. A testing program is under way at SERI to determine the tracking accuracy of this new generation of sun trackers. This paper defines the three major types of trackers, describes some recent sun tracker developments and outlines the testing that is under way.

1397

Sun Tracking Controller for Multi-KW Photovoltaic Concentrator System : Semma R P & Imamura M S, "Proceedings 3rd Photovoltaic Solar Energy Conference ", Cannes (France), (1980).

This paper describes a two axis sun tracking controller designed to achieve low cost, high performance reliability and operational flexibility for multi-KW to multi hundred KW photovoltaic concentrator applications. The controller is based on an active sun sensing approach and allows an array assembly to track the sun within five arc minutes. It has the capability to stow the array at night or during inclement weather conditions, communicate with an external computer and provide autonomous operation and diagnostic monitoring. These functions along with the basic design goals were achieved by using a dedicated microprocessor. The prototype of the third generation sun tracker design has been fabricated and is now in the verification stage. It represents a significant improvement over the two earlier generations which were based on hard wired logic. The sun tracking unit can be readily adapted for use on any application that requires sun pointing.

Materials Problems in Solar Concentrators

1398

Environmental Effects on Solar Concentrator Mirror : Bethea R M, Barriger M T, Williams P F & Chin S, " Solar Energy ", 27, 497 (1981).

Multiple samples of eight candidate solar concentrator mirrors, including both first and second surfaces types, were deployed 2.13 m above ground level. This exposure included the majority of the 1977, 1978 and 1979 dust storm seasons in western Texas. Additional samples were deployed at 28.96 – 30.48 m elevation on a radio transmitter tower. The samples were evaluated at regular intervals and after every major dust and hail storm and after most thunder storms in the area. All materials tested failed due to the disintegration of the reflective (second) surface; one of the first surface materials failed due to hail stone pittings. Other materials failed due to blistering of the reflective surface or their inability to withstand 1.27 – 2.54 cm hailstones. These results are confirmed by the reflectivity and topographic studies (by scanning electron microscopy) reported here. The effects of different methods used to clean naturally occuring dust deposits from the mirrors is reported in terms of changes in reflectivity.

1399

Reflectivity of Silver and Aluminium Based Alloys for Solar Reflectors : Hummel R E, "Solar Energy", 27, 449 (1981).

Numerous attempts have been made over the years to improve the corrosion resistance of silver and the reflectivity of aluminium by using alloying techniques. This paper describes the current understanding of the limitations of the attempts, based on the electron theory of metals. Experimental optical data on alloys that utilize silver and aluminium as base metals are summarized. It is shown that (i) alloying of silver and aluminium reduces the solar weighted reflectances (ii) solute additions may slightly improve the corrosion resistance of silver but in turn reduce the reflectance (iii) the parameters which affect the reflectance of pure metal include kind and number of solute atoms, surface reactions, surface conditions, kind of mirror preparation and interband transitions. Suggestions for future investigations are given.

Miscellaneous Concentrator Designs / Patents Etc.

1400

Solar Concentrators Using A Vacuum Contoured Surface for Tracking : Alexander W and Howell J R, " Proceedings 18th AIAA Aerospace Science Meeting ", Pasadena, California (USA), (1980).

Aluminized thin film can be formed into cylindrical troughs or spherical segments by applying a pressure differential across their surfaces. The pressure differential can be easily obtained by evacuating a plenum behind the film. If the film is segmented into two or more segments, the concentration can be applied across each segment. This paper reports on an analysis of such a trough collector, & on the experimental confirmation of the analysis. It is shown that the concentration ratios varying from 31 to 16 are obtained for incidence angles varying from normal to the trough to 40 from normal at solar noon for two segment collectors. It is found that deviations between the idealized used in the analysis and the experimental results are small, so that the computer predictions of performance can be used with confidence.

can be used with confidence. Experimental results showing reflected beam profiles from the collector onto an absorber tube for trough collectors with 'f' numbers (focal/aperture ratio) of 1.16 and 1.32 are presented and compared with predictions.

1401

<u>Assessments of Generic Solar Thermal Systems for Large Power Applications</u> : Apley W J and Bird S P, "ASME Paper No. 81 - WA/Sol - 25 ", (1981).

A comparative analysis of generic solar thermal conversion configuration was performed to evaluate and rank the principal concepts under consideration in the DOE Solar Thermal Power Program. Year long performance simulations were conducted for the 50 to 200 MWe systems using Barstow, California, meteorological and insolation data. Multi-attributable methodology was used to rank the eleven concepts.

1402

<u>Luminescent Solar Concentrators -2. Experimental and Theoretical Analysis of their Possible Efficiencies</u> : Batchelder J S,Zewail A H and Cole T, "Applied Optics", 20, 3733 (1981).

Experimental techniques are develop to determine the applicability of a particular luminescing center for use in a luminescent solar concentrator (LSC). The relevant steady state characteristics of eighteen common organic laser dyes are given. The relative spectral homogeneity of such dyes are shown to depend upon the surrounding material using narrow band laser excitation. The authors developed three independent techniques for measuring self absorption rates; these are time resolved emission, steady state polarization anistropy and spectral convolution. Preliminary dye degradation and prototype efficiency measurements are included. Finally, the authors give simple relationships relating the efficiency and gain of an LSC to key spectroscopic parameters of its constituents.

1403

<u>La Production de Chaleur et d'Electriate par Des Collecteurs Solaires Distribues A Concentration/(Heat and Electricity Production Using Distributed Solar Collectors with Focussing)</u> : Boy - Marcotte J L, "Entropie", 16 (96), 66 (1980).

Technical and economical elements are given to establish a compromise price performances for solar focussing collectors intended for industrial use with a peak power of about one thermal megawatt and in the 100 - 300 deg C temperature range. The transformer of this solar thermal energy into mechanical or electrical energy is defined according to system requirements. CEA COSS collectors and Bertin & Cie. organic fluid turbine which are used by SOFRETES under a COMES contract in the 100 KWe solar power station with distributed collectors in Corsica, are presented.

1404

<u>Effect of Receiver Optical Properties on Solar Thermal Electric Systems</u> : Call P J, Jorgensen G J and Pitts J R, "Journal of Solar Engineering (Transactions ASME) 103, 207 (1981).

The importance of reducing the thermal emittance of the receiver surface on the cost effective operation of intermediate and high temperature ($\geqslant$400 deg C) solar thermal electric power plants is discussed. Computer codes for seven systems (point and line focus) are used to independently determine optimum operating conditions for selective (low emittance) and nonselective receiver surfaces.

1405

<u>Solar Collectors</u> : Cassiday V M, "Specif. Eng. ", 46 (6), 113 (1981).

Practical applications of solar energy in commercial, industrial and institutional buildings are considered. Two main types of solar collectors are described; flat plate collectors and concentrating collectors. Efficiency of air and hydronic collectors among the flat plate types are compared. Also several concentrators are described, including their suntracking mechanisms. Descriptions of some recent solar installations are presented and a list representing the cross section of solar collector manufacturers is furnished.

1406

<u>Simple Tracking Strategies for Solar Concentrations</u> : Cope A W G & Tully N, " Solar Energy " 27, 361 (1981).

Consideration is given to the validity of the single axis tracking systems for solar concentrators of low to medium concentration ratios having moderate acceptance angles. If the misalignment between the sun and reflector normals is within the acceptance angle perfect tracking can be assumed. Relation about a fixed polar axis gives a constant misalignment equal to the sun's declination angle on that day. Rotation about a declination axis gives perfect alignment of noon, but increasing misalignment towards each end of the day varying with the time from the equinoxes. Data is given for monthly adjustment of the declination axis. All the results are independent of latitude.

1407

<u>Elastic Bent Plate Concentrators</u> : Dame R E, " Proceedings ISES (American Section) Conference ", Phoenix, Arizona (USA), 3.1, 496 (1980).

This paper describes some theoretical and experimental work under way at Mega Engineering (USA) to develop elastically deformed reflectors to become extremely low cost and optically accurate solar collector concentrator reflectors.

1408

<u>Control System Development for a 1 MWe Solar Thermal Power Plant</u>: Daubert E R, Begthold Jr. F M and Fulton D G, " Proceedings 16th IECE Conference ", Atlanta, Georgia (USA), 2, 1770 (1981).

A point focussing distributed receiver solar thermal power plant was developed for use in small community and industrial applications. The control requirements and the mechanisation used to implement these requirements are described. The power plant, rated at 1 MWe consists of approximately 55 power modules delivering power to a central collection point where the power is appropriately converted and delivered to the utility interface.

1409

<u>Progress on the Development of Luminescent Solar Concentrators</u> : Friedman P S, " Optical Engineering ", 20, 887, (1981).

Various planer configurations of the luminescent solar concentrator (LSC) are discussed including; the uniformly doped, the stacked plate, the thin film, and the multilayered film LSC. Radiation which is lost from the luminescent plate by falling within the critical angle for total internal reflection, is examined in terms of the above configuration. Several optical distribution factors are defined and experimentally evaluated for the luminescent plate and associated photovoltaic cell. These factors are used to calculate a plate - to - cell optical coupling coefficient for two commercially available solar cells. LSC's based on organic and inorganic phosphor systems are compared and the problem of dye stability discussed.

1410

Monte Carlo Modelling of Non Tracking Concentrator Using Light Trapping Techniques : Gordon B A, Knasel T M, Houghton A & Bain C N, "Proceedings ISES (Am. Sec.) Conference", Phoenix, Arizona (USA), 3.1, 615 (1980).

Monte Carlo methods have been used to model the performance of a non tracking solar collector which uses light trapping to provide concentration. Light trapping results from total internal reflection of light diffused from the bottom surface of the cover sheet of the collector. Gains of 20% to 30% are readily achievable and have been measured on a simple prototype.

1411

Canali A Pareti Riflettenti da Impiegare Come Concentratori e Come Riflettori di Radiazioni - I/(Reflecting Wall Channels for use as Concentrators and Radiation Reflectors - I) : Grillo S, "Energy Alternative Habitat Territ Energy ", 2 (4), 164 (1980).

The design of special channels with reflecting walls for the concentrator of diffuse radiation and direct radiation whose direction varies with time, is described. The collected radiations have directions that form angles with respect to the central axis of the channel. These angles are not greater than a certain fixed value. These channels are shorter than the usual fixed concentrators. They have two apertures, one larger and one smaller and can function either as concentrators or radiation reflectors.

1412

Canali A Pareti Riflettenti da Impiegare Come Concentratori e Come Riflettori di Radiazioni - II /(Reflecting Wall Channels to Use as Radiation Concentrators and Reflectors - II): Grillo S, "Energy Alternative Habitat Territ Energy ", 2(5), 267 (1980).

Models of the proposed channels are compared with a composite parabolic channel (CPC) for $\theta = 30°$. It is concluded that multiple reflecting wall channels (MRWC) are less cumbersome than the conventional concentrators and therefore more economic. Their use can be reversed i.e. they can be used as fixed reflectors. They also offer great possibilities of variations and choice.

1413

Focal Plane Flux Distributions Produced by Solar Concentrating Reflectors : Harris J H and Duff W S , " Solar Energy " 27, 403 (1981).

Efficient numeric equation based methods of computing focal plane target flux distributions are developed for four different types of solar concentrating reflectors. Reflecting surface quality and concentrator contour accuracy are taken into account. Several approximations yield computational short - cuts so that fast running computer programs to characterize focal plane flux distributions can be written.

1414

Studies With High Temperature Solar Collectors : Hayward K, "IEE Conference Publication No. 192 ", p. 20 (1981).

A description of two types of collectors together with some preliminary analysis of their thermal performance characteristics is presented. In addition, some points of interest regarding the installation and operation of the collectors are discussed.

1415

Optimization of a Hybrid Solar Thermal Power System : Irwin R E, Rodriguez D A & Pons R L, "ASME Paper 81 - WA/Sol-28 ",(1981).

A detailed study is carried out for a hybrid solar thermal power system which incorporates both storage and fuel burning operation. The solar plant is a Point Focussing Distributed Receiver (P F D R) system with distributed (electrical) generation. Energy storage is provided by batteries; both conventional (lead acid) and advanced technology (sodium sulfur) batteries are considered.

1416

Device for Regulating Radiant Flux in Solar Installations : Khakimov R A, Zakhidov R A & Sizov Yu. M , "Applied Solar Energy", 16 (3), 22 (1980).

Devices for regulating luminous flux in solar installations that concentrate solar radiation are considered. Shutter type luminous flux regulators located on heliostats are considered, as well as devices that permit automatic variation in orientation of each concentrator facet in accordance with the prespecified program. The advantages and drawbacks of the devices described are cited.

1417

Optical and Thermal Analysis of Linear Solar Receiver for Process Industries : Kumar R, " Proceedings 2nd International Alternative Energy Sources Conference", Miami (USA), (1979).

In this paper, the performance of an eccentric cylindrical annulus receiver is projected for industrial applications. In section II, a two stage CPC, one over the other, is discussed for photovoltaic application where high cost demands that the absorber area be as small as possible.

1418

Effects of Regional Insolation Differences upon Advanced Solar Thermal Electric Power Plant Performance and Energy Costs : Latta A F, Bowyer J M & Fujita T, "Journal of Solar Energy Engineering (Transactions ASME)", 103, 213 (1981).

This paper presents the performance and cost of the four 10 MWe advanced solar thermal electric power plants situated in various regions of the continental U S. It contains a discussion of the regional insolation data base, a description of the solar system's performances and costs and a presentation of a range for the forecast cost of conventional electricity by region and nationality over the next several decades.

1419

Use of the Thermal Lens Technique to Measure the Luminiscent Quantum Yields of Dyes in PMMA for Lumi-

<u>nescent Solar Concentrators</u> : Lesiecki M & Drake J M, "Applied Optics ", 21, 557 (1982).

The thermal lens technique has been used to measure absolute luminescent quantum yields of dyes in liquid and solid (PMMA) solutions. The validity of approach has been established by measuring quantum yields of known compounds, and its application to dye/polymer systems that would be suitable for luminescent solar concentrators has been stressed. The results of the authors for Rhodamine 6 G & Eluoral 555 give 0.93±0.04 and 0.88±0.03 respectively, for their quantum yield in PMMA.

1420

<u>New Concepts for Static Concentration of Direct and Diffuse Radiation</u> : Luque A, Eguren J, Ruiz J M, Cuevas A, Del Alamo J, Gomez J M, Acuna M and Sala G, "Proceedings 3rd Photovoltaic Solar Energy Conference ", Cannes (France), (1980).

The rules for casting maximum energy on a cell placed in a static concentrator of minimum entry aperture are derived. A concentration of 9.13 for colecting the direct sunbeam throughout the year of 4.5 for collecting diffuse light are upper bounds when using practical materials. Practical bifacial solar cells required to achieve those figures are presented. A prototype of concentrator with bifacial cells has been fabricated and its results are also presented. Based on the possible improvements of such a concentrator the authors arrived at a cost estimate of $ 3.19 W/peak.

1421

<u>Finite Lambertian Source Analysis of Concentrators : Application to Solar Reflectors</u>: Luque A & Gomez J M, "Applied Optics ", 20, 4193 (1981).

The maximum power cast on a collector from a source of finite angular extension by a concentrator of fixed position occurs when the collector sees the concentrator as a Lambertian source. Concentrators not fully meeting this requirement are evaluated using a parameter called the shape quality factor.Lambertian concentrators can be obtained with mirrors but not with lenses of finite 'n'. In many cases they are not ideal and some rays are cast outside the collector leading to an intercept factor below unity. Among those mirrors with the highest intercept factor, shape inaccuracies reduce this parameter and the shape quality factor to the extent which is analyzed. Rules for effective cost design are presented, leading to the conclusion that Lambertian concentrators should be used if the concentrator cost is high.

1422

<u>Radiation Testing of Spectrally Selective Reflectors for Solar Array Concentrators</u> : Layons J W (III) & Meulenberg Jr.A, "Proceedings 16th IECE Conference ", Atlanta, Georgia (USA), 1, 393 (1981).

Solar arrays which utilize light concentration have the potential advantages of lower cost and higher specific power (W/kg) than convention planar arrays. A study has been performed to evaluate the use of spectrally selective reflectors in light concentrating solar arrays.

1423

<u>Dynamic Analysis and Control of a Solar Power Plant II: Control System Design and Simulation</u> : Maffezzoni C and Parigi F, "Solar Energy", 28, 117 (1981).

The paper describes the main results of the study concerning dynamic analysis and control of the E E C Solar Power Plant, coupled to the grid for the first time in April 1981. Based on the process analysis developed in the first part of the present study and on frequency and time domain methods, the design of the steam generator control system of the EEC solar power plant is dealt with in this paper. More precisely, two different solutions are considered the one of quite simple and traditional implementation, using only Proportional Integral Differential (PID) regulators; the other including more sophisticated control functions, using adaptive dynamic compensators of special form.

1424

<u>Dynamic Analysis and Control of a Power Plant - I: Dynamic Analysis and Operation Criteria</u> : Maffezzoni C and Parigi F, "Solar Energy", 28, 105 (1982)

The paper describes the main results of the study concerning dynamic analysis and control of the EEC solar power plant, coupled to grid for the first time in April 1981. In this Part I of the work it is shown how a careful dynamic analysis of the process is necessary for the verification and the assessment of the receiver design, the precise formulation of the plant operation procedures and safety conditions, the specifications of the control system requirements. In particular, it is recognized that such kinds of processes can be adequately simulated by means of accurate partial derivative models based on suitable simulation codes.

1425

<u>Medium Temperature Solar Systems: More than Prototypes</u>: Menzo J P, "Specific. Eng. ", 46(3), 72 (1981).

Four prototype solar systems are described, (i) distributed power generation, (ii) industrial process heat, (iii) production of shaft work and (iv) total energy systems. A method for estimating the performance of these systems is described. The economic feasibility of solar heat storage is also discussed.

1426

<u>On the Angular Error of a Cylindrical Concentrator of Sheet Material</u> : Markman M A, "Applied Solar Energy", 16(5), 17 (1980).

An original method is proposed for obtaining concentrators of quite high accuracy by mechanical loading of an aluminium sheet. Experiment confirms the calculation results.

1427

<u>Design and Development of a Solar Conical Concentrator</u> : Mathur S S, Sharma J K, Dang A and Garg H P, " International Journal of Energy Research", 6, 73 (1981).

A portable solar conical concentrator of 1 meter aperture and consisting of 63 triangular plane mirror strips has been designed and fabricated. The device has been tested using cylindrical receivers having radii of 7.5 cm and 3 cm respectively. It required 2 hours for 8 kg of water in a large cylinder to reach boiling point, whereas it required only 20 minutes for 1.25 kg of water to boil in a smaller cylinder. The steam formed can beutilized to sterilize medical instruments or for other purpose as desired. The paper includes a detailed analytical treatment of heat exchange at the surface of the absorber, an evaluation of the appropriate heat transfer coefficients, and solutions of the energy balance equation for cylindrical absorbers of the sizes noted. Efficiencies as high as 60 per cent have been observed.

1428

Optical Evaluation of Cylindrical Elastical Concentrators : Mc Cormick P G," Solar Energy ",26, 519 (1981).

A study of elastical profiles formed by the elastic buckling of sheet reflectors and the application of such profiles to cylindrical concentrating collectors has been carried out using a ray trace analysis. Maximum concentration ratios of approximately 6 are obtained using untruncated elastical profiles. Edge truncation of the reflector substantially increases the maximum concentration ratio truncation of the center of the reflector is shown to provide a method for obtaining a uniform flux distribution on t h e receiver.

1429

Regime Characteristics of a Solar Thermoelectric Generator and Comparison of Experimental and Calculated Data : Movsumov E A and Bairamov A M, " Applied Solar Energy ", 16 (6), 64 (1980).

A thermoelectric generator simple & conventional in design is described; it employs an inexpensive concentrator.

1430

Luminescent Solar Concentrators and the Reabsorption Problem: Olson R W, Loring R F and Fayer M D, "Aplied Optics ", 20, 2934 (1981).

The problem of reabsorption in luminescent solar concentartors (LSC) is examined. A mathematical development is presented which enables the LSC gain to be calculated based on the optical properties of the materials and a random walk formalism. Two and three dimensional analyses are used. A detailed set of calculations for a common dye (rhodamine) is used to examine the practicability of employing a single dye. The effects of diameter, thickness and quantum yield on the LSC output are presented. The spectrum of the LSC output as a function of concentration is calculated. It is suggested that LSC's can be made more efficient with a system which utilizes radiationless electronic excited state transport and trappings as intermediate steps between absorption and re-emission.

1431

Performance of an Air Heating Receiver for Line Focussing Solar Collectors : Parker B.F, "Transactions ASME ", 24, 743 (1981).

An air heating solar collector with a parabolic trough reflector and linear receiver was conducted and tested. The results are presented in a modified form of the Hottel – Whillier Bliss generalised performance equation by plotting efficiency versus $\Delta T/I$. The test unit, which has some imperfections in constructing the receiver core, operated at an efficiency of 48 to 57% when heating ambient air depending on flow rate.

1432

Effect of Concentrator Field Layout on the EE – 1 Small Community Solar Power System : Pons R L & Irwin R E, "Proceedings 16th IECE Conference ", Atlanta, Georgia (USA), 2, 1759 (1981).

This paper presents an evaluation of the effect of concentrator field layout on the technical/economic performance of a point focussing distributed receiver (PFDR) solar thermal power plant rated at 1 MWe. The plant design is based on the EE – 1 prototype currently under development by the FACC for JPL/DOE.

1433

Yearly Average Performance of the Principal Solar Collector Types : Rabl A, " Solar Energy ", 27,215 (1981).

The results of hour by hour simulations for 26 meteorological stations are used to derive universal correlations for the yearly total energy that can be delivered by the principal solar collector types : flat plate, evacuated tubes CPC, collectors that track about one axis, collectors that track about two axes and central receiver.
The correlations are polynomials of the first and second order in yearly average insolation latitude & threshold (= ratio of heat loss and optical efficiency). With these correlations the yearly collectible energy can be found by reading a single graph and multiplying the coordinates by the collector parameters. This simple method reproduces the results of hour by hour computer calculations with an accuracy (rms ± error) of 2 % for flat plates and 24 % for concentrators. This method can be applied to any system where the collectors operate year round in such a way that no collected energy is discarded. This includes photovoltaic systems, solar augmented industrial process heat systems, and solar thermal power systems.
In addition, the method is recommended for rating collectors of different types or different manufacturers on the basis of yearly average performance. The method is also useful for evaluating the effects of collector degradation, the benefits of collector cleaning, and the gains from collector improvements (due to enhanced optical efficiency or decreased heat loss per absorber surface). For most of these applications, the method is accurate enough to replace a system simulation.

1434

Wind Load Definition for Line Focus Concentrating Solar Collectors : Randall D E, " Proceedings Line Focus Solar Thermal Energy Technology Development Semina ", Albuquerque, New Mexico (USA), (1980).

In this paper, a procedure is outlined to enable the designer to define peak wind loads which might be anticipated during the operational life time of a solar collector array. This procedure is intended to provide the information necessary to the preparation of economical collector and foundation design characteristics while simultaneously meeting the necessary safety and reliability constraints for line focus concentrating collector arrays.

1435

A Bielement Reflecting System for Focussing of Parallel Rays :Ray P, " Proceedings National Solar Energy Convention

A bielement reflecting system is contemplated in this paper. It is shown that the system which comprises two reflectors, is able to bring to a point focus or a line focus incident parallel rays. It is suggested that the system may be incorporated advantageously in solar energy conversion systems.

1436

Investigations of Optical Characteristics of Individual Reflector for Solar Power Plant (SPP) : Rekant N B, Aparisi R R and Kokhova I I, "Applied Solar Energy ", 16 (3), 25 (1980).

Results are reported for an investigation of a model SPP reflector. An individual SPP reflector measuring 3 X 5 m is described, together with the construction of a flat 5 X 9 m screen that serves to receive the reflected radiation. A method is presented for measuring the reflected radiation in field tests of the reflector.

1437

Thermal Damage in Luminescent Solar Concentrators (LDC) for Photovoltaic Systems : Rico F J M, Jaque F and Cusso F, "Journal of Power Sources", 6, 383 (1981).

An evaluation of the new type of solar concentrator is made. The temperature dependence of the optical efficiency of luminescent solar concentrators (LSC) has been studied. It has been found that in the range room temperature (RT) - 100 deg C an important thermal degradation is produced. The damage depends not only on the temperature but also on the nature of the dye.

1438

Transformation du Rayonnment Solaire par Fluorescence : Application a L'encapsulation des Cellules/(Transformation of Solar Radiation by Fluoresence : Application of Cell Encapsulation) : Sarti D, Le Poull F & Gravisse Ph., "Solar Cells", 4 (1), 25 (1981).

The presence of fluorescent material in the polymethacrylate (PMMA) which coats silicon solar cells, allows an increase of 25 % in the conversion efficiency of photovoltaic modules. This is due to two effects: a better match between the incident radiation and the solar cell spectrum; and concentration of the radiation because of the total internal reflections in the PMMA matrix.

1439

Computational Techniques for Ray Tracing and Tolerancing Solar Collectors : Vanderwerf J, Irving B, King C and Wientzen R, " Proceedings SPIE Conference", 237, 189 (1980).

Software that has been developed for the analysis and tolerancing of solar concentrator optical systems is described. The software models the conventional and the non-imaging components used in high energy solar concentrators.

1440

Chimie et Concentration due Rayonnement Solair/(Chemistry and Concentration of Solar Radiation) : Vilaron A, " Entropie ", 17 (97), 134 (1981).

The author recalls to mind some data concerning solar radiation and chemical thermodynamics, taking into account technological considerations related to industrial reactors.

1441

Analytical Dynamic Modelling of Line Focus Solar Collectors ; Wright J D, "Journal of Solar Engineering (Transactions ASME)", 103, 244 (1981).

Solar thermal power and industrial process heat systems may require a constant outlet temperature from the collector field. This constant temperature is most efficiently maintained by adjusting the circulating fluid flow rate. Models relating deviations in outlet temperature to changes in inlet temperature, insolation, and fluid flow rate illustrate the basic response and the distributed parameter nature of line focus collectors.

1442

Evaluation of Concentrator Characteristics : Zakhidov R A and Klychev Sh.I, "Applied Solar Energy", 16 (5), 36 (1980).

The cost and weight indexes of paraboloidal one piece and faceted concentrators with diameters from 1.5 to 5 m having different degrees of concentration are examined.

1443

On the Conversion of Solar Radiation with Fluorescent Planar Concentrators (FPCs) : Zastrow A, Heidler K, Sah R E, Wittwer V and Goetzberger A, "Proceedings 3rd Photovoltaic Solar Energy Conference ", Cannes, (France), (1980).

Though experimental results obtained with FPCs in the visible region are encouraging, the utilization of the infrared region of the solar spectrum is still involving great problems due to low fluorescence quantum yields of dyes, strongly overlapping bands and bad stability. An investigation of 30 infrared laser dyes shows that quantum yields decrease statistically with increasing wavelength and depend strongly on solvent viscosity and temperatures in several cases. The suppression of radiationless processes by deterioration of NH_2 groups in oxazine dyes enhances quantum yields substantially. Efficiency, concentration ratio and stability of collectors absorbing in the visible region could be improved.

Author Index

Name	References
Bond J A	0687
Boone J L	0841
Borania C H	0848
Bordoloi K C	0971
Borgese D	0613
Borner S	0612
Borton D (N)	0485,1193, 1194,1195, 1204
Borukhov M Yu.	0373,0972, 0973
Bos P B	0785
Bosnjakovic F	0974
Bouquet F L	0871
Bowyer J M	1120,1418
Boy-Marcotte J	0975,1147, 1403
Boyaner M R	0872
Boyd D A	0230
Boyes J D	083
Bracewell R N	0778,1333
Bradford A P	0873
Bradley J O	007,0963, 0977
Branch G P	009
Brander A I	1107
Brantley Jr L W	0150
Braslavskya M V	0231,0976
Breadley J O	0977
Bregg G N	1001
Brenner J L	0382
Brent C R	0978
Broadbent S	0786
Brosens P J	0979,0980
Brown C T	0594,0595, 0614,0615, 0702,0708, 0720,1357
Brown C I	1357
Brown G L	0616
Brown J	0807
Brown J E	0981
Brown K C	1253
Brumleve T D	0682,1349
Brune J M	0927
Brzuskiewicz J	0884
Buchholz R L	0419
Buckwatter C Q	0894
Buehl W H	0303
Burgess E L	079,0779, 0780,0781, 0782,0788, 1373
Burkhard D G	0179,0180, 0232,0233, 0982,1218, 1219,1234
Burkhard D L	0179
Burleson J	0787
Burr A A	1194
Buschow A G	0874
Butler B L	0900
Butler C P	0983
Butz J R	0080
Buzin E I	0737,0738, 0984
Byfield D A	011
Byrne R E	1154
Bystrov V V	0374
Byxbee R C	0837

C

Name	References
Cafaratti G	0613,0617
Call P J	1404
Call R L	1376
Cannady M D	1390
Canning J S	0234
Cape J A	0618,0783, 1374
Caputo R S	1033,1211
Carden P O	0375,0838
Carito J A	0981
Carley W J	1334
Carlson J A	0875
Carlton R J	012
Carratelli E P	0943
Cassiday V M	1405
Castellazzi P	042
Castle C H	0876
Castle J N 11	1110
Chambers B	0769
Champion R L	013
Chao B T	0985
Charron F	0452
Charters W W S	1344
Chauhan R S	0986,1203
Chaves E J	0971
Chawla O P	0298
Cheema L S	057,081, 0197,1312
Chen C W	0987
Chen F M	0727
Chen M M	0988
Chering J C	044
Cherne J	014
Chiam H F	1311
Chilcoat M	0201
Chin S	1398
Chubb T A	0376
Clausing A M	0151,0152, 0181,0183, 1350
Claverie M	1375
Claxton R J	082
Clements L D	0153,0154, 0816
Clemot M	0619
Clevett M L	0863
Cline H E	0377
Cluff C B	1376
Cobble M H	083,0378, 0739,0740, 0989,0990, 0991,0992, 0993,0994, 0995,1377
Coello-Vera A	1125
Coeytaux M	0628
Coffelt J	1199
Cole R	0235,0236, 0237,0242, 0243,0321
Cole T	1402
Coleman G	0714
Coleman M G	0800
Coleman W J	0877
Collares-Pareira M	0238,0239, 0240,0241, 0242,0243, 0299,0741, 0742,0743, 0996
Collaros G J	056
Collier R K	0182
Collon A	1354
Combesure C	0360
Conn W M	0998
Conti M	0944
Cope A W G	1406
Cope N A	0847,1395
Cosby R M	084,085, 086,094, 096
Costello D	0977
Costigue E	0799,
Costogue E N	1382
Cottingham J G	0620
Cotton E S	0379
Courrege Ph.	0999
Courtot P	0570
Coutures J P	0368,0400, 0467,0505, 1030,1342
Cowman C D	1000
Coxon De W	1001
Crabtree W L	1390
Craig G T	0248,1036
Craig J I	1086
Crawford R	0839
Cuevas A	0806,1420
Curtner K L	015,0639
Cusso F	1437

D

Name	References
Dahan G	1147
Daly J C	1002
Dame R E	1407
Damescu A L	1170
Dang A	1427
Daniel J L	0894
Darnell J R	1003
Darsey D M	0621,0622, 0623,0624
Dasgupta S	0380,0381, 0654
Dascher R E	056
Daubert E R	1408
David J P	1378
Davies J M	0379,1163, 1164
Davis B I	0397
Davis D B	0656
Davis W J	1370
Davison J H	010,1004
Day F D	1276
Deaver F K	1379
Deb S	0784
De Dufour W F	0853
De Geus A M	1005
De Gezzele A	0840
Del Alamo J	0806,1420
De Larue Jr R E	0382
Delano E	087,088, 089
Delfolie B	0452
Dellin T A	0625,0626, 0637,1288, 1351
De Meo E A	0785
Demichelis F	1006,1380, 1381
Denisenko E T	0491,0492,
De Plomb E P	0132
Derrick G H	0226,0227, 0228,0229
Dersus B	0619
De Socio L	008
Desai P	1007
Desautel J.	1008,1167
Deyhmy I	0618
Diaz A	001
Dimarogonas A D	0842
Dimitriu S	1170
Divakaruni S M	0589
Djalalov Z	0548,0549
Djalilov B (N)	0358
Dobie J	1199
Donovan R I	0787
Donovan R L	0786
Dorman J	1009
Drake J M	1419
Drenker G	0447,1113
Drew G G	1010
Drost M K	1352
Duban N	0147
Dubrovskii L A	0913,1285
Dudko O A	0386
Dudko Yu. A	0383,0384, 0385,0386, 0585,0586, 1285
Dudley V E	016,090, 0145,0387, 1011,1012, 1013,1014, 1015,1016, 1017,1018, 1019,1214, 1413
Duff W S	(see Dudley V E references)
Duffie J A	040,041, 0388
Dundev V	0627
Dupas A	1375
Duren J C	1151
Durbin R C	0841
D'utruy B	0628
Duwez P	0389,0390
Dvering G	1020
Dvernyakov V S	0391,0403, 0474,0475, 0476,0491, 0492,1021

E

Name	References
Eason E D	1022,1023
Easton C R	0629,0630, 0631,0632, 0633,0634, 0652,0698, 0709
Eckert E R G	0050
Eddy C S	0750
Edenburn M W	017,0776, 0777,0788, 0789,0790, 0824,0825, 1379
Edgecombe A L	0181,0183, 1253
Edlin F E	0392
Edwards D K	1024
Efimovskaya T V	1227
Eggers G H	0126,0127, 0133,0135, 0136,0139, 0140
Egorov A V	0843
Egorov V D	1107
Eguren J	0806,1420
Eicker P J	0927
Eldifrawi A A	0184
Eldighidy S M	0214
Eldridge B G	1060
Elevich G V	0843
Elfner P	0569
El Gabalawa N	1033
Elgabalawi N	1088
Elliott D G	0173
El Wakil M M	1313,1321
Emrich W	0133
Engel R K	1025
Engira R M	0395
Enileev H H	0732
Enoch J M	0337

Name	References
Shulmeister L F	0843
Sibers D O	0696
Silverman C G	0981
Simanovskii L M	1292
Simmons D E	0376
Simmons H	0239,0241
Simons J	0664
Simon A W	0501,0502, 0503,0504, 1222
Simon M	0826,0827, 1394
Singh M	1202
Singh P	057
Singh R	1223
Singh R N	058,059, 060,061, 062,0115, 0116,0745, 0746,0755, 0756,1293, 1296,1310
Singhal A K	0116,1293, 1296,1310
Sizov Yu.M	1416
Skaggs R L	075
Skinrood A C	0683,0697, 0701
Skripkar L N	0470
Sletton C J	1224
Smith P R	083,092, 0994
Smith R H	1225,1226
Smokovdina G S	1227
Snail K	1324,1325
Snyder H E	0405,0406, 1041
Sobin A	0698
Sobirov O Yu.	063,0903
Soderstrand M A	0623,0624
Soderstrom K G	0301
Sodha M S	1228,1229
Sokolov E V	1106
Sokolov S B	1135
Sokolova Yu.B	0761,1260
Solodovnikova L A	1106
Solodynikov Yu.A	0543,0914, 1230
Sommerlad R E	1363
Sootha G D	006
Soundarnayagam S	1251
Sparrow E M	050,064
Spiga M	0457
Spitzberg L A	0117
Springer T	0714
Srinivasan M	0506
Sriramulu V	055
Stadjuhar S A	1231
Stahl D	1348
Stamboltsyan S G	0496
Starodubtsev S V	0507,0508
Steele H L	1340
Steele K	0869
Stehle H	1348
Stein C	0241
Stein M L	0279
Stephens J B	1143,1144
Stephens W H	0699
Sterne K E	1232
Steward W G	0162
Stewart D E	0509
Stickley R A	0925
Stojanovic M S	0828
Straatman J T H	1233
Straub A	0700
Strobel G L	0179,0180, 1234
Stromberg R T	0968,0969,
Stromberg R P	1235
Strong J D	0951
Strong S	0807
Stultz J W	0904
Suaginyan L R	0403
Sucherpakov Yu.K	0902
Sukhatme S P	0212
Suleimanova F N	0546
Suleimanov S Kh.	0734
Surek T	0772
Suresh D	1341,1342, 1343,1344
Sutyagine V M	0542,0571, 0868
Suyarov Kh.	0921
Suyushkaliev N H	0553
Swafford C J	0702
Swanson P	0905
Swanson R M	0778,0829
Swanson S R	0202
Swartz G A	0813,0814
Swarup G	1236
Swet C J	1237,1238, 1239
Szulmayer W	0118,1240

T

Name	References
Tabor H	0169,0213, 0324,1241
Taha I S	0214
Tahakoff W	0510
Tairova L P	0902
Taketani H	0906,0907
Tallerico L N	0701,1242
Tamehiro H	1291
Tamutus D	0813
Tan J F	1243
Tanaka T	054,065, 066,067, 1244,1245
Tani T	054,065, 066,067
Tarcici A	1246
Tarnizhevskii B V	0584,0511 0512
Tate R E	051,1294
Taylor H D	1247
Teagan W P	0236
Teague H L	0595,0702
Tenukest C J	0372
Teplyakov D I	076,0353, 0462,0500, 0513,0514, 0515,0516, 0517,0518, 0519,0520, 0521,0522, 0523,0524, 0525,0526, 0527,0579, 0580,0581, 0582,0583, 0703,1248 1249,1346
Tester J W	1250
Tewary V K	1316
Thacher E I	0740
Thacher E F	0995
Thalhammer E D	0704
Thanvi K P	0200
Thatree N	0217
Thodos G	0325
Thomas A	1251,1297, 1298
Thomas M T	0894
Thomas W C	1252
Thyagrajan K	1228,1229
Thompson R	0528
Thomson W B	0660
Thornhill J W	1382
Thornton J P	0705,0706, 0863,1253
Tietz T F	0423,0456
Tikush V L	0403
Timar T	0405,1041
Tobin R D	0688
Togami H	0825
Toiliev K	0361
Tome G	0936
Tornielli G P	1289
Tornvall T	0529
Toth G	0700
Tracey J R	0612,0707, 0708
Trarieux P	0147
Treadwell G W	045,068, 069,070, 1254,1299, 1300
Triglia A	0252
Tripathi T C	006
Trombe F	0359,0452, 0530,0531, 0532,0533, 0534,0535, 1255
Troong T Y	0908
Truesdal K L	0454
Trukhov V S	072,0557, 0881,0882, 0883
Truong H V	0215,0216
Truscello V C	0536,0537, 0538,0539, 0636,1211, 1345
Tsukada K	0865
Tully N	1406
Tunnell H A	1256
Turfler R M	0770
Tveryanovich E V	0103,0540, 1257,1258
Tyagi R C	0119

U

Name	References
Udagawa M	031,071
Umarov G Ya.	072,0349, 0489,0499, 0500,0507, 0508,0541, 0542,0543, 0544,0545, 0546,0547, 0548,0549, 0550,0551, 0552,0553, 0554,0555, 0556,0557, 0558,0559, 0560,0561, 0562,0563, 0564,0732, 0757,0758, 0759,0760, 0761,0762, 0763,0903, 0909,0911, 0912,0914, 1259,1260, 1261,1262, 1345A
Umemoto Y	0120
Umesh G	1229
Uselton R B	1032,1138
Usmanov M	0560

V

Name	References
Vainer A A	0164,0561, 0584,0763, 0765
Vaishya J S	006
Vallee E C	0473
Valleva I A	0896
Van Wieringen J S	1233
Vanderhoff J	1439
Vanderwerf D F	0121
Vannucci S N	0565
Vant-Hull L L	0593,0597, 0609,0631, 0632,0652, 0653,0670, 0671,0673, 0675,0709, 0710,0711, 0712,0713, 0714,0721, 0722,0890, 1087,1364
Vartanyan A V	0566,0567, 0715,0917, 0918,1258, 1263,1264
Vasilyeva L V	0754
Vasagam R M	1108
Veinberg V B	0568
Veneruso A F	1265
Venkataraman S T	1266
Verma V K	091
Venis J	1233
Vermeulen P J	0569
Vicariot H	0570
Vignola F	0192
Vilaron A	1440
Vij S K	031
Vigourex F E	1267
Vilkova S N	0542,0571, 0903,0910, 0911,0912,
Villanueva J	0215,0216
Visentin R	0944
Vittitoe C N	0603,0604, 0605,0606, 0607,0716, 0717,0718, 0958,0959, 0960,1171
Vladimirova L N	0572
Vorotnikov V I	1230

W

Name	References
Wabrek R M	1377
Wachtell G P	1157
Wachtler W J	1268
Waddington D	0706
Wahlig M	0657,1056
Wagner W	0698
Walker R A	1309
Walker W E	0133,0142, 0143

Publishers of Sources

Journals

Acta Electronica, Laboratoires d'Electro-
nique et de Physique Appliquee, 3 Avenue
Descartes, 94450 Limeil-Brevannes, France .

Advances in Instrumentation, Instrument
Society of America, 400 Stanwix Street,
Pittsburgh, PA 15222, U S A.

AIAA Monograph, American Institute of Aeronautics & Astronautics, 1290 Avenue of the
Americas, New York, NY 10019, U S A.

American Journal of Physics, American Institute of Physics, 335 E. 45th Street, New York,
N Y 10017, U S A.

Applied Energy, Applied Science Publishers
Ltd., Ripple Road, Barking, Essex, England
IG11 0SA, U K.

Applied Optics, Optical Society of America,
at American Institute of Physics, 335 E.45th
Street, New York, N Y 10017, U S A.

Applied Physics, Springer - Verlag, 175 Fifth
Avenue, New York, N Y 10010, U S A .

Applied Solar Energy, (Akademiya Nauk Uzbek-
skoi S S R) Allerton Press Inc., 150 Fifth Ave.,
New York, N Y 10011, U S A.

A S H R A E Transactions, American Society of
Heating, Refrigerating & Air-Conditioning
Engineers Inc., 345 E. 47th Street, New York,
N Y 10017, U S A.

A S M E Transactions, American Society of
Mechanical Engineers, 345 E. 47th Street, New
York, N Y 10017, U S A.

Astronautics & Aerospace Engineering,
American Institute of Aeronautics & Astro-
nautics, 1290 Avenue of the Americas, New
York, N Y 10019, U S A .

Astronautik, Hermann Obertch Gessellschaft
E.V., Bremen D - 2800, West Germany.

Brennstoff - Waerme - Kraft (now published as
B W K), Verein Deutscher Ingenieure Verlag
GmbH, Graf - Recke - Strasse 84, Postfach 1139,
4000 Duesseldorf 1, West Germany.

Building Systems Design (now published as
Energy Engineering), Fairmont Press, Box 411,
Brooklyn, NY 11202, U S A.

Bulletin of Electrotechnical Laboratory ,
Electrotechnical Laboratory, Tanashi, Tokyo,
Japan .

Chemical Engineering Progress (CEP), American
Institute of Chemical Engineers, 345 E. 47th
Street, New York, N Y 10017, U S A .

Chemical Engineering World, Industrial Publications, Taj Building, 3rd Floor, 210 Dr Dada-
bhoy Nauroji Road, Bombay, India 40001.

Co-operation Mediterraneane pour L'Energie
Solaire (COMPLES) Bulletin, University of
Marseilles, France.

Deutsche Luft und Raumfahrt Forschungsber -
icht, Deutsche Forschungs und Versuchsanst
fur Luft und Raumfahrt E.V., Linder Hoehe ,
Postfach 906058, 5000 Köln 90, West Germany.

Elektronikpraxis, Vogel Verlag KG, Max -
Planck Strasse 7, Postfach 6740, 8700 Würz-
burg 1, West Germany.

Elektrotechnik und Maschinenbau (now pub -
lished as E und M), Springer - Verlag, 175
Fifth Avenue, New York, N Y 10010, U S A.

Entropie/Revue Internationale de Thermody-
namique et de Techniques Avancees, Editions
Bartheye et Cie, 54 Ave. Marceau, 7 5 0 0 8
Paris, France.

Heating/Piping/Air Conditioning, Penton-IPC,
Penton Plaza, 1111 Chester Ave., Cleveland,
Ohio 44114, U S A .

Heat Transfer - Japanese Research, Scripta
Publishing Co., 7961 Eastern Ave., Silver
Spring, Maryland 20910, U S A.

High Temperature, (Teplofizika Vysokikh Tem-
peratur) Consultants Bureau, 233 Spring St.,
New York, N Y 10013, U S A .

IEEE Spectrum, Institute of Electrical and
Electronics Engineers Inc., 445 Hoes Lane,
Piscataway, New Jersey 08854, U S A .

IEEE Transactions Aerospace and Electronic
Systems, Institute of Electrical and Electronic Engineers Inc., 445 Hoes Lane, Piscataway, New Jersey 08854, U S A .

Industrial Diamond Review, De Beers Industrial Diamond Division Pty. Ltd., Charters,
Sunninghill, Ascot, Berks, SL5 9PX, U K.

Industrial Technology /In Tech , Korea Scientific & Technological Information Center,
C.P.O. Box 1229, Seoul, South Korea.

International Journal of Control, Taylor &
Francis Ltd., 4 John St, London, WC1N 2ET, UK.

International Journal of Energy Research ,
John Wiley & Sons Ltd., Baffins Lane, Chi-
chester, Sussex, England, PO19 1UD, U K.

Institut Technique du Batiment et des Trevaux
Publics Annales, 9 Rue la Perouse 7 5 7 8 4 ,
Paris 16, France.

Iron Age, Chilton Co. Inc., Chilton W a y ,
Radnor, Pennsylvania 19089, U S A .

ISA Transactions, Instrument S o c i e t y of
America, 400 Stanwix St., Pittsburgh, Pennsylvania, 15222, U S A .

Israel Journal of Technology, Weizman Science Science Press of Israel, B o x 8 0 1 ,
Jerusalem 91000, Israel.

Journal of Energy, American Institute o f
Aeronautics & Astronautics, 1290 Ave. of the
Americas, New York, N Y 10019, U S A .

Journal of Engineering for Power (Trans -
actions of ASME - Series A), American Society of Mechanical Engineers, 345 East 47th
Street, New York, N Y 10017, U S A .

Journal of Environmental Sciences, Insti -
tute of Environmental Studies, 940 E a s t,
Northwest Highway, Mt. Prospect, Illinois
60056, U S A .

Journal of Heat Transfer (Trans ASME), American Society of Mechanical E n g i n e e r s,
345 East 47th St., New York, N Y 10017, U S A .

Journal of Optics/Nouvelle Revue d'Optique,
Masson, 120 Blvd. Saint - Germain, 7 5 2 8 0
Paris Cedex 06, France.

Journal of Physics Education, Ministry of
Education & Social Welfare, Shastri Bhawan,
New Delhi, India 110001

Journal of Pressure Vessel Technology, American Society of Mechanical Engineers, 345 E ,
47th St., New York, N Y 10017, U S A .

Journal Rech. Cent. Nat.Rech. Sci./ Courrier du C N R S Centre National, de la Recherche Scientifique, 15 Quai Anatole France, 75700 Paris, France.

Journal of Solar Energy Engineering (Trans ASME), American Society of Mechanical Engineers, 345 East 47th St., New York, N Y, 10017, U S A.

Journal of Spacecraft and Rockets, American Institute of Aeronautics and Astronautics, 1290 Ave.of the Americas, New York, N Y, 10019, U S A.

Journal of Vacuum Science and Technology, American Institute of Physics, 335 E 45th St., New York, N Y, 10017, U S A.

Mechanical Engineering, (Also pub.from Korea & Japan) American Society of Mechanical Engineers, 345 East 47th St., New York, N Y, 10017, U S A.

Memoires of Government Industrial Research Institute, Nagoya, Japan.

Nature, Macmillan Journals Ltd., 4 Little Essex St., London WC2R 3LF, England, U K.

Optica Acta (International Journal of Optics) Taylor & Francis Ltd., 4 John St., London WC1N 2ET, England, U K.

Optica Applicata, Politechnika Wroclawska, Wybrzeze Wyspianskiego 27, 50-370 Wroclaw, Poland. (Sub. to Ars Polona-Ruch, Krakowskie Przedmiescie 7, Warsaw, Poland.).

Optical Engineering, Society of Photo-Optical Instrumentation Engineers, 405 Fieldston Rd., Box 10, Bellingham, WA 98225, U S A.

Optical & Quantum Electronics, Chapman and Hall Ltd., 11 New Fetter Lane, London, EC4P 4EE, England, U K.

Optical Society of America Journal, American Institute of Physics, 335 East 45th Street, New York, N Y, 10017, U S A.

Optics Letters, Optical Society of America, American Institute of Physics, 335 E 45th, Street, New York, N Y, 10017, U S A.

Optik, Zeitschrift für Licht und Elektronenoptik, (Deutsche Gesellschaft für Elektronenmikroskopie e.V) Wissenschaftliche Verlagsgesellschaft mbH, Postfach 40, 7000 Stuttgart 1, West Germany.

Photogrammetric Engineering and Remote Sensing, (Previously Photogrammetric Engineering) American Society of Photogrammetry, 105 N, Virginia Ave., Falls Church, VA 22046, U S A.

Power Engineering, Technical Publishing Co., 1301 S.Grove Ave., Barrington, Illinois 60010, USA. (Foreign Subs.to: J B Tratsart Ltd. 154a Greenford Rd., Harrow, Middlesex HA1 3QT, England, U K.

Progress in Astronautics & Aeronautics, American Institute of Aeronautics & Astronautics, 1290 Ave.of the Americas, New York, N Y 10019, U S A.

Raumfahrtforschung, Deutsche Forschungs-und Versuch-Sanstalt für Luft Und Raumfahrt e.V., Leibigweg 10, 8012 Ottobrunn, West Germany.

RCA Review, R C A Corporation, Princeton, New Jersey 08540, U S A.

Revue de l'Aluminium, Societe d'Edition et de Documentation des Alli ges Legers, 5 Rue Saint Phillipe-Du-Roule, Paris, France.

Revue de l'Energie, Editions Techniques et Economiques, 3 Rue Soufflot, 75005 Paris, France.

Revue Generale de Thermique, Editions Europeenns Thermique et Industrie, 30 Rue de la Source, 75016 Paris, France.

Revue Internationale des Hautes Temperatures et des Refractaires, Masson, 120 Blvd. Saint Germain, 75280 Paris Cedex 06, France.

Revue de Physique Appliquee, Editions de Physique, Zone Industrielle de Courtaboeuf, B.P. 112, 91402 Orsay, France.

Review of Scientific Instruments, American Institute of Physics, 335 East 45th Street, New York, N Y 10017, U S A.

SAMPE Quarterly, Society for the Advancement of Material and Process Engineering, Box 613, Azusa, California 91702, U S A.

Science, American Association for Advancement of Science, 1515 Massachusets Avenue, N.W., Washington D.C. 20005, U S A.

Seismological Society of America Bulletin, Seismological Society of America, 2620 Telegraph Avenue, Berkeley, California 94704, USA.

Solar Cells, Elsevier Sequoia S.A., Box 851, CH-1001 Lausanne 1, Switzerland.

Solar Energy, International Solar Energy Society, Pergamon Press Inc., Journals Div., Maxwell House, Fairview Park, Elmsford, New York 10523, U S A. (Also at Headington Hill Hall, Oxford OX3 0BW, England, U K.)

Solar Engineering Magazine, (Previously Solar Engineering), Solar Engineering Publishers Inc., 8435 N.Stemmons Freeway, Suite 880, Dallas, Texas 75247, U S A.

Soviet Journal of Optical Technology / Optiko-Mekhanicheskaya Promyshlennost, Optical Society of America, American Institute of Physics, 335 East 45th Street, New York, N Y 10017, U S A.

Sunworld, International Solar Energy Society, Pergamon Press Inc., Journals Divn., Maxwell House, Fairview Park, Elmsford, N Y 10523, U S A. (Also at Headington Hill Hall, Oxford OX3 0BW, England, U K).

Teknisk Tidskrift, (Sveriges Civilingenjoers-foerbund), Ingenjoersfoerlaget A.B., Box 5703, 11487 Stockholm, Sweden.

Teploenergetika, Izdatel'stovo Energiya, Slyuzovayanab 10, Moscow Z-114, U S S R.

Termotecnica, Associazione Termotecnica Italiana, Viale Premuda 2, 20129-Milan, Italy.

Tohoku University Research Institute for Scientific Measurements Bulletin, Kenkyusho, 19-1 Sanjomachi, Sendai 980, Japan.

Publishers Note :

The Journal " Jet Propulsion " was published by the American Rocket Society until it merged with the American Institute of Aeronautics and Astronautics in 1963.

Full or complete names of journals and of their publishers with addresses in respect of the following journals appearing in this bibliography are not available :

* Associacao Brasilleria de Metais Boletim (Brazil).
* Regional Journal of Energy Heat and Mass Transfer.
* Air Fleet Equipment and Aircraft Design.
* C.R.Hebd.Sean.Acad. Science Bulletin (France
* Dopovidi Akad. Nauk. (U S S R).
* Izvestiya Otdelenie Tekhrucheskikh Nauk Energetika Aotomatika (U S S R).
* Policy Analysis and Information Systems.

Publishers of Conference Proceedings

Academic Press Inc., 111 5th Avenue, New York, N.Y.,USA

AIAA/American Institute of Aeronautics & Astronautics, 1290 Ave.of the Americas, N.Y., N.Y.10104,USA

AIChE/ American Institute of Chemical Engineers, 345 East 47th Street, N.Y., N.Y. 10017, USA

American Power Conference, c/o I.I.T, Technology Center, Chicago, IL 60616, USA

ASTM/ American Society for Testing Materials, 1916 Race Street, Philadelphia, PA 19103, USA

ASAE/ American Society of Agricultural Engineers, P.O.Box 410, St. Joseph, MI 49085, USA

ASME/ American Society of Mechanical Engineers, 345 East 47th Street, New York, N.Y. 10017, USA

American Society of Photogrammetry, 210 Little Falls Street, Falls Church, VA 22046, USA

Applied Research Laboratories, Univ. of Texas at Austin, c/o A.J.Tucker, 1000 Burnet Road, Austin, TX 78758, USA

Applied Science Publishers Ltd., Ripple Road, Barking, Essex, England IG11 OSA, U.K.

Central Salt & Marine Chemicals Research Institute, Waghawadi Road, Bhavnagar, India 364002

Clean Energy Research Institute, University of Miami, P.O.Box 248294, Coral Gables, FL 33124, USA

Consultants Bureau, c/o Plenum Publishing Co., 227 West 17th Street, New York, N.Y. 10011, USA

CNRS/ Centre Nationale D'Etudes Spatiales, 18 Av. Edouard-Belin, F-31055 Toulouse Cedex, France

Duetsche Gesellschaft Fur Sonnenenergie, Postfach 200604, D-8000 Munich-2, West Germany

Development Analysis Associates Inc., 675 Massachusetts Avenue, Cambridge, MA 02139, USA

Edison Electric Institute, 1111 19th Street N.W., Washington D.C. 20036, USA

Electric Power Research Institute, Research Reports Center, P.O.Box 10090, Palo Alto, CA 94303, USA

Electrochemical Society, 10 South Main Street, Pennington, N.J. 08534, USA

Elsevier Sequoia S.A., Box 851, CH-1001 Lausanne 1, Switzerland

European Community Information Service, 2100 M Street N W, Washington D.C. 20037, USA

Florida Solar Energy Center, 300 State Road #401, Cape Canaveral, FL 32920, USA

Franklin Institute Press, P.O.Box 2266, Philadelphia, PA 19103

Gordon & Breach Science Publishers Inc., 1 Park Avenue, New York, N.Y. 10016, USA

Gulf Publishing Company, P.O.Box 2608, Houston, TX 77001, USA

Halstead Press, One Wiley Drive, Somerset, N.J. 08873, USA

Hemisphere Publishing Corporation, 19 West 44th Street, New York, N.Y. 10036, USA

IEEE Publications, 445 Hoes Lane, Piscataway, N.J. 08854, USA

IES/ Institute of Environmental Sciences, 940 East Northwest Highway, Mount Prospect, IL 60056, USA

Institute of Measurement & Control, 20 Peel Street, London W8 7PD, England, U.K.

Institute of Electrical Engineers, P.O.Box 8, Southgate House, Stevenage, Hetrs, SG1 1HQ, England,U.K

ISA/ Instruments Society of America, P.O.Box 12277, Research Triangle Park, N.C.27709, USA

ISES(Am. Section)(Now American Solar Energy Society), 27 Harrison Street, Bridgeport, CT 06604, USA

ISES(UK Section) c/o Royal Institution of Great Britain, 19 Albemarle Street, London W1X 3HA,England

IPC Science and Technology Press, P.O.Box 63, Westbury House, Bury St., Guildford, Surrey GU2 5BH, England, U.K.

Los Alamos Scientific Laboratory, P.O.Box 1663, Los Alamos, NM 87545, USA

Masson S.A., 120 Blvd.St.Germain, F-75280 Paris Cedex 06, France

Mid Atlantic Solar Energy Association, 2233 Gray's Ferry Avenue. Philadelphia, PA 19146, USA

M.I.T.Press, 28 Carleton Street, Cambridge, MA 02142, USA

North Carolina State Univ. at Raleigh, Raleigh, N C 27650, USA

National Engineering Consortium Inc., 1211 West 22nd Street, Oak Brook, IL 61521, USA

New Mexico State Univ., Physical Sciences Lab., P.O.Box 3-PSL, Las Cruces, N M 88003

NTIS/ National Technical Information Service, Springfield, VA 22161

Oak Ridge National Laboratory, P.O.Box Y, Oak Ridge, T N 37830, USA

Oklahoma State Univ., Engineering Extension, 301 Engg. North, Stillwater, OK 74078, USA

Optical Society of America, 1816 Jefferson Place N W, Washington D.C. 20036, USA

Pergamom Press, Headington Hill Hall, Oxford OX3 OBW, England, U.K.

Pergamon Press, Fairview Park, Elmford, N.Y. 10523, USA

SAMPE/ Society for the Advancement of Material & Process Engineering, P.O.Box 613, Azusa, CA 91702,USA

Society of Automotive Engineers, 400 Commonwealth Drive, Warrendale, PA 15096, USA

SPIE (Now International Society for Optical Engineering), P.O.Box 10, Bellingham, WA 98227, USA

Technical Publishing Co., 1301 South Grove Avenue, Barrington, IL 60010, USA

UNIPUB, 345 Park Avenue South, New York, N.Y. 10010, USA

United Nations Publications,Sales Section, Rm LX-2300, New York, N.Y. 10017, USA

DOE/ US Dept. of Energy, Office of Energy Research, Washington D.C. 20545, USA

UAH/ University of Alabama Press, P.O.Box 1247, Huntsville, AL 35807, USA

Univ. of Chicago Press, 1150 South Langley Avenue, Chicago, IL 60628, USA

Univ. of Illinois Press, P.O.Box 5081 STA A, 54 East Gregory Drive, Champaign, IL 61820, USA

Univ. of Maryland, Dept. of Mechanical Engineering, College Park, MD 20740, USA

Univ. of Miami, Conference Services, School of Continuing Studies, P.O.Box 248095, Coral Gables, FL 33124, USA

UMR/ Univ. of Missouri at Rolla, Continuing Education, 111 Engineering Research, Rolla MO 65401, USA

Wiley John & Sons. Inc., One Wiley Drive, Somerset, N.J. 08873, USA

Western Periodicals Co., 13000 Raymer Street, North Hollywood, CA 91605, USA

World Energy Conference/Dunya Enerji Konferansi, Turk Milli Komitesi, P.K. 32, Bakanliklar, Ankara, Turkey